Interdisciplinary Evolution Research

Volume 3

About the Series

The time when only biologists studied evolution has long since passed. Accepting evolution requires us to come to terms with the fact that everything that exists must be the outcome of evolutionary processes. Today, a wide variety of academic disciplines are therefore confronted with evolutionary problems, ranging from physics and medicine, to linguistics, anthropology and sociology. Solving evolutionary problems also necessitates an inter- and transdisciplinary approach, which is why the Modern Synthesis is currently extended to include drift theory, symbiogenesis, lateral gene transfer, hybridization, epigenetics and punctuated equilibria theory.

The series Interdisciplinary Evolution Research aims to provide a scholarly platform for the growing demand to examine specific evolutionary problems from the perspectives of multiple disciplines. It does not adhere to one specific academic field, one specific school of thought, or one specific evolutionary theory. Rather, books in the series thematically analyze how a variety of evolutionary fields and evolutionary theories provide insights into specific, well-defined evolutionary problems of life and the socio-cultural domain.

Editors-in-chief of the series are Nathalie Gontier and Olga Pombo. The Series is edited from within the Applied Evolutionary Epistemology Lab, more information on the lab is available at http://appeel.fc.ul.pt.

More information about this series at http://www.springer.com/series/13109

Nathalie Gontier
Editor

Reticulate Evolution

Symbiogenesis, Lateral Gene Transfer, Hybridization and Infectious Heredity

Springer

Editor
Nathalie Gontier
AppEEL—Applied Evolutionary
Epistemology Lab
University of Lisbon
Lisbon
Portugal

ISSN 2199-3068 ISSN 2199-3076 (electronic)
Interdisciplinary Evolution Research
ISBN 978-3-319-16344-4 ISBN 978-3-319-16345-1 (eBook)
DOI 10.1007/978-3-319-16345-1

Library of Congress Control Number: 2015939804

Springer Cham Heidelberg New York Dordrecht London

Printed on acid-free paper

Springer International Publishing AG Switzerland is part of Springer Science+Business Media (www.springer.com)

Preface and Acknowledgments

Although originally a systematic term associated with speciation by hybridization, this book exemplifies "reticulate evolution" as it occurs by mechanisms and processes of symbiosis, symbiogenesis, lateral gene transfer, hybridization or divergence with gene flow, and infectious heredity. These phenomena are currently taking up a prominent role in the evolutionary sciences. Almost on a daily basis, new and fascinating data are presented that prove that evolution can result not merely from natural selection which brings forth a vertical pattern of linear descent with modification, but also from various types of reticulate evolution whereby evolutionary lineages connect, merge, and dissolve into one another.

A vast majority of these data are coming in from molecular biology, molecular systematics, and molecular phylogenetics. These fields developed after the foundation of the Modern Synthesis, and they currently enable a unification of bacteriology, virology, medicine, cell biology, exo-, and astrobiology. Together, these fields are building the new and fascinating research area of reticulate evolution. Many of the means whereby reticulate evolution can occur, however, were already identified before the foundation of the Modern Synthesis. Research on hybridization first developed in pre-evolutionary times and reaches as far back as the seventeenth century. Symbiosis and symbiogenesis has mostly been associated with ecological disciplines as well as with research on (a) biogenesis, developmental biology, and cell cytology. These disciplines originated in the late nineteenth century, during a period that has been characterized by Julian Huxley as the "eclipse of Darwin" because the then rising fields favored non-selectionist evolutionary explanations. Infectious heredity and the mechanisms of lateral gene transfer in prokaryotic organisms were first identified in the beginning of the twentieth century, within the bacterial and biomedical sciences, and today associated with virology, bacteriology, and genetic engineering and with attempts to personalize medicine. None of these fields belong to the classic Modern Synthesis that combined theoretical population genetics with Mendelian hereditary laws and aspects of diverse mutation theories. Today, scholars are therefore more and more pleading for an Extended Synthesis that integrates these research fields and their important data into a larger and richer theoretical framework whereby we can understand the evolution of life.

There is a general recognition that reticulate evolutionary mechanisms are stirring up and overthrowing many tenets of the standard neo-Darwinian framework, but how the new findings can become integrated with the standard paradigm, and how a new evolutionary biology might look like, is nonetheless still very much in the open. Natural selection, for example, is traditionally defined as a slow and gradual process, while symbiogenesis and lateral gene transfer are argued to occur rapidly in time. Research today is adding to the complexity, by demonstrating that the various evolutionary mechanisms are often simultaneously active within the various domains of life. Symbiotically acquired organelles, for example, or ontogenetically acquired viruses, can exchange genes laterally with the nuclear genes. Horizontally transferred genes in turn become integrated into the germ line where they evolve via natural and sexual selection mechanisms. When both vertical and reticulate mechanisms are active within the same individual or population, then how do we define the pace(s) and mode(s) of overall evolution? At present, we do not have the epistemic frameworks that synthesize vertical with reticulate evolutionary theories. The mechanisms of reticulate evolution are currently (1) awaiting synthesis into a standard reticulate evolutionary paradigm, and (2) that paradigm in turn is awaiting integration with the classic, neo-Darwinian one.

Though knowledge on the various means by which reticulate evolution occurs has never been so vast, a wider recognition and dissemination of these ideas toward the public sphere as well as within and across the general evolutionary sciences continues to lag behind. Almost all founders of the Modern Synthesis, and many leading neo-Darwinian scholars, brought natural selection to a wider academic and non-academic audience by publishing accessible books on their findings. It brought forth a general recognition that, regardless of our private or public institutional affiliations, the study of evolution by means of natural selection concerns us all and therefore needs integration in all scientific and public domains. Whether we are biologists or not, and academics or not, reticulate evolution equally impacts how we are to understand life's natural history, and it has a vast array of applications in the study of human health and disease, genetic engineering, or agriculture. Today, few experts unfortunately feel chosen to bring their work outside their field-specific disciplines, and instead, they report their results in technical journals. Field-specific handbooks that have as goal to train future expert-scientists are equally inaccessible to non-experts because of the necessary technical language.

The goal of this book is to provide an introduction on reticulate evolution to scholars and students working outside the fields, but it is not an encyclopedia of the numerous advocates. Such encyclopedias will eventually be edited by the biologists themselves. For this book, I chose to focus on the mechanisms that underlie reticulate evolution. Here too, choices had to be made, because there are many ways in which reticulate evolutionary mechanisms can be explained and exemplified. One can choose to provide a dictionary or glossary of new scientific jargon; one can give a state of the art on a selection of current data; one can dive into history and examine the original context of discovery wherein the mechanisms were first identified and perform an epistemic, philosophical analysis on how these ideas were and are received as well as what their consequences are for scientific theory in general; one can exemplify one reticulate evolutionary mechanism by providing examples of how it occurs within diverse lineages; or one can exemplify the relation between the various reticulate evolutionary mechanisms by exemplifying how they together bring forth the evolution of one or a couple of species. In agreement with the theme, the authors in this work have merged these various approaches, and their works focus either on theory formation or on particular case studies, or they either provide historical and philosophical contextualization or introductions to the state of the art.

This book also draws the line under a one-year long project funded by the Templeton Foundation that set out to investigate how reticulate evolutionary theories, and also macroevolutionary theories, are currently expanding evolutionary thought and how they can be implemented into the sociocultural domain. With the project, my collaborators, Marcia Belchior, Francisco Carrapiço, Luís Correia, Larissa Mendoza Straffon, Marco Pina, Olga Pombo, and Emanuele Serrelli, and I organized evolution schools for pre- and postdoctoral research scholars that featured courses on reticulate evolution taught by Michael Arnold, Frédéric Bouchard, Francisco Dionisio, Luis Villarreal, and Douglas Zook (http://evolutionschool.fc.ul.pt/videos); a symposium session on how symbiogenesis, lateral gene transfer, and virolution extend the synthesis for the 2013 meeting of the American Association for the Advancement of Science (http://appeel.fc.ul.pt/sub/eve/dir/aaas/aaas2013.html); a public conference on evolution that featured talks for a general audience and classes for teenagers on topics of reticulate evolution (http://evolutionconference.fc.ul.pt); and a conference on *Evolutionary Patterns: Horizontal and Vertical, Micro- and Macroevolutionary Patterns of Biological and Sociocultural Sciences* (http://evolutionarypatterns.fc.ul.pt/videos). Along the way, we captured several video interviews with the scholars involved, which can be viewed either on the respective video channels mentioned above, or directly on Appeel's YouTube channel (https://www.youtube.com/user/appeellisboa).

Many of the scholars that partook in the project and its associated events are also featured in this volume. I am very grateful to the authors who were willing to disseminate their and other people's work on reticulate evolution to a larger academic community, and I am quite confident that this book will provide an excellent entry point to interested scholars outside the field. Cordial thanks furthermore go out to Andreas Bohn, Frédéric Bouchard, Jorge Carneiro, Maurizio Casiraghi, Claudine Chaouiya, Eveline Kolijn, Alan Cooper, Cristina Cruz, Tal Dagan, Lee Dugatkin, Ricardo Guerrero, Frank Kressing, Paulo Madruga, William Martin, Ana Noronha, Frank Ryan, Jan Sapp, Rosalia Vargas, Davide Vecchi, Luis Villarreal, Tyler Volk, Richard Watson, and Slava V.I. Yukalov. Finally, my gratitude, as always, goes out to the entire Springer team, especially Anette Lindqvist, Sabine Schwarz, and the team in India.

Nathalie Gontier

Contents

Contributors

Michael L. Arnold Department of Genetics, Fred C. Davison Life Sciences Complex, University of Georgia, Athens, GA, USA

Amanda N. Brothers Department of Genetics, Fred C. Davison Life Sciences Complex, University of Georgia, Athens, GA, USA

Francisco Carrapiço Centre for Ecology Evolution and Environmental Change (CE3C); Centre for Philosophy of Science, Department of Plant Biology, Faculty of Science, University of Lisbon, Lisbon, Portugal

André F.P. Carvalho Centre for Environmental Biology (CBA); Centre for Ecology, Evolution and Environmental Change (CE3C), University of Lisbon, Lisbon, Portugal; Gulbenkian Institute for Science (IGC), Oeiras, Portugal

Luís Correia BioISI—Biosystems and Integrative Sciences Institute, Faculty of Science, University of Lisbon, Lisbon, Portugal

Francisco Dionisio Centre for Environmental Biology (CBA); Centre for Ecology, Evolution and Environmental Change (CE3C), Department of Plant Biology, Faculty of Science, University of Lisbon, Lisbon, Portugal; Gulbenkian Institute for Science (IGC), Oeiras, Portugal

Vitor G. Faria Gulbenkian Institute for Science (IGC), Oeiras, Portugal

João Alves Gama Centre for Environmental Biology (CBA), Centre for Ecology, Evolution and Environmental Change (CE3C), Faculty of Science, University of Lisbon, Lisbon, Portugal; Gulbenkian Institute for Science (IGC), Oeiras, Portugal

Nathalie Gontier AppEEL—Applied Evolutionary Epistemology Lab, University of Lisbon, Lisbon, Portugal

Jennafer A.P. Hamlin Department of Genetics, Fred C. Davison Life Sciences Complex, University of Georgia, Athens, GA, USA

António Manso BioISI—Biosystems and Integrative Sciences Institute, Faculty of Science, University of Lisbon, Lisbon, Portugal; Polytechnic Institute of Tomar, Quinta do Contador, Tomar, Portugal

Noland H. Martin Department of Biology, Texas State University, San Marcos, TX, USA

Caetano Souto-Maior Gulbenkian Institute for Science (IGC), Oeiras, Portugal

Élio Sucena Gulbenkian Institute for Science (IGC), Oeiras, Portugal; Department of Animal Biology, Faculty of Science, University of Lisbon, Lisbon, Portugal

William C. Summers Program on History of Science and Medicine, Yale University, New Haven, CT, USA

Sunni J. Taylor Department of Biology, Texas State University, San Marcos, TX, USA

Laura S. Weyrich Australian Centre for Ancient DNA (ACAD), University of Adelaide, Adelaide, Australia

Douglas Zook Global Ecology and Science Education, Boston University, Boston, MA, USA

Reticulate Evolution Everywhere

Nathalie Gontier

Abstract Reticulation is a recurring evolutionary pattern found in phylogenetic reconstructions of life. The pattern results from how species interact and evolve by mechanisms and processes including symbiosis; symbiogenesis; lateral gene transfer (that occurs via bacterial conjugation, transformation, transduction, Gene Transfer Agents, or the movements of transposons, retrotransposons, and other mobile genetic elements); hybridization or divergence with gene flow; and infectious heredity (induced either directly by bacteria, bacteriophages, viruses, prions, protozoa and fungi, or via vectors that transmit these pathogens). Research on reticulate evolution today takes on inter- and transdisciplinary proportions and is able to unite distinct research fields ranging from microbiology and molecular genetics to evolutionary biology and the biomedical sciences. This chapter summarizes the main principles of the diverse reticulate evolutionary mechanisms and situates them into the chapters that make up this volume.

Keywords Reticulate evolution · Symbiosis · Symbiogenesis · Lateral Gene Transfer · Infectious agents · Microbiome · Viriome · Virolution · Hybridization · Divergence with gene flow · Evolutionary patterns · Extended Synthesis

1 Reticulate Evolution: Patterns, Processes, Mechanisms

According to the Online Etymology Dictionary (http://www.etymonline.com), the word *reticulate* is an adjective that stems from the Latin words "*rēticulātus*" (having a net-like pattern) and *rēticulum* (little net). When scholars identify the evolution of life as being "reticulated," they first and foremost refer to a recurring *evolutionary pattern.*

N. Gontier (✉)
AppEEL—Applied Evolutionary Epistemology Lab, University of Lisbon, Lisbon, Portugal
e-mail: nlgontier@fc.ul.pt

N. Gontier (ed.), *Reticulate Evolution*, Interdisciplinary Evolution Research 3,
DOI 10.1007/978-3-319-16345-1_1

Net(work)-like patterns can be found in the way organisms belonging to distinct groups, species, or higher ranks of life *interact* with other such entities and exchange material and energy at a biochemical, behavioral, sexual or ecological level; as well as in the *phylogenetic reconstructions* of life's evolved lineages that scholars obtain by comparing the genes, proteins, and overall morphological and behavioral features of organisms and species.

Reticulate evolution brings forth *rapid evolutionary change* characterized by a network-like pattern of *horizontal crossings and mergings* that often precede a pattern of vertical descent with modification. This contradicts standard neo-Darwinian evolutionary theory that understands life to evolve gradually by means of natural selection that brings forth a bifurcating or ramificating pattern.

To understand why and how evolution is reticulate in mode and pattern, and why the tempo of reticulate evolution is often fast and non-gradual, scholars have to determine the *processes and mechanisms* that bring forth these reticulate evolutionary patterns. From the nineteenth century onwards, and mostly from outside or within the margins of the Darwinian and neo-Darwinian paradigm, botanists, microbiologists, bacteriologists, cytologists, and molecular geneticists have been increasingly able to identify these mechanisms and processes. Reticulate evolution today is a *vernacular concept* for evolutionary change induced by mechanisms and processes of *symbiosis, symbiogenesis, lateral gene transfer, hybridization or divergence with gene flow*, and *infectious heredity*.

1.1 Symbiosis

The concept of symbiosis was introduced in botany by de Bary (1878) who defined symbiosis as "the living together of unlike-named organisms." de Bary was inspired by the zoologist Van Beneden (1873, 1875), who a couple of years earlier had distinguished between "commensalism," "mutualism," and "parasitism" to characterize the "social lives" of animals.

Symbiosis thus refers to species interactions, and symbiotic associations have been mostly studied from within ecological research fields (Buchner 1921, 1939; Paracer and Ahmadjian 1986; Sapp 1994). Distinct organisms interact by providing a habitat or ecological niche for one another, by serving as a nutritional source, by enabling reproduction (in the case of pollination, for example), or by providing metabolic functions, morphological traits, and behavioral features neither of the partners are able to develop on their own. When organisms engage in a symbiotic association, both the *host* (the larger partner in the association) and the *symbiont(s)* (the smaller partner) often develop new features and sometimes form new individuals with characteristics not found in the individual organisms (Margulis 1991, 1998). Lichens, for example, result from a conjunctive symbiosis between a fungus (the mycobiont) and algae or cyanobacteria (the phycobiont).

Symbiosis can be *temporary* and *facultative* or extend prolonged periods of time, sometimes resulting in *obligate* and *hereditary* symbiosis (Buchner 1921,

1939; Wallin 1927; Lederberg 1952; Sagan 1967). When distinct organisms live on the surface of other organisms (on their skin, the leafs or roots of plants, the gill of fish, the outer membrane of cells), the symbiosis is called *ectosymbiosis*; and when organisms live inside other living organisms (inside the cells, leaf cavities or roots of legumes, or inside the vascular, lymphatic or gastro-intestinal systems), it is called *endosymbiosis*. Symbiotic associations are also differentiated based upon actual (penetrating) physical contact between the host and symbiont (*conjunctive symbiosis*), or the mere living inside each other's vicinity (*disjunctive symbiosis*) (Albany 1998).

Symbiotic associations can be acquired by *horizontal* or *vertical* transmission (outside or via the germline) (Archibald 2014; Douglas 2010; Gontier forthcoming). And although symbiosis per definition defines symbiotic relations to occur between *living* organisms, also viruses (genetic agents) and prions (infectious proteins) can be understood as symbiotic partners although neither are considered basic units of life (Lederberg 1952, 2003; Roossinck 2012).

The original symbiosis concept does not specify the nature of the living arrangement that exists between distinct organisms. The exact nature of the symbiotic relation between distinct individuals can be characterized further as *neutralism* (when the symbiosis neither harms nor benefits either of the partners), *commensalism* (when one partner benefits from the symbiosis and the other is unaffected), *mutualism* (when both partners benefit), *parasitism* (when one organism benefits and the other is harmed), *amensalism* (where one organism is harmed or killed and the other is unaffected), and *synnecrosis* (where both partners are harmed or killed by the symbiotic association) (Table 1).

Symbiotic interactions are numerous and diverse. One organism can *simultaneously* entertain different kinds of symbioses with a variety of organisms. So far, scholars have not been able to delineate a limit on how many organisms can simultaneously engage in a symbiotic association. What is becoming increasingly clear though, is that many commensal and mutual symbiotic associations are often *necessary* to obtain and maintain normal development, successful survival, and reproduction. Because symbiosis impacts adaptation, reproduction, and fitness, symbiosis can affect speciation and, in cases such as synnecrosis or amensalism, extinction (Brucker and Bordenstein 2012; Gontier forthcoming; Pound 1893; Schneider 1897).

Table 1 Possible symbiotic associations

Types of symbioses	Effects on species 1	Effects on species 2
Neutralism	0	0
Commensalism	+	0
Amensalism	−	0
Mutualism	+	+
Parasitism	+	−
Synnecrosis	−	−

+ beneficial; − harmful; 0 indifferent

Commensal, mutual, parasitic, or amensal symbiotic associations with microorganisms also impact organismal health, which is something we return to in the part on infectious heredity.

The *effects* of symbiotic associations extend the organisms that engage in the living arrangement because symbiosis can significantly alter biotic and abiotic ecological systems from the lowest to the highest hierarchical level. The evolution of photosynthesizing cyanobacteria, for example, which is estimated to have occurred between 2.7 and 2.4 billion years ago, is known to have severely impacted the earth's atmosphere and climate (Carrapiço 2006; Dole 1965; Flannery and Walter 2012; Holland 2006; Melezhik 2006; Pentecost and Franke 2010; Robert et al. 2005). The origin of photosynthetic life forms (organisms that produce oxygen as a waste product), led to the great oxygenation event which commenced somewhat 2.4 billion years ago. The transition from a reducing atmosphere to an oxygen-rich atmosphere led to the oxygen catastrophe, i.e., the first major extinction event where obligate anaerobe life forms that evolved in the Hadean and Archean became severely threatened. The great oxygenation event was a precondition for oxygen-respiring life forms to evolve, and it triggered the Huronian glaciation (the first ice age). These environmental changes were also one of the triggers for the evolution of symbiogenesis out of permanent symbiosis, where, as an adaptive environmental response, various life forms increased in size and sought permanent shelter in one another to find protection against the devastating environmental conditions.

In this volume, **Zook** provides us with a current state of the art as well as a new definition of symbiosis, **Carrapiço** reviews the history of symbiosis research, and **Faria and Sucena** exemplify how endosymbiosis can induce rapid speciation.

1.2 Symbiogenesis

Symbiogenesis is an evolutionary mechanism that occurs through "long-term hereditary symbiosis" (Margulis and Dolan 2000: 157). The fact that symbiotic associations can become hereditary was first acknowledged by von Faber (1912) who attested that bacteria found inside tropical plants engaged in a form of "erbliche Zusammenleben." The latter term was translated as "hereditary symbiosis" by Cowles (1915) and was later adopted by scholars such as Buchner (1921, 1939), Wallin (1927) and Lederberg (1952).

The concept of symbiogenesis was first introduced by the Russians Constantin Merezhkowsky, Andrey Famintsyn, and Boris Kozo-Polyanski (Sapp 1994). By building on earlier work of Andreas Schimper, the Russians pointed out that chloroplasts, organelles found in algae and plant cells, had evolved from cyanobacteria that engaged in long-term symbioses. The permanent endosymbiosis resulted in symbiogenesis: the cyanobacteria evolved into organelles, cellular structures that permanently reside inside the cells. With the Russians, hereditary symbiosis became understood as a causal agent in the evolution of new morphological features, and symbiogenesis was identified as an evolutionary mechanism whereby species evolve. A symbiogenetic origin for mitochondria, which are also eukaryotic organelles, was conjectured by Paul Portier in France (who identified them as "cellular symbiotes"), and in America by Ivan Wallin who adopted the notion of hereditary symbiosis and introduced the concept of "symbionticism" (Sapp 2003).

Our modern notions on symbiogenesis come from Lynn Margulis (Sagan 1967; Margulis 1970, 1998), who from the late 1960s onwards reintroduced, systematized, and expanded these ideas into the encompassing Serial Endosymbiotic Theory (Fig. 1). SET-theory,

> … presents a theory of the origin of … discontinuity between eukaryotic (mitosing or 'higher') and prokaryotic cells. Specifically, the mitochondria, … and the photosynthetic plastids can all be considered to have derived from free-living cells, and the eukaryotic cell is the result of the evolution of ancient symbioses. Although these ideas are not new [Merechowsky (1910) and Minchin (1915) in Wilson (1925), Wallin (1927), Lederberg (1952), Haldane (1954), Ris and Plaut (1962)], in this paper they have been synthesized in such a way as to be consistent with recent data on the biochemistry and cytology of subcellular organelles. (Sagan 1967: 226)

Contrary to Woese (Woese et al. 1978, 1990; Woese and Fox 1977) Woese and Fox 1977) who divides life into three domains, namely Archaea, Bacteria, and Eukaryota, Margulis endorsed a 5-kingdom classification of life and understood symbiogenesis as the distinguishing feature that separates prokaryotic organisms such as Archaebacteria and Eubacteria belonging to the Monera kingdom from all eukaryotic organisms, i.e., the Protoctists or protists (for the difference see Rothschild 1989), Fungi, Animal and Plant Kingdoms (Whitaker and Margulis 1978). For Margulis (1998: 42):

> symbiogenesis is the factor that distinguishes all nucleated-cell life from all bacterial life. No middle ground exists - either a group of organisms evolved by symbiogenesis or it did not. My claim is that all nucleated organisms (protoctists, plants, fungi, and animals) arose by symbiogenesis …

SET provides a theory for the origin of the four eukaryotic kingdoms which have evolved by three symbiogenetic mergings (Fig. 1).

The first merger is still controversial among scientists and involves the origin of the eukaryotic cell. According to SET, the eukaryotic cell evolved from a permanent hereditary symbiosis between different prokaryotes, namely *Archaeplasma*-like archaebacteria (*Thermoplasma acidophilium*) and *Spirochete*-like eubacteria. *Archaeplasma* bacteria are anaerobe and fermenting microorganisms that today are classified as a genus in the Archaea domain. *Spirochetes* are a phylum of double-membraned, corkscrew-shaped, mobile bacteria, today classified as belonging to the domain of Bacteria. The symbiotic merger between these distinct individuals, for Margulis, enabled the origin of the first nucleated cells, overall cell movement, and the formation of the mitotic spindle (Margulis et al. 2000, 2006).

In eukaryotic cells, the nucleated genes are organized on separate chromosomes. The mitotic spindle is a microtubule-rich organellar structure found outside the nucleus that helps in pulling apart the chromosomes during mitosis. Mitosis involves a series of complex movements of compartmentalized genes, and for

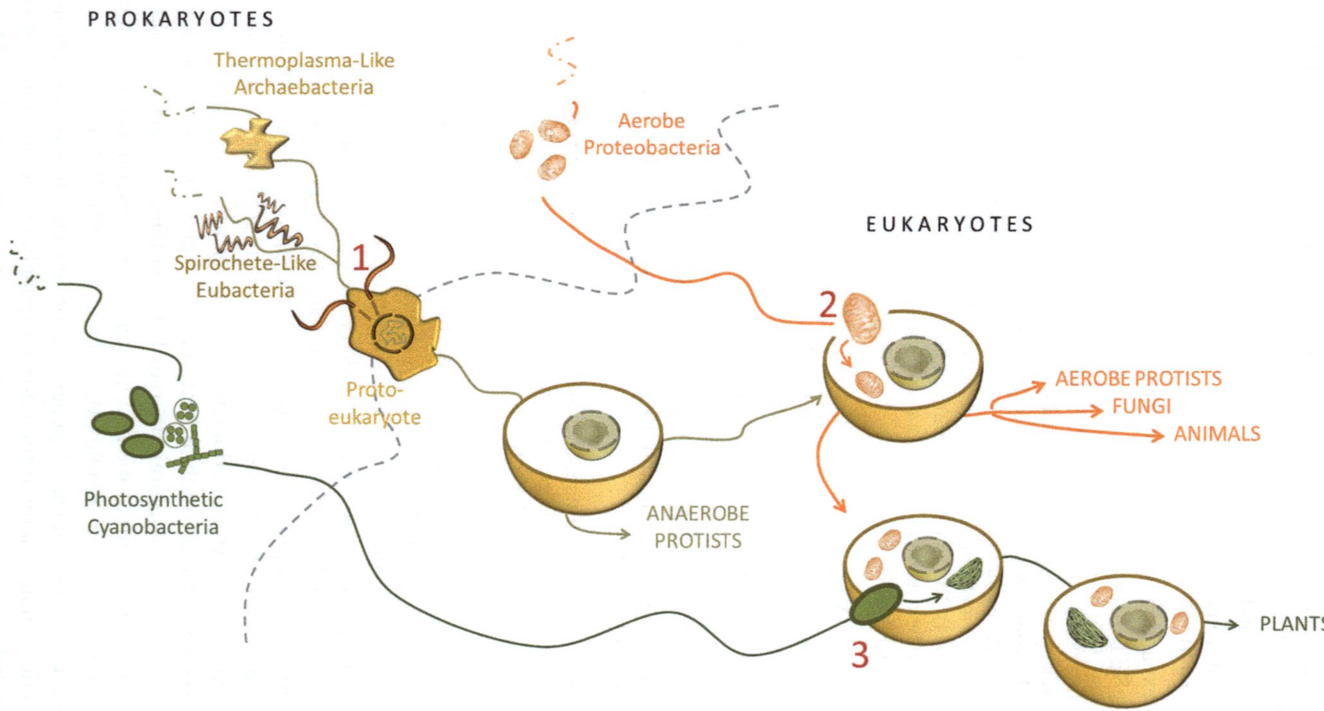

Fig. 1 Schematic of the three symbiogenetic mergings that underlie the evolution of the four eukaryotic kingdoms. In a first merger, *Thermoplasma*-like archaebacteria and *Spirochete*-like eubacteria merged and evolved into the first eukaryotic cells. In a second merger, anaerobe eukaryotic protoctists engulfed aerobe *proteobacteria* and evolved into the mitochondria that are found in aerobe protoctist, fungi, plant and animal cells. In a third merger, the products of the second merger engulfed cyanobacteria that evolved into the chloroplasts found in plant cells

Margulis (1998: 40–43), spirochetes are what enabled this internal movement or "dance of the chromosomes." The first merger also enabled external movement, because it gave rise to undulipodia ("waving feet"), including cilia, the "tails" and "hairs" of eukaryotic cells. For that reason, the first merger is also called "motility symbiosis."

Evidence for motility symbiosis is found in the structure of centrioli, undulipodia and cilia. Centrioli make up the centrosome, i.e., the microtubules-organizing center important for mitosis. Centrosomes also lie at the formation of kinetosomes, the basal bodies wherefrom moving organelles (undulipodia and cilia) extend. Undulipodia and cilia are made up of microtubular structures that in their shaft (the axoneme) have microtubules arranged according to a [9(2) + 2] pattern and in their basal bodies (kinetosomes) they all have microtubules arranged according to a [9(3) + 0] pattern. This latter pattern is identical to the microtubular organization found in centrioli (Fig. 2), and the centrioli are responsible for the formation of the kinetosome as well as the mitotic spindle. Based upon their morphological similarity, in SET theory, centrioli, undulipodia, and cilia are conjectured to have evolved from once-free-living spirochetes because free-living spirochetes often contain cytoplasmic tubules that resemble microtubules (Margulis and Dolan 2001: 89–96).

Contrary to SET theory that explains the origin of all eukaryotic cell types as resulting from permanent symbioses between different prokaryotes, several scholars (Livingston Bell 2001; Villarreal and Witzany 2010: 699) have suggested a viral origin for the eukaryotic nucleus. In this scenario, archaea-like organisms symbiogenetically integrated double-stranded DNA virus(es) which enabled the origin of hypercyclic DNA compartmentalization. Both scenarios need not be mutually exclusive, but so far, no scholar has tried to integrate both views into an overall tripartite chronological sequence.

The second and third merger involve the origin of mitochondria and chloroplasts, two eukaryotic cell organelles. The second merger of SET theory describes the evolution of mitochondria from aerobe proteobacteria that started to entertain permanent symbiotic relations with some of the first eukaryotic beings (possibly in response to the oxygen crisis); and in a third merger, chloroplasts evolved from the intracellular incorporation by phagocytosis (eating or engulfing) of cyanobacteria. In both cases, these once-free-living bacteria were engulfed by the first eukaryotic life forms, the endosymbiosis with the intracellular guests became permanent and hereditary, and this hereditary symbiosis led to the evolution of the respective organelles. Not all cyanobacteria and proteobacteria (which both encompass large taxonomic groups) engaged in symbiosis, and to this today, both cyanobacteria and proteobacteria continue to live independently of eukaryotic organisms.

Mitochondria and chloroplasts contain their own DNA and their endosymbiogenetic, bacterial origin is today undisputed because there is proof coming from comparative molecular phylogenetics (Bonen and Doolittle 1975, 1976; Bonen et al. 1977). The DNA found in these cellular organelles still relates more closely to the free-living bacteria where they presumably evolved from than it does to the nuclear genes of the cells they belong to.

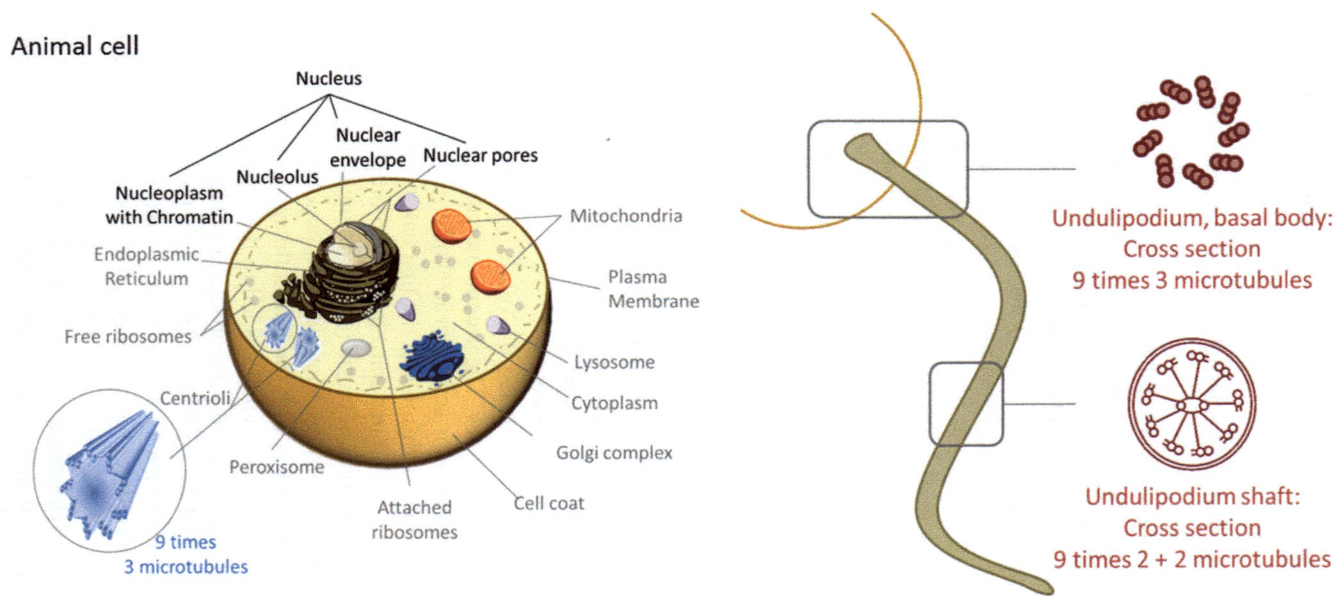

Fig. 2 *Left* An illustration of the basic structure of an eukaryotic animal cell. *Right* A pictorial cross section of an undulipodium. The microtubules-structure found in the basal bodies of undulipodia (kinetosomes) is identical to the microtubules-structure found in centrioli, and Margulis' SET theory therefore assumes an evolutionary homologues relationship

All chloroplasts thus appear to be related by common descent from cyanobacteria. Nonetheless, these plastids have been acquired repeatedly, often as *primary*, *secondary,* and *tertiary endosymbiosis* events. Eukaryotic organisms with chloroplasts in place and forming a primary endosymbiosis were completely engulfed by other eukaryotes where the engulfed organism, as a whole, started to function as a chloroplast. Products of this secondary endosymbiosis in turn have also been engulfed by other eukaryotes, a process called tertiary endosymbiosis (Archibald 2011, 2014). Symbiogenesis as an evolutionary mechanism therefore not merely evidences a pattern of reticulation, it also demonstrates a pattern of increased embedding, comparable to Russian dolls, though the dolls have different morphologies rather than being identical. Mitochondria, in turn, have all evolved from proteobacteria with which they still share a high genetic similarity.

Evidence is furthermore accumulating that proves that these organelles have undergone considerable gene loss after their symbiogenetic acquisition (Archibald 2014), and they have engaged in lateral gene transfer with the nuclei of eukaryotic cells, in both directions (Archibald and Richards 2010; Blanchard and Lynch 2000; Martin and Herrmann 1998). Finally, Margulis also associated SET theory with the Gaia hypothesis which was first introduced in its modern form by Lovelock (1972) and later elaborated by both authors (Lovelock and Margulis 1974).

Besides chloroplasts that are found in all plant cells, and mitochondria, found in all aerobe protist, plant, fungal and animal cells, eukaryotic cells contain many more organelles, and their evolutionary origin remains obscure. The Belgian cytologist, de Duve et al. (1974), who first discovered lysosomes (eukaryotic organelles found in animal cells and involved in housekeeping), also suggested a symbiogenetic, bacterial origin for these organelles.

In this volume, **Zook** elaborates upon *primary*, *secondary,* and *tertiary endosymbiosis*, which is especially relevant for understanding the origin of green and red algae as well as dinoflagellates (marine plankton that often combines photosynthesis with phagotrophy: the engulfment and eating of prey). And both **Zook** and **Carrapiço** explain how symbiogenesis is to be understood as an evolutionary mechanism in and of itself that complements the mechanism of natural selection. **Correia and Manso** provide a computational model to simulate symbiosis, symbiogenesis, and lateral gene transfer.

1.3 Horizontal or Lateral Gene Transfer

Lateral gene transfer is the process whereby genes are exchanged horizontally, either between distinct organisms with different genealogical histories, or between distinct genomes present in the same organism (e.g. between gene-containing organelles and the nucleus; or between the bacterial genome and plasmids residing inside the bacterial cell). In prokaryotes, lateral gene transfer occurs mainly by mechanisms of *transformation*, *transduction,* and *bacterial conjugation* (Fig. 3).

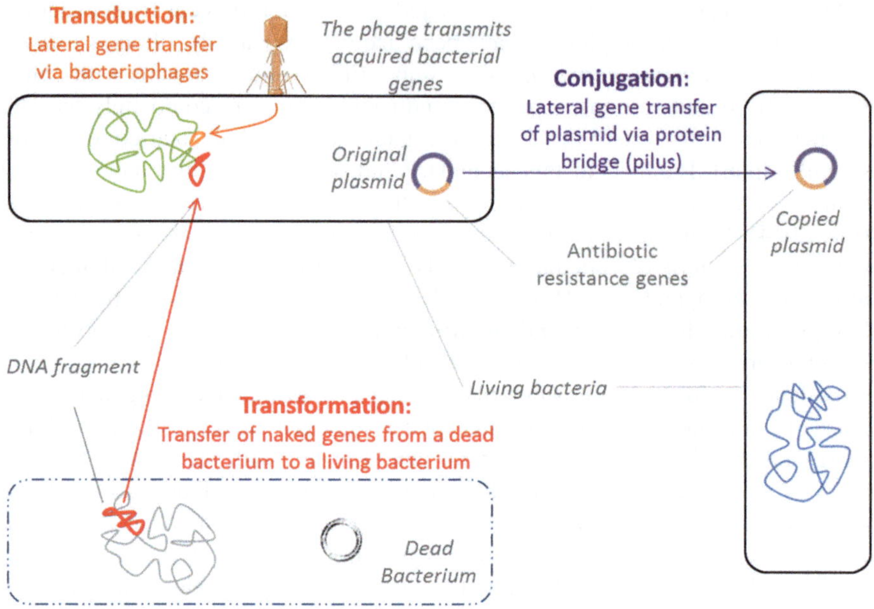

Fig. 3 Schematic of the three main mechanisms of lateral gene transfer in prokaryotes

Transformation involves the uptake of naked DNA from the surroundings, and the process was first described by Frederick Griffith and later confirmed by Oswald Avery, Colin MacLeod, and Maclyn McCarthy. Transduction was first described by Joshua Lederberg and Norton Zinder and involves the transfer of bacterial genes via bacteriophages, i.e., bacterial viruses. Bacterial conjugation, or bacterial mating as it is often called, was discovered by again Joshua Lederberg, in collaboration with Edward Tatum, and involves the transfer of plasmids.

A *plasmid* is an extrachromosomal (Lederberg 1952), mobile genetic element (Shapiro 1983), often made up of circular DNA. Plasmids are central agents for lateral gene transfer by means of bacterial conjugation whereby a single strand of the double-stranded plasmid is laterally transferred from a donor bacterium to a recipient. Plasmids often carry antibiotic resistance genes, and via bacterial conjugation, these resistance genes are exchanged between bacterial populations. Such "extra genes" are not necessary for the bacterium to survive, but they can nonetheless increase the bacterium's chances of survival and therefore also its fitness.

Several bacteria also contain *Gene Transfer Agents* (GTAs) in their genome. GTAs are bacteriophage-like elements that are horizontally exchanged (Maxmen 2010; Stanton 2007), and they present a fourth form of lateral gene transfer among bacteria.

Prokaryotes and eukaryotes alike contain "jumping genes" (McClintock 1950, 1953) or *transposons*. These are mobile genetic elements that can change their position in the genome and move to another location. They can switch their position inside the genome they belong to, or they can travel horizontally from the bacterial genome to a bacterial plasmid or vice versa, or from organellar DNA to nuclear DNA and vice versa. Retrotransposons are a subclass of transposable elements found in eukaryotes (Engels and Preston 1981; Frost et al. 2005; Kazazian et al. 1988; SanMiguel et al. 1996; Shapiro 1969; Singer 1982; Taylor 1963). Retrotransposons are alternatively known as *transposons via RNA intermediates*, because they move about by *copying and inserting* themselves via RNA intermediates. Transposons are always made up of DNA, and they *cut and paste* themselves into genetic sequences (Finnegan 1989).

Transposons leave gaps at the places where they cut themselves and often interrupt the gene sequence where they insert themselves, while retrotransposons enable genome growth by duplication of gene sequences, and both therefore enable "genetic transformation" (Rubin and Spradling 1982) of the organismal genome they belong to. In other words, they change the genetic make-up of organisms and are therefore key players in evolution.

Another type of mobile genetic elements are *retroviruses*. Retroviruses can insert their genes into the host's genome, and they can become transmitted vertically. Retroviruses furthermore resemble certain retrotransposons, making some scholars believe they are evolutionary related (Flavell 1981; Nelson and Hooley 2004; Ryan 2009; Temin 1980).

Scientists are currently mapping the various mobile genetic elements there exist in order to find recurring structures, elements, patterns, and mechanisms whereby these elements are transmitted. These efforts are designated as the *mobilome* projects (Frost et al. 2005; Siefert 2009).

The abundant occurance of lateral gene transfer in all three domains of life has only been recognized in recent years. Molecular phylogenetic reconstructions (Doolittle 2000; Gogarten 2000; Bapteste 2014; Sapp 2009) now provide conclusive evidence for "alien" or exogenous DNA uptake, which has greatly contributed to the general academic reception and recognition of the phenomena. Nonetheless, the existence of jumping genes and many of the mobile genetic elements, as well as the basic mechanisms whereby prokaryotes exchange genetic material horizontally, were already identified in the beginning of the twentieth century, mostly under artificial laboratory conditions.

In this volume, **Summers** provides a history of plasmids, and **Dionisio et al.** provide a symbiotic account of non-transferrable plasmids. **Gontier** sketches the discoveries of lateral gene transfer mechanisms in history and relates it to current epistemic debates on the "web" versus "tree" of life.

1.4 Hybridization

Originally, the neo-Darwinian framework mainly provided a theory on animal evolution, and both natural and sexual selection theories rely heavily on eukaryotic reproduction systems such as sex that enable the differential vertical descent of (mutated) genes over generations through time (Gontier forthcoming). Prokaryotes, however, reproduce by division and also many plants and flowers reproduce asexually by division or "cloning." When plants and flowers do reproduce sexually, they do so by means of cross-fertilization (where the gametes of sexually different individuals belonging to the same species join—similar to animal sex), self-fertilization (many flowers have both male and female sex cells that recombine during reproduction within the same, bisexual individual), pollination (the transfer of pollen from anther to stigma often mediated by insect species such as wasps and bees that live in symbiotic association), or hybridization (López-Caamal and Tovar-Sánchez 2014).

Hybridization occurs when two genetically distinct individuals (that in turn can belong to different subspecies, species, genera, and even families) reproduce offspring. The offspring can be infertile, but most of the time they are fertile, and the hybrid can reproduce either with its parental lineages (backcrossing or introgression) or only with similar hybrids. In both cases, hybridization can lead to the introduction of novel features as well as new species altogether (Arnold 1997, 2004, 2006; Harrison 1990; Mallet 2005, 2007; Rieseberg 1995, 2001).

In many ways, the Modern Synthesis has prohibited hybridization to become recognized as an evolutionary mechanism that can, and often does, induce speciation. Hybridization poses a problem for the neo-Darwinian paradigm. Mayr's (1942) biological species concept, for example, defines species based upon sexual compatibility and geographical accessibility. Per definition, individuals that can produce fertile offspring belong to the same species and such a definition logically excludes speciation to occur *because* of sexual exchange between individuals belonging to distinct species. But this is exactly what happens during hybridization, when individuals of distinct species mate and produce offspring.

Hybridization of animals was already recognized in ancient societies. Mules, for example, were deliberately bred. The word "mule" stems from "mulato" or "half-breed," and it was also used to designate humans with multiple-ethnic origins from the Middle Ages onwards. At the time, scholars falsely divided the human species into separate races. Colonization led to many mixed marriages and "bastard children," leading naturalists and clergyman of the seventeenth, eighteenth, and nineteenth century to speculate on the long-term consequences of mixing. John Ray, for example, in a paper presented at the Royal Society of London in 1684, argued that hybridization violates the divine order in the world for God had created the species in a fixed form, and he speculated that hybridization would have devastating influences on the "pure breeds" (Kingsbury 2009). In short, debates on the consequences of hybridization ran high in pre-evolutionary societies and are very much comparable to current debates on the long-term consequences of genetic engineering that artificially combines hybridization with endosymbiosis and LGT techniques.

The mechanisms of pollination in flowering plants and the recognition that also plants have sexes and reproductive organs was only recognized in 1694, by the German scholar Camerarius or Rudolph Jakob Camerer (Roberts 1929; Zirkle 1934, 1935). Thomas Fairchild in London and Josef Gottlieb Kölreuter in Germany would attempt to produce deliberate crosses of various plant species (López-Caamal and Tovar-Sánchez 2014). In 1720, Fairchild presented to the British academic community what would become known as "Fairchild's mule," a deliberate cross he produced in 1717 between two plant species belonging to the *Dianthus* genus, known as *Dianthus barbatus* and *Dianthus caryophyllus*. Kölreuter mixed various species of tobacco plants in the 1760s, but many of those turned out sterile, making Kölreuter agree with Ray and conclude that hybridization was against divine creation and that it would eventually lead to sterility in the offspring of all crosses.

Carolus von Linnée or Linnaeus (1753) also applied the incoming knowledge of the existence of sexual organs in plants. In his double-volumed work on the taxonomy of plants, he provided the first systematic classification of round and about 9000 different plant species.

Linnaeus classified plants based upon a sexual system which he dubbed "Clavis Systematis Sexualis," a system later incorporated in the 10th edition of his *Systemae Naturae*. Plants were considered to undergo "public" or "clandestine marriages," and plant species were further differentiated based upon whether or not the marriage between the sexual partners lasted in time, whether they had different means to engage in sexual reproduction (e.g., by pollination, self-fertilization), and whether or not the species were monogamous or endorsed promiscuous relations with multiple partners. Though his system made use of logical and binary oppositions, and thus remained based upon artificial classifications, he first attempted to classify the naturally occurring hybrids of different plant species and he also came to recognize that hybridization challenges the idea that species are fixed entities that undergo no significant change through time.

The incoming results on the rather "promiscuous" intercourse and "marriages" between various plant species thus first facilitated evolutionary thinking. For Christian Konrad Sprengel, who would later inspire Darwin, hybridization led him to understand that species are not fixed but in constant flux, and also Karl Friedrich von Gärtner, who was able to produce fertile crosses, recognized the potential hybridization had for agriculture and the production of more nutritional crops (for reviews, see Kingsbury 2009; Camaal and Sanchez 2014). Darwin himself endorsed ideas on genetic blending and recognized hybridization to occur, but nonetheless, the neo-Darwinians focused on genetic recombination as it occurs by cross-fertilization between distinct sexual members of the same species.

At the turn of the twentieth century, Erich von Tschermak von Seyssenegg (one of the rediscovers of Mendel) in Austria also studied hybridization, as did the Danish scholar Øjvind Winge who was able to produce stable hybrids, and the Swedish geneticist Arne Müntzing who discovered chromosomal recombinations (Camaal and Sanchez 2014). Nonetheless, plant hybridization and introgression (the backcrossing of diverging species with the parental stock) (Fig. 4) was especially brought to the attention of the modern scientific evolutionary community by Anderson (1949) and Stebbins.

Stebbins was responsible for integrating plant studies into the Modern Synthesis by introducing the first "botanical synthesis" (Smocovitis 1997, Smocovitis and Ayala 2000) in his major 1950 work on "Variation and Evolution in Plants" (Stebbins 1950).

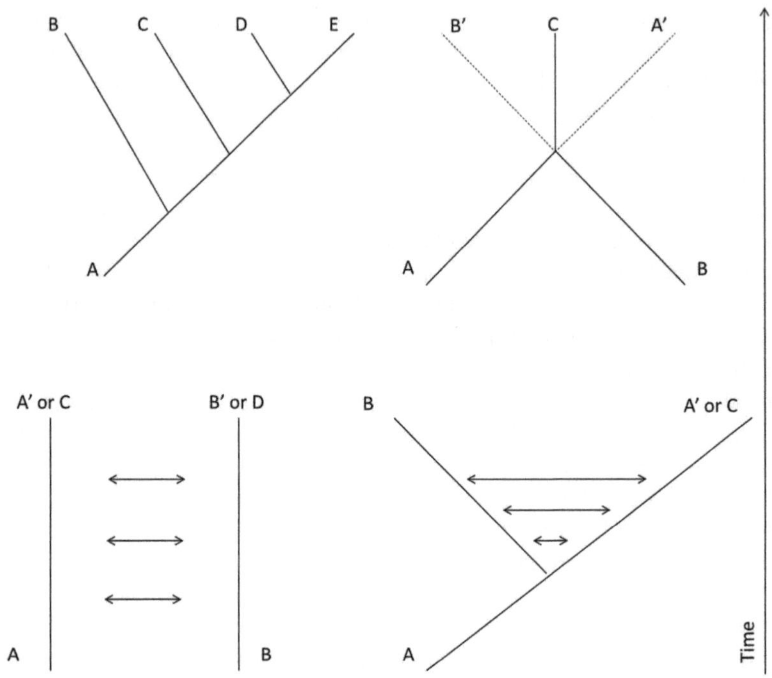

Fig. 4 Different modes of speciation, *Top left* Speciation by natural selection and drift: Species *B*, *C*, *D* evolve by splitting off from species *A* (cladenogenesis), while Species *A* gradually evolves into a new species *E* (anagenesis). *Top right* Speciation by hybridization: Members of species *A* and *B* cross and form a new species *C*, while species *A* and *B* either seize to exist due to the crossings or continue to evolve independently. *Bottom left* Speciation by symbiosis, symbiogenesis, lateral gene transfer, or hybridization: species *A* and *B* maintain symbiotic relations, acquire symbionts, or exchange genes horizontally, or they regularly hybridize, while they remain distinct species. Species A' and B' are nonetheless genetically, morphologically, or behaviorally altered by the various crossings, transfers and symbiotic associations in time, possibly up to the point that they evolved into new species (species *C* and *D*). *Bottom right* Divergence by gene flow or introgression: During its divergence from species *A*, species *B* either regularly backcrosses with its parental species (introgression), or exchanges genes laterally (directly or via symbiosis), thereby causing both species to diverge in time. This leads to the evolution of a new species *B*, and also the parental species is genetically altered (species A'), possibly up to the point that it evolved into a new species (species *C*)

Stebbins began his career by studying the *Crepis* genus, a genus of flowering plants popularly known as Hawks Beard that contains around 200 different species and that belong to the Cichoriaea tribe that also includes common lettuce, chicory, and other plants. Babcock and Stebbins (1938) discovered that many *Crepis* species regularly hybridize, that hybridization leads to polyploidy (chromosome doubling), and they pointed out that hybridization maximizes both variation and the potential to occupy diverse ecological niches. For Stebbins (1940), polyploidy in particular was important to understand the evolution of new plant genera. With

Stebbins, *Crespis* species soon became what *Drosophila* provides for geneticists, *Escherichia coli* for bacteriologists, and *Wolbachia* for scholars studying lateral gene transfer and symbiosis: a model organism.

By invitation of Ernst Mayr and Theodosius Dobzhansky, both major founders of the Modern Synthesis, Stebbins (1959) combined hybridization with natural selection theory and theoretical population genetics, systematics, and taxonomy. With the 1959 book, he launched the new field of evolutionary plant biology and he dedicated full chapters to hybridization and polyploidy which he understood to be targets of natural selection.

Anderson (1949), who coined the term introgression, pointed out the creative role hybridization can play because hybrids may backcross with their parental species, thereby increasing genetic diversity, adaptation, and fitness of both populations. Anderson was also a member of the Society for Evolution that gave way to the foundation of the Modern Synthesis (Gontier forthcoming; Smocovitis 1997). Together with Stebbins, he emphasized that hybridization plays a significant role in evolution because hybridization introduces new variation and enables a wider occupancy of ecological space (Anderson and Stebbins 1954).

Stebbins and Anderson's ideas on hybridization as adaptive for individual organisms and long-term beneficial for species are today proven by numerous scholars (e.g., Arnold 2004, 2006; Harrison 1990; Mallet 2005, 2007; Riesenberg 1995, 2001), who furthermore add that hybridization facilitates speciation and extinction, as well as provides a means to enter the genome of foreign species (Mallet 2005, 2007).

Because plant hybridization and introgression is well-documented and well-recognized to occur, in this volume, **Arnold et al.** focus on animal hybridization and introgression, or as the authors prefer to call it, "divergence with gene flow", in mammalian lineages.

1.5 Infectious Heredity in Health, Disease and Evolution

Many diseases are caused by the body's own (mutated) genes (e.g., following radiation), or by the malfunctioning of the individual's own metabolism and auto-immune system (e.g., systemic, auto-immune deficiencies), but the majority of diseases are caused by infectious agents that an organism haphazardly acquires during its lifetime. Infections can cause abnormal growth associated with diseases such as cancer, or benign but nonetheless obstructive tumor formation.

All three domains of life are prone to viral infections, or "viral colonization" as Villarreal calls it (Villareal and Defilipps 2000; Villarreal and Witzany 2010). There are around 50 known double-stranded, and two single-stranded DNA viruses that infect Archaea (Pietilä et al. 2014), bacteria are vulnerable to infections by bacteriophages (bacterial viruses), and eukaryotes can become infected by numerous DNA and RNA viruses as well as bacteria, fungi, worms, and small protozoan organisms.

In multicellular organisms, parasitic bacteria such as *pneumococci*, for example, enter their eukaryotic host and start multiplying inside the organism. They can block vital airways such as the lungs which can lead to respiratory problems; or they can start competing with the body's own cells for resources, thereby inducing cell mortality in their host.

To enable the formation of new viruses, viruses make use of the host metabolism and upon release, they kill the host cell. Many viruses can also copy their genetic material into the genome of the host. *Endogenous retroviruses* or *ERVs* (Gifford and Tristem 2003; Löwer et al. 1996; Ryan 2009) are viruses that upon infection can horizontally insert their genetic material into their host genome. ERVs resemble retrotransposable elements (Nelson and Hooley 2004), and they are often classified as a subtype of the latter. ERVs make use of the genetic apparatus of the somatic cells, but they can also integrate in the genomes of the sex cells and nestle inside the germ line. Once they become part of the germ line, the genes become the subject of vertical transmission where they are passed on to future generations in a Mendelian fashion. It is now well established that the genomes of mammals contain bits and pieces of these viruses in regions that were previously designated as "junk DNA." The genomes that acquire retroviral genes, however, not merely serve as containers for the latter. On the contrary, the acquired retroviral genes often play crucial functional roles in developmental pathways. It has been proven that endogenous retroviruses played a significant role in the formation of the female placenta (Knerr et al. 2002; Sugimoto and Schust 2009). Evidence furthermore suggests that our human ancestors caught endogenous retroviruses from Neanderthals (Marchi et al. 2013). At least theoretically, it is likely that Neanderthals reciprocally caught some of our infectious diseases, which might have eventually contributed to their decline.

Our specific human history is also filled with pandemics such as the plague, cholera, tuberculosis, Ebola, SARS, HIV, and child diseases such as the measles or rubella. These diseases often spread nation- and worldwide. Travel induced by war, colonization, or commerce enables the spread via various modes of human contact and as such these epidemics and pandemics can influence human life history as well as human evolution (Gontier 2006, 2007). In this regard, Ryan (2005, 2006; 2009) has introduced the term *plague culling*. When infectious diseases populate biological groups, species, or higher taxa, or when they make their way into the germ line, then over evolutionary time, they can introduce new features, cause bottle necks, or induce speciation events, and as such play a creative role in evolution.

Research on neurodegenerative diseases has led to the identification of prions by Prusiner (1982, 1991). Prions are infectious proteins that underlie mammalian neurodegenerative diseases such as Creutzfeldt-Jakob disease in humans and Bovine Spongiform Encephalopathy (BSE) or mad cow disease in bovines; as well Kuru in humans, and Scrapie in sheep. Creutzfeldt-Jacob disease and BSE are also related in etiology and caused by similar prions. Prions are proteins that undergo post-translational, epigenetic changes in their three-dimensional folding structure. Thus, after the genetic code is transcribed and translated into proteins, the protein that makes up the prion undergoes further, non-genetically encoded alterations in form. What exactly causes the proteins to change form is still uncertain, but *Spiroplasma* bacteria have been implicated (Bastian et al. 2007). Once the proteins flip into prions and take on the altered morphological form, these prions can bind to the regular proteins and make them change form as well. The prion-induced disease is able to spread across the brain and causes neurodegenerative, spongiform diseases where the brain starts to shrink in size and morphologically starts to resemble a sponge.

Prions cannot only become spread intraspecifically, they can also spread interspecifically by horizontal transmission. When humans, for example, eat with BSE-infected cow meat, it can induce the development of Creutzfeldt-Jacob disease which is exactly what happened in the early 2000s, in the UK and other European countries (see, e.g., the European Parliament and Council Regulation (EC) No 999/2001 on the "TSE-regulation" or the laws and decrees enacted against the spread of Transmissible Spongiform Encephalopathies at http://ec.europa.eu/food/food/biosafety/tse_bse/legisl_en.htm).

In sum, viruses, bacteriophages, bacteria, fungi, worms, protozoa, and also prions (Fig. 5) can function as pathogens or infectious agents. They are horizontally acquired, and they can become intra- and interspecifically transmitted in both vertical and horizontal fashion, via the germ line, or via the blood, milk, mucus, or other bodily fluids; they are ingested via food resources; or caught via inhaling infected air. Infectious agents can also become horizontally transmitted via vectors, i.e., symbiotic organisms that themselves carry microorganisms which are transmitted from the symbiont to the host.

All infectious agents are also symbionts, and not all infectious agents are pathogens that cause disease. Our gastro-intestinal tract, for example, provides an

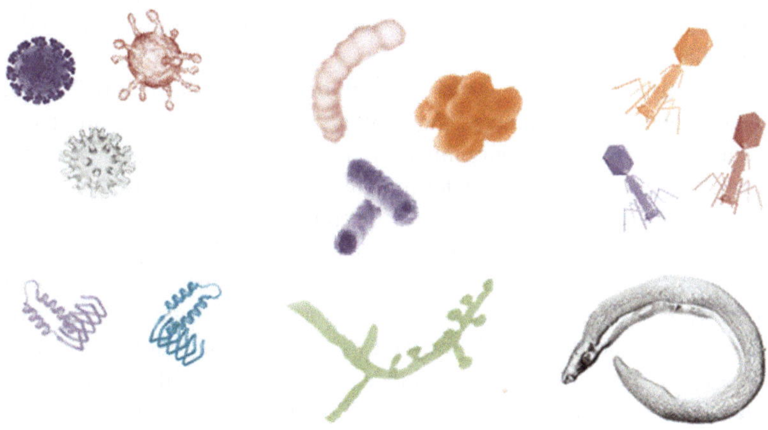

Fig. 5 Examples of infectious agents (not to scale). From *left to right* and *top to bottom:* viruses, bacteria, bacteriophages, prions (infectious proteins), fungi, worms. The prions are based on https://microbewiki.kenyon.edu/index.php/File:R7_prion.jpg

oxygen-low environment, and is therefore a suitable niche for the anaerobe organisms that first evolved under a reduced atmosphere. Over millions of years, these anaerobes have found shelter in multicellular organisms, and in return, the symbionts often provide the host with traits and biochemical substances that the host can neither produce nor establish on its own. Anaerobe gut flora is known to contribute to digestion of certain food substances, they help build the colon walls, and they often protect their host against infections with less beneficial microbes (Backhed et al. 2005; Turnbaugh et al. 2007, 2009; Ley et al. 2006). Current studies are even pointing towards the various compositions of microbiomes to explain body weight, sexual attraction, stress responses, temperament, and personality (Foster and McVey Neufeld 2013; Ley 2010; Bravo et al. 2012; Venu et al. 2014).

Multicellular, eukaryotic organisms have evolved complex anatomical forms and their various bodily organs and systems are populated by numerous microorganisms with which the eukaryotic hosts entertain symbiotic relationships. Scholars are increasingly demonstrating that besides parasitic symbiotic associations, also mutual and commensal associations between infectious agents and their hosts contribute to acquiring and maintaining normal development and overall health. Scholars are currently engaged in mapping the various *microbiota*, i.e., protozoan, microbial, and viral communities, that symbiotically live inside or onside eukaryotic organisms. These endeavors are known as *microbiome* and *viriome* projects and include the *Human Microbiome Project* that was launched by the American National Institute of Health (The NIH HMP Working Group 2009; the official website of the Human Microbiome Project can be found at http://www.hmpdacc.org/).

The complex symbiotic associations with the microbiomes, viriomes, and other microbiota furthermore need to be understood in terms of coevolution and lateral gene transfer (Dunning Hotopp et al. 2007). The host often provides environmental and ecological conditions suitable for microbial or viral growth, and (parasitic) symbionts can exchange genes laterally with their host, leading to altered genetic codes and altered metabolism.

Lederberg first coined the terms "microbiota" and "microbiome" in the early twenty-first century, to delineate "the ecological community of commensal, symbiotic, and pathogenic microorganisms that literally share our body space and have been all but ignored as determinants of health and disease" (Lederberg and McCay 2001). Earlier in time, and to emphasize the association there exists between infectious agents and symbiosis, he popularized concepts such as "hereditary symbiosis" (Lederberg 1952), and "infective transmission" (Lederberg 1998: 1) to delineate "DNA-mediated transformation, or virus-mediated transduction" (Lederberg and Lederberg 1956), as well as "infective heredity"—a concept first used by one of his collaborators, Zinder (1953), to describe Lederberg's work on bacterial transduction and conjugation. Writing in a time before "lateral gene transfer" as a notion was coined, he associated both infectious agents as well as the various means of prokaryotic horizontal gene exchange with symbiosis theory. For Lederberg (2003: 287), "We should think of each host and its parasites as a superorganism with the respective genomes yoked into a chimera of sorts." This introduces a "sociological development," or, in other words, a coevolutionary and social epistemic dimension to research on chimeric organisms.

In short, infectious heredity deals broadly with the horizontal acquisition of infectious agents, as well as the impact infectious agents have on health and disease of their host, both ontogenetically and phylogenetically. Classically, infection is understood as an ontogenetically acquired trait. Because the founders of the Modern Synthesis adhered to the Weismann barrier that rejected any type of Lamarckian inheritance and evolution, ontogeny was not considered to influence phylogeny. Research today on the contrary demonstrates that ontogenetically acquired infectious agents can most certainly influence phylogeny.

Infectious heredity therefore links the biomedical sciences with the evolutionary and ecological sciences. The acknowledgment that disease and health is induced by microorganisms was first put forward in the bacteriological and biomedical sciences when scholars like Joseph Lister in Great Britain, Louis Pasteur in France, and Robert Koch in Germany advanced the *germ theory of disease* (for reviews, see Sapp 1994, 2003, and for a timeline, see Campbell's 2007–2015 Germ Theory Calendar at http://germtheorycalendar.com/). Disease, in turn, became correlated to research on immunity and medicinal therapies by British scholars such as Edward Jenner and Alexander Fleming, the Russian zoologist Ilya Mechnikov, and the German scholars Paul Ehrlich and Emil von Behring (for reviews, see Gaudillière and Löwly 2001).

As early as 1949, J.B.S. Haldane, one of the population geneticists, linked the advances in bacteriology, microbiology, and the overall biomedical sciences, with evolutionary theory in an article on "Disease and Evolution" wherein he discussed various infectious diseases as agents of natural selection (Lederberg 1999). Diseases like malaria, for example, which is caused by parasitic protozoans, alter the successful survival rates of infected individuals, and certain hemoglobin disorders in turn protect against malaria.

For Haldane (1949), infectious agents can be understood as a medium through which natural selection becomes expressed. But as said, the relation between organismal diseases and the infectious agents that cause them additionally needs to be understood in terms of coevolution, symbiosis, and lateral gene transfer. Many bacteria and viruses "know" how to infect organisms, and many organisms "know" how to fight against or collaborate with the infectious agents. In other words, over the course of evolutionary history, infectious agents and their hosts have coevolved behavioral and biochemical repertoires to recognize and respond to one another. Many possess the biochemical "keys" of our bodies' "locks," and our bodies have evolved intricate immune responses that enable the identification, limitation, and even eradication of unwanted foreign agents, as well as means to recognize and use beneficial traits provided by these foreign bodies. Reticulate evolution is therefore pivotal in understanding the epidemiology of infectious disease as well as immunity.

In this volume, examples of infectious heredity, microbiome and viriome studies are discussed by **Souto Maior, Weyrich, Zook, Arnold,** and **Gontier.**

2 Introduction to the Chapters

Authors in this volume provide a state of the art on current data and theory. They exemplify the mechanisms and processes by reviewing case studies of reticulate evolution as they occur in various ranks of life; by delineating the historical context of discovery wherein reticulate evolutionary mechanisms were first recognized to occur; and by explaining how reticulate evolution challenges some of the classic tenets of the standard evolutionary, neo-Darwinian paradigm.

Douglas Zook, the decade-long former president of the International Symbiosis Society (http://iss-symbiosis.org/) and inheritor Lynn Margulis' first course on symbiosis that she developed at Boston University, provides a current state of the art of symbiosis research. He provides a new definition of symbiosis as "the acquisition of an organism(s) by another unlike organism(s), and through subsequent long-term integration, new structures, and metabolism(s) emerge."

In his chapter "Symbiosis: Evolution's Co-Author," the author takes on an overall ecological approach and details how the biosphere (the global ecosystem that encompasses the habitable zones of life) is the outgrowth of intimate symbiotic interactions between living organisms and the abiotic environment. Organisms have from the very beginning and continuing over billions of years played crucial roles in the evolution of the earth's atmosphere and its biomes via processes of biomineralization, lithification (the formation of rocks), and by aiding and sustaining crucial biochemical cycles such as the nitrogen, oxygen, and carbon cycle.

From stromatolites onwards, Zook reconstructs life's early origins and explains how primary, secondary, and tertiary symbiosis events have molded early eukaryotic life. The author explains why concepts such as "mutualism," "parasitism," and "commensalism" are outdated. The concepts imply a "compartmentalization," while Zook understands symbiosis as evolution's coauthor. Both symbiosis and natural selection are the primary mechanisms whereby life's immense biodiversity evolves, and he explains why symbiosis and natural selection are not mutually exclusive concepts.

The author adopts Margulis' "holobiont" concept that designates the new entity that forms as a result of a symbiotic association as a new unit of evolution. The holobiont concept also plays a crucial role in the Rosenberg's Hologenome theory that explains how holobionts are new units of *selection*, and Zook details how holobiont selection results in rapid adaptation and increased fitness.

Zook ends his chapter with extracts from an unreleased video interview he conducted with Lynn Margulis on how she understood the relation between the standard neo-Darwinian paradigm and the evolutionary symbiogenetic view of life.

In his chapter Can We Understand Evolution Without Symbiogenesis?, **Francisco Carrapiço**, one of the former secretaries of the International Symbiosis Society, understands symbiogenesis as an evolutionary mechanism crucial for understanding biodiversity as well as speciation events.

Carrapiço shares his truly encyclopedic knowledge on the rich history that precedes symbiosis and symbiogenesis research, and reviews when concepts such as "consortia," "commensalism," "parasitism," "mutualism," "symbiosis," and "symbiogenesis" were first introduced in time as well as how they got redefined over the ages. He lines up numerous pioneering scholars, including Simon Schwendener, Heinrich Anton de Bary, Pierre-Joseph Van Beneden, Albert Bernhard Frank, Andreas Schimper, Constantin Merezhkowsky, Andrey Famintsyn, Hermann Reinheimer, Paul Portier, Ivan Wallin, Boris Kozo-Polyansky, Lynn Margulis and many many others.

Carrapiço systematically demonstrates the difficult epistemic relations there have been between symbiologists, Darwinians, and neo-Darwinians, causing symbiosis research to have developed parallel and mostly outside the standard evolutionary paradigm. The malreception of symbiosis theory by neo-Darwinian scholars is explained as resulting from different notions both paradigms entertain on the nature of the organism, species-specificity, cooperation and interaction, and the overall role ecology plays in understanding the evolution of life. The author tracks the rise of these ideas and situates them in opposing sociopolitical ideologies of the nineteenth century.

Carrapiço ends by providing guidelines on how the evolutionary paradigm can be re-conceptualized to include the important results brought forth by research on symbiosis and symbiogenesis, and the author avers for a fuller and richer understanding of the evolution of life. Symbiosis leads to "*synergies*" and enables the evolution of "*consortia*" that can be characterized as "symbiogenic *superorganisms,*" which the author defines as "new entities or consortia formed by the integration of individual organisms, that possess characteristics that go beyond the sum of the individual properties of each element of the association, resulting in the development of new attributes and capacities as an integrated whole." He exemplifies the concept by reviewing his own work on *Azolla*, an aquatic fern that entertains symbiotic relations with the microorganisms that inhabit its leaf cavities.

(Endo)symbiosis is not a phenomenon confined to the evolution of organismal cell types associated with the four eukaryotic kingdoms. Rather, symbiologists agree that symbiosis continues to impact speciation. Speciation events, however, are rarely witnessed in nature, and neo-Darwinian scholars or symbiologists alike therefore have to combine observational knowledge with theory to explain how either natural selection or symbiosis, or both, can enable speciation. Symbiologists are rapidly catching up in providing new species concepts as well as theoretical scenarios on how symbiosis can lead to speciation.

The evolutionary-developmental biologists **Vitor Faria and Élio Sucena** detail how endosymbiosis influences and facilitates speciation of both hosts and symbionts. They explain how in particular intracellular coevolution between facultative endobacteria and their insect hosts can contribute to rapid phenotypic change and speciation of the host's progeny. In their chapter Novel Endosymbioses as a Catalyst of Fast Speciation, the authors provide a five-step scenario for the appearance of novel host lineages. Facultative bacterial endosymbionts of eukaryotic organisms are not only transmitted horizontally, they are often transmitted vertically among members of the host species. As such, they become a defining feature of the host lineage's phenotype, and they impact the fitness of their host.

A prototypical example is *Wolbachia,* a genus of bacteria that entertains parasitic and mutual symbiotic associations with many insect species. *Wolbachia* is often vertically transmitted via the female eggs and impacts the reproductive success of both the males and females of the host insect, by disabling or enabling sexual compatibility. Faria and Sucena detail how endosymbionts like *Wolbachia* either impact "directional" or "disruptive selection" of their host among its conspecifics. In disruptive selection, the host's possibility to mate with its conspecifics is reduced by the presence of the endosymbiont; in directional selection, the host's ability to mate with conspecifics is increased by the presence of the endosymbiont. In both cases, the host's symbionts introduce barriers that facilitate rapid speciation by symbiosis.

Besides *Wolbachia,* Faria and Sucena exemplify their proposal for how speciation-by-endosymbiosis possibly occurs with numerous real-life case studies on coevolution between host and endosymbiont, fitness impacts of facultative endosymbionts, horizontal and vertical transmission of symbionts between hosts, and endosymbiont-induced phenotypic and genotypic novelties.

In the chapter on Historical and Epistemological Perspectives on What Lateral Gene Transfer Mechanisms Contribute to our Understanding of Evolution, **Nathalie Gontier** first reviews how lateral gene transfer has been brought to the attention of the larger academic community by results coming in from molecular phylogenetics. In the beginning of the 1990s, species-genome sequencing techniques as well as ribosomal RNA comparisons of various taxa led to the introduction of Carl Woese' three-domain classification of life. Such research also made it obvious that lateral gene transfer and symbioses occur abundantly, and scholars such as Gogarten (2000), Doolittle (2000) and Bapteste et al. (2005), among others, subsequently started to question the standard neo-Darwinian tree of life iconographies. The scholars introduced new metaphors, such as the "web of life" and "net of life" metaphor, which in turn upset neo-Darwinians such as Richard Dawkins, Jerry Coyne, and Daniel Dennett. Polemic debates followed in various journals and media. Gontier reviews these polemics and places them in historical and epistemological context.

In the second part of her chapter, she reviews the basic mechanisms according to which horizontal gene transfer occurs in both pro- and eukaryotes. Gontier traces the identification of bacterial transformation, transduction, and bacterial conjugation to pre-synthetic times where discoveries made by Frederick Griffith, Oswald Avery, Colin MacLeod, Maclyn McCarty, Norton Zinder, Joshua Lederberg, Edward Tatum, Barbara McClintock, François Jacob, and many others improved knowledge on bacteriology, immunology, and disease. She investigates why these phenomena were long considered biomedical peculiarities rather than genuine evolutionary mechanisms relevant to understanding the evolution of life.

That it is beyond reasonable doubt that horizontal gene transfer occurs abundantly, but many of the mechanisms by which genes are transferred between eukaryotic species remain obscure. It is becoming increasingly obvious though that symbioses, symbiogenesis, and hybridization act as facilitators of lateral gene transfer, and Gontier investigates how scholars today are trying to identify recurring patterns and mechanisms. Along the way, she identifies where and how the incoming results conflict with specific tenets put forward by the founders of the Modern Synthesis.

In the chapter Plasmids: Histories of a Concept, the historian of science and molecular biologist **William C. Summers** provides the context of discovery of plasmids and reviews how definitions of plasmids and associated concepts such a episomes have changed over the last decennia.

It has taken biologists some time to determine what the exact nature of hereditary particles is, and where such hereditary material is stored inside the cell. Summers details how the first observations of mitosis and meiosis led to the formulation of the chromosome theory and the gene theory, and how both became combined, making scholars assume that the "Mendelian factors" or "genes" are located on chromosomes and transmitted vertically from parents to offspring. Nonetheless, cytologists also observed the cytoplasmic (lateral) transfer of non-chromosomal biochemical substances, "plasmagenes," which made them introduce theories on cytoplasmic inheritance. One such cytoplasmic biochemical substance that can be transferred laterally is the bacterial plasmid, which today we know is made up of circular DNA. Bacterial conjugation involves the lateral transfer of a plasmid-strand from a donor to a recipient bacterium.

Summers details how work on *E. coli* bacteria by Joshua and Esther Lederberg, Edward Tatum, Luigi Luca Cavalli-Sforza, William Hayes, and Allan Campbell led to the identification of plasmids and their lateral transmission by bacterial conjugation. The plasmid concept was first introduced by Joshua Lederberg in 1952, a year before the unravelling of the double helix, to designate "extrachromosomal hereditary particles." Lederberg's paper carries the title "Cell genetics and hereditary symbiosis," and he understood bacterial conjugation as one type of hereditary symbiosis.

Summers reviews how studies on plasmid transfer gave way to the discovery of the fertility factor (F-factor) necessary to induce bacterial conjugation, and how it became clear that genetic exchange can also occur between plasmid DNA and chromosomal DNA (e.g., in Hfr strains where the F-factor becomes part of the genome of *E. coli*).

In 1958, François Jacob and Elie Wollman introduced the episome concept, to identify "genetic elements which were optionally associated with the chromosomes of the cell" and Summers describes how, because of advanced knowledge into the biochemical nature of plasmids and episomes, the plasmid concept was favored over the episome concept.

Bacteria, and to a lesser extend Archaea, house many plasmids that are non-transferrable. Some of these can nonetheless become mobilized by other conjugative plasmids that reside inside the host cell, but around 48 % of proteobacterial plasmids are neither conjugative, nor mobilizable, and thus always non-transferrable. These non-transferable plasmids contain genes that are not essential for the bacterial host, and the bacteria sometimes lose these plasmids over time or the plasmids undergo considerable gene loss. Many are nonetheless able to maintain their position. This poses an interesting scientific riddle: Are plasmids "selfish genes" that entail a fitness cost for their bacterial hosts and if so, does there exist selection against the presence of these non-transferrable plasmids towards plasmid-free cells; or is there instead selection towards the maintenance of a symbiotic relationship between the host and the plasmid?

In their chapter Symbiosis Between Non-Transferable Plasmids and Prokaryotic Cells, **Francisco Dionisio, João Alves Gama, and André F.P. Carvalho** detail how the prokaryotic organisms entertains a symbiotic relationship with the non-conjugative plasmids and how there can be selection for the maintenance of such relationship. By examining the selective mechanisms that underlie stable symbiotic associations between the host and the non-transferrable plasmid, the biologists provide a new means to understand the intricate interaction between symbiosis and natural selection.

From the chapter, we learn that there are more connections to be drawn between how neo-Darwinists and symbiologists approach their research subject. Dionisio, Alves Gama, and Carvalho apply sociobiology, especially "the public goods theory" to bacteria. Metaphorically speaking, a bacterium harboring plasmids can be considered a public entity or public space where different plasmid individuals (the goods of the public entity) compete over resources (the goods of the host) as well as the occupation of that space provided by the host. Most importantly, that public entity itself also sets rules on who can inhabit the niche and how the space is occupied. The reader is introduced to several trade-off scenarios and cost-benefit equations that help conceptualize the symbiotic association between plasmids and their host at the micro-organismal level.

The disease ecologist **Caetano Souto-Maior** explains how both symbiosis and lateral gene transfer provide innovative ways in which we can understand (1) host–symbiont relations, (2) symbiont–pathogen relations, and (3) pathogen–host relations as crucial for the transmission of infective disease. Many diseases are transmitted by vectors, i.e., symbiotic organisms that harbor pathogens which in turn infect the host of the symbiont. When these vectors endure long-lasting symbiotic associations with a host, both the vector (the symbiont) as well as the pathogens (residing inside the symbionts and affecting the host) can become vertically transmitted in the host lineage.

In his chapter, Host-Symbiont-Pathogen-Host Interactions: Wolbachia, Vector-Transmitted Human Pathogens, and the Importance of Quantitative Models of Multipartite Coevolution, the author highlights several case studies. Different species of mosquitos and worms that parasitize humans often themselves carry various bacterial strains such as *Wolbachia*, or viruses such as the dengue virus, and both can cause disease in humans. *Wolbachia* infections have been implicated in various human diseases, including river blindness (van den Hurk et al. 2012) and elephantiasis (a lymphatic disease characterized by swellings of the lower limbs). The dengue virus causes dengue fever, a tropical blood disease that induces rashes, gastro-enteritis, muscle and joint pains, and potentially lethal fevers as well as potentially lethal hemorrhagic shock (the uncontrollable release of blood from the veins leading to severe bleedings).

Wolbachia provide their host mosquito with protection against infection with the dengue virus, but *Wolbachia* also harm mosquitos by reducing their fitness and intervening in the sexual maturation of the female mosquito eggs (Kozek and Ramakrishna 2007; Hurst et al. 1999). With *Wolbachia*-infected mosquitos reduce the chance that the dengue virus infects mosquitos and that they in turn infect humans with the dengue disease via mosquito bites.

Experimental projects have been introduced whereby scholars intentionally infect the mosquitos prone to dengue infection with *Wolbachia* strains that are harmless to humans (see e.g., http://www.eliminatedengue.com). These mosquitos have subsequently been released in nature, with the hope to eradicate dengue fever infections in humans. Similar experiments have also been conducted with the hope to reduce the spread of yellow fever and the chikungunya virus (van den Hurk et al. 2012), as well as the West Nile virus (Hussain et al. 2013), which are all transmitted by mosquitos. Such genetic engineering can help eliminate infective disease.

Souto Maior furthermore details several cases of horizontal gene transfer between the *Wolbachia* genome and the host's nuclear genome. To understand these intricate and complex horizontal interactions between hosts, symbionts, and pathogens, the author illustrates how important it is to develop tri- and multipartite population dynamics. Evolutionary models, for the author, should include ecological, immunological, and epidemiological accounts on the interactions hosts, symbionts, and pathogens entertain. He furthermore emphasizes that most infections occur stochastically, and drift, more than natural selection theory should underlie population genetics and symbiology, as well as the epidemiology or spread of disease.

In the chapter Evolution of the Human Microbiome and Impacts on Human Health, Infectious Disease, and Hominid Evolution, the anthropologist **Laura Weyrich** exemplifies studies on the evolution of the human microbiome. Ancient feces (coprolite) provide insight into the evolution of the human gut microbiome, and calcified dental plaque gives knowledge on the various microorganisms that have populated the oral cavity. Weyrich demonstrates an intricate coevolution between lifestyle, microbiome, health, and disease.

She starts her chapter by comparing the human microbiome with the microbiome of our closest living relatives, the chimpanzees and bonobos, in order to reconstruct the microbiome of our last common ancestors. She subsequently compares incoming data on the microbiomes of Western urbanized, and Indigenous populations. Her overall conclusion is that the microbiome is ecologically determined: when populations share the same environment and thus the same food resources, they share the same microbiome.

The author then turns to more ancient human lineages. Using next-generation sequencing techniques (especially meta-barcoding), Weyrich, together with her colleagues at the Australian Centre for Ancient DNA, was able to identify the changes in the oral human microbiome over the past 8000 years. They found that especially the Neolithic Revolution (the onset of agriculture some 7500 years ago) and the Industrial Revolution (which occurred around 200 years ago) severely changed the human oral microbiome, mostly in negative ways.

The introduction of agriculture marks a transition from a hunter-gatherer lifestyle to a more sedentary lifestyle characterized by the domestication and cultivation of crops. While the introduction of agriculture is often characterized as a "great leap forward," Weyrich demonstrates that hunter gatherers fared much better healthwise than the early agriculturalists did. The ancient biofilms even enable Weyrich to infer when bacterial pathogens entered the human microbiome in time, and thus to infer when certain diseases started to plague humankind; and she can backtrack the coevolutionary process the microbiome has undergone with the human immune system.

The Industrial Revolution was characterized by the invention of the machine which in turn enabled the production of manufactured foods as well as the preservation of food products by pasteurization, sterilization, or canning. Polluted air from factories and metal poisoning are some of the negative consequences, while on the other hand, the industrialization also marks an end to famine that characterized Western societies for centuries. These events are also evidenced in shifts in the composition of the human microbiome, and the data show an intricate, commensalist coevolution between human hosts and microbial communities.

In the final parts of her chapter, Weyrich reviews how the hybridization that took place between early *Homo sapiens* species and Neanderthals and Denisovans (a sister taxa of Neanderthals), have impacted the evolution of the human microbiome, and how the microbiomes of the various species has in turn contributed to successful hybridization. In short, the human microbiome contributes to physical health, infectious disease, successful adaptation, hybridization, and possibly also extinction and speciation.

Michael Arnold, Amanda Brothers, Jennafer Hamlin, Sunni Taylor, and Noland Martin also take us to the Animal Kingdom and write on Divergence-With-Gene-Flow—What Humans and Other Mammals Got up to. The authors define divergence with gene flow as "evolution of diverging populations with some amount of continued genetic exchange between them," and understand the concept as an alternative to the notion of hybridization that historically invokes negative connotations and assumptions on hybrid sterility or assumptions that hybrid genotypes induce a genetic burden on their carrier. The authors demonstrate that these assumptions are untenable. "Divergence with gene flow" furthermore enables the inclusion of incoming research on symbiosis and lateral gene transfer.

Arnold and coauthors prove that divergence with gene flow occurs abundantly in animal life. To make their case, the authors have chosen to baffle us with numerous case studies and scientific evidence of divergence with gene flow as it has been reported in scientific works since 2008. They in particular focus on the mammalian lineage and include data on our own species, *Homo sapiens*.

The authors give an impressive lineup that starts with the cooptation of retroviral DNA in early mammalian lineages and the role these viruses play in the formation of the placenta. Making their way through the mammalian tree, they illustrate divergence with gene flow in marsupials, mice, rats, chipmunks, hares and rabbits, shrews, minks, pole cats, polar and brown bears, panthers, wild cats, boars and domesticated pigs, wildebeest, chamois, deer species, marine mammals, horses, and bats. When turning to the primates, Arnold and coauthors note that the "clade in general is a rich source of examples of reticulate evolution." From Lemurs to Old World monkeys, numerous proofs exists for inter-taxa mating. Within our own *Homo* lineage, several subtaxa have mixed: there was admixture between *Homo sapiens* and *H. neanderthalensis*, and also various human sub-populations have introgressed with more archaic species, thereby incorporating Denisovan genes, as well as currently unidentified Melanesian, African, and European lineages of archaic *H. sapiens*. The authors emphasize that "these data falsify the hypothesis of simple replacement of archaic forms by our species and instead favor a scenario of mutual attraction and genetic exchange leading to a human genome that is a mosaic of recent and ancient DNA sequences."

Their case studies demonstrate that divergence with gene flow occurs abundantly and rapidly. Repeated divergence with gene flow does not, as a rule, lead to sterility. Most of the time, rather than pose a genetic burden on the mixing species, divergence with gene flow increases successful survival as well as speciation, it occurs more than sympatric or parapatric speciation, and divergence with gene flow contributes to biodiversity.

The authors conclude that reticulate evolution does not confine itself to lateral gene transfer between prokaryotes and hybridization between plants, it also occurs abundantly in animals, by both divergence with gene flow as well as lateral gene transfer. Along the way, the authors also introduce the reader to new scientific jargon as well as a series of innovative techniques and methodologies by which scholars can, beyond any reasonable doubt, make the case for understanding genetic exchange not as linear but reticulate and "web-like."

Evolutionary biology has greatly advanced by adopting bioinformatics and overall computational approaches that help test evolutionary hypotheses as well as model evolutionary scenarios. In their chapter A Multiset Model of Multi-Species Evolution to Solve Big Deceptive Problems, **Luís Correia and António Manso** demonstrate how reticulate evolution can be modelled artificially.

In previous work, the authors have developed a Multiset Genetic Algorithm (MuGA) that enables to model competitive multiple species evolution. Instead of depicting populations as a collection of individuals, in MuGA, the populations are represented as multisets (multi-populations), and the operators explore the multisets in order to optimize problems. Such multisets are not found in the natural world, but the models are interesting to examine engineering problems.

In this chapter, they present a variant of their model, SMuGA, which is a novel approach to artificial symbiosis. The model integrates symbiosis and lateral gene transfer with MuGA to model cooperative coevolutionary and symbiotic relations between hosts and parasites.

Their model is able to simulate symbiotic collaborations between a single host and multiple symbionts. More specifically, they model how a single host receives genetic material from multiple parasites with varying genome length, and they model the interaction between the multiple parasites and the host. They can investigate how artificial symbiogenetic evolution enables optimization of fitness calls, thereby accelerating optimization of deceptive problems.

The model has two phases: in the first phase, symbiotic interactions are generated and competition exists over the composition of the next generation of host population, and in the second phase, hosts and parasites first evolve independently, but the parasites compute their own fitness based upon the host's fitness which enables a computing of successful collaborations instead of actually generating them. Symbionts are thus enabled to "evaluate" and "explore" their host.

In general, it is hard to simulate real-life events because of the complexity involved. Symbiosis, symbiogenesis, and lateral gene transfer pose additional problems and challenges to be overcome by modelers, not in the least because of the numerous additional relations that need to be brought into the system. The authors present an innovative model as well as new techniques and methodologies, to model the complex interactions and integrations of symbionts and their genes into the host.

3 Reticulate Evolution, the Modern, and the Extended Synthesis

As this introduction makes clear, there are merely fine lines to be drawn between the various mechanisms whereby reticulate evolution can occur, and most of the time, the various mechanisms are simultaneously active within the same organisms.

Both symbiosis and symbiogenesis can impact the future course of evolution. Symbionts can become horizontally and vertically transmitted without inducing symbiogenesis. The major difference is that in symbiosis, the individuals maintain some form of individuality although both partners, and at a higher level also the populations they belong to, are affected by the symbiotic relation (which is the case with *Wolbachia* and their insect hosts, for example). Symbiogenesis occurs through a permanent form of hereditary and obligate symbiosis, whereby the partners start to become dependent upon one another, up to the point that they become a single new individual.

The easiest way to distinguish between lateral gene transfer and symbiosis or symbiogenesis is by following Margulis (1998) differentiation: lateral gene transfer is characterized by "gene fusions," while endosymbiosis is characterized by "cell fusions" or "body fusions." During horizontal gene transfer, the genes are not literally fused, but they are horizontally exchanged between distinctly evolved organisms, an exchange that leads to the insertion of foreign DNA into the recipient's genome. During symbiogenesis, not genes but whole cells or multicellular organismal bodies fuse, literally, one organism engulfs the other in its totality, and such a fusion leads to symbiogenesis.

According to this distinction, also any type of meiotic, eukaryotic sex is primarily based upon endosymbiosis, and such a characterization in turn makes the line between hybridization and symbiogenesis or symbiosis more fluent. Per definition, hybridization always requires a form of sex. But sexual contact can be understood as a form of symbiosis or symbiogenesis, where the sex cells and genes come together into a new and stable individual. An example is human sexual reproduction where the male and female temporarily engage in a facultative form of conjunctive symbiosis; and upon fertilization, the head of the sperm cell permanently enters the egg cell. The haploid chromosomes of both cells form diploid pairs, and the zygote starts to differentiate into the various structures that make up the newly formed multicellular organism. Or as Margulis (1998: 40–42) put it: "Sex, too, is the coming together, the merging of cells of different histories and abilities. In sex the cells that fuse are closely related and the fusion is reversible; in serial endosymbiosis the cells that fuse are only distantly related, and the fusion is permanent."

Hybridization, by necessity, only occurs in sexual and thus eukaryotic organisms, while symbiogenesis is not confined to eukaryotic life forms, it also occurs in asexual individuals. The same goes for lateral gene transfer. It crosses all domains of life, and it occurs by asexual means.

Infectious heredity blurs the divide between the living and the non-living. Prions and viruses are not considered to be living entities or basic units of life. Nonetheless, they evolve by means of reticulate evolution. They affect the evolution of life, and they might also be the outcome of reticulate mechanisms themselves. The origin of viruses or genomes in general imply a combination of various genes into a hypercyclic structure. Prions obtain their structure from interactions between proteins and possibly also certain bacteria. Infectious heredity occurs through all known media of reticulate evolution and was introduced as a separate form to emphasize the important role it plays in health and disease, which in turn impacts the future course of evolution.

It is important to note that until recently, the various means whereby reticulate evolution occurs were studied from within varied disciplines. Just as communication was lacking between the founders of the Modern Synthesis and scholars who studied reticulate evolution, communication was also lacking between the scientists who studied hybridization, symbiosis, symbiogenesis, lateral gene transfer, and infectious heredity.

Studies on hybridization and symbiosis first arose in *botany* and *zoology*. From the very onset, symbiosis research has developed in close contact with *ecological* research fields, where the symbiotic association was interpreted as a *behavioral* phenomenon displayed by different organisms that entertain various contact modes (commensalism, parasitism, or mutualism).

With the introduction of symbiogenesis as an evolutionary mechanism, Merezhkowsky introduced symbiogenesis into *evolutionary biology*. Merezhkowsky also linked symbiosis and symbiogenesis with the then-rising fields of *bacteriology* and research on the origin of life, *abiogenesis,* and *astrobiology*. But his work was by and large ignored.

Bacteriology and *virology* have from its very beginning been intricately related to the *biomedical sciences*, especially *immunology* and *epidemiology*. It was in this context that the modes of lateral gene transfer were first described. But until recently, the biomedical sciences did not engage in evolutionary studies because, rather than focusing on the past, they focused on the present (the ontogeny and etiology of disease) and the future (by finding cures that eradicate diseases).

Bacteriology and virology were the first fields that defined *microbiology* as a separate area of research. Microbiology also forms a bridge between *evolutionary biology*, *(an)organic chemistry*, and *abiogenesis*, because Archaea provide insight into the first life forms, as do viruses, that might have played a significant role in the (pre-)RNA world as well as the formation of the eukaryotic nucleus, in a symbiogenetic fashion. In fact, it was the study of bacterial transformation that first evidenced that genes are the seats of heredity, insights that contributed to the rise of *molecular genetics*.

Cytoplasmic biology has brought to light that extrachromosomal structures such as plasmids and organelles exist and that extrachromosomal heredity plays a significant role in the evolution of life.

Nonetheless, at the turn of the twentieth century, botany, zoology, ecology, ethology, bacteriology, virology, astrobiology, cytoplasmic biology, developmental biology, epigenetics, and the biomedical sciences, were distinct research areas with little interdisciplinary contact. Besides zoology and to a lesser extend botany, these epistemic fields evolved separately from overall evolutionary theory. Symbiosis and symbiogenesis, or ecology, epigenetics, and developmental biology find their historical beginnings in a period designated by Julian Huxley as the "eclipse of Darwin." Research on cytoplasmic heredity, the mechanisms of lateral gene transfer, and the impacts of infectious heredity date back to the beginnings of the twentieth century, but the disciplines matured their theoretical and evidential frameworks outside or in the margins of the standard neo-Darwinian paradigm.

It is only in recent years that recognition of their significant data became well-received and that scholars are developing inter- and transdisciplinary practices that enable them to cross field-specific boundaries. The main reason for this is that the *molecular phylogenetic reconstructions* of the tree of life, that were based exclusively on neo-Darwinian frameworks, have led to anomalies that can only be explained by accepting reticulate evolution as a fact of life. Molecular phylogenetics in turn combines *bioinformatics* and *computational evolutionary approaches*.

The current challenges we are faced with are (1) to combine these emerging reticulate theories into encompassing reticulate evolutionary paradigms and (2) to integrate reticulate evolutionary theories with the existing theories on natural selection and drift into a more encompassing evolutionary synthesis.

Today, reticulate evolutionary mechanisms themselves are becoming combined into unifying frameworks, and such unification in turn provides a means to *unify* zoological and botanical evolutionary biology with molecular genetics, cell biology, microbiology, virology, mycology, ecology, developmental biology and epigenetics, and the biomedical sciences. Reticulate evolution also provides new methodologies and theoretical frameworks to investigate and understand old evolutionary problems, it enables innovative means for *biochemical and genetic engineering*, and it opens up intriguing ways to personalize medicine.

Even the *sociocultural and linguistic sciences* are applying key concepts of reticulate evolution to understand complex behavioral and sociocultural phenomena, and in turn, reticulate scholars are beginning to integrate sociocultural studies to understand the behavioral and biochemical communication and interaction that underlies symbiosis, symbiogenesis, hybridization, lateral gene transfer, and infectious heredity.

Neo-Darwinian theory has made significant progress by understanding not only anatomical form, but also the behavior of animals as outcomes of natural selection. Beginning with sociobiology, scholars have been able to extend the evolutionary framework towards the sociocultural and behavioral sciences, by understanding differential phenotypic behavior as the outcome of social or cultural learning. Sociobiological and behavioral theories are today applied within bacteriology and microbiology. Microorganisms do not have a brain, but they nonetheless display differential phenotypic behavior that is relevant from an ecological point of view. Communication need not involve spoken or signed language, it can also be of a biochemical kind.

Reticulate evolution also raises fascinating questions on units and levels of selection as well as cooperation that extend the individual towards higher ranks such as the group, bacterial types, colonies, or species. This necessitates an ecological and overall *hierarchical* approach to evolution that enables scholars to conceptualize how individual and group behavior, higher and lower-level evolution, as well as higher- and lower-level interactions occur.

Turning to the second challenge, the neo-Darwinian synthesis combines Darwin's mechanism of natural selection with Mendelian hereditary laws, chromosome and gene theories, aspects of mutation theories, and insights from theoretical and experimental population genetics. This theoretical effort has brought forth a standard paradigm according to which we can understand vertical evolution: the *Modern Synthesis*. The Modern Synthesis has helped explain why the tree of life, and especially the evolution of eukaryotic animal and plant life, takes on a *vertical pattern of descent with modification*, a splitting pattern characterized by the bifurcation and ramification of evolutionary lineages.

When we compare insights on reticulate evolution with the standard neo-Darwinian text books, it reads very much as science fiction. Nonetheless, reticulate evolution has and continues to be a determining factor in the evolution of life. It brings forth a pattern of intricate mergings in the tree of life that takes on net and web-like shapes when we cartography the crossings.

Reticulate evolution and vertical evolution induced by mutation, drift, natural selection, and migration are often theorized to be complementary principles, where natural selection and drift are hypothesized to follow after symbiosis or symbiogenesis took place. Scholars such as Merezhkowsky, Wallin, Kozo-Polyanski, or Margulis understood symbiosis as the primary source of evolutionary novelty, and natural selection was a secondary principle that acted upon the novel variation introduced by symbiogenesis. For them, natural selection was (merely) a weeding-out mechanism.

Both neo-Darwinians and early symbiologists alike have also often opposed themselves in “either/or” debates and have understood vertical and reticulate mechanisms as mutually exclusive principles. In practice, however, life evolves according to numerous evolutionary mechanisms, and they simultaneously influence the organism and higher ranks of life at multiple levels. Eukaryotic organisms incorporate organelles that evolved by means of symbiogenesis, which was the result of an intricate symbiosis of the original merging individuals. These eukaryotic organisms also evolved according to selectionist principles that underlie the vast biodiversity that characterizes the tree of life. Nonetheless, the eukaryotic organisms can become infected by microbiota, and during ontogeny, numerous symbiotic associations are entertained by all living organisms.

The future therefore consists of finding out how these various evolutionary mechanisms simultaneously bring forth the evolution of life. At a meta-level, we therefore need to ask how these mechanisms interact, and whether or not there is a higher-order sorting of evolutionary mechanisms. Does evolution sometimes favor selection over symbiogenesis, hybridization over symbiosis, or infectious heredity over lateral gene transfer? Or is there sometimes selection for reticulate evolution, or does reticulate evolution induce selection? What would induce such higher-order sorting? Is it the nature of the organism, the type of group it belongs to, or the environments the various taxa inhabit? These are questions that need to be tackled by a future generation of researchers. At present, we do not know and we also lack the epistemic frameworks to adequately frame the questions.

We live in an age of fascinating new discoveries and data collection, similar to the exiting times the early naturalists lived through when they first started to detail the adaptive behaviors and anatomical traits of animal life.

Data on reticulate evolution is currently ahead of theory and understanding. Integrating reticulate evolution into the overall existing evolutionary framework will undo many of the assumptions the latter once made. How we define organisms, groups, species, genera, or higher taxa requires reconceptualization that takes the numerous interactions that exist between organisms into account. It requires a basic reformulation of notions such as behavior, communication, fitness, adaptation, speciation, and extinction. Reticulate evolution has identified new units of evolution (such as hybrids, mobile genetic elements, symbionts, and holobionts), as well as levels of evolution. The “environment” is both abiotic as well as biotic. A multicellular organism is itself an entire community, from the intragenetic and intracellular level all the way up to the outer layers that bound it.

Whether it is possible to synthesize reticulate with vertical evolution into a revised evolutionary synthesis remains an unanswered question. Some scholars plead for an integration and a revision of the synthesis, others deny the possibility and call out for a rupture with the Modern Synthesis. Only the future will tell and though this volume is (merely) of an introductory level, we do hope it will inspire scholars to engage in finding the answers to these fascinating questions.

Acknowlegements This work was written with the support of the Portuguese Fund for Scientific Research (grant ID SFRH/BPD/89195/2012 and project number UID/FIL/00678/2013) and the John Templeton Foundation (grant ID 36288).

Glossary

Aerobe organisms Organisms that require gaseous oxygen to metabolize. *Compare to anaerobes*

Algae (Aquatic) eukaryotic organisms that photosynthesize

Anaerobe organisms Organisms that are poisoned by gaseous oxygen and that live in oxygen-low or oxygen-free environments. *Compare to aerobes*

Archaea First domain of life, previously designated as Archaeabacteria in the kingdom of Monera

Axoneme Shaft of undulipodia

Bacteria Second domain of life, previously designated as Eubacteria in the kingdom of Monera

Bacteriophage Virus that infects bacteria

Centrioli Cylindrical cell organelle, found in pairs (together called the centrosome) in many eukaryotic organisms, built up from microtubules (tubulin protein structures) structured according to a [9(3) + 0] pattern. They help build the mitotic spindle that separates the chromosomes during division. *Compare to undulipodia and cilia*

Chloroplasts Organelles found in plant cells that have evolved by symbiogenesis from photosynthesizing cyanobacteria, currently enabling cells to photosynthesize

Cilium/Cilia Type of undulipodium that visually appears as hairs on the cell and functions as sensory organelles, often enabling motility. Their basal body has a [9(3) + 0] microtubular structure, and their shaft a [9(2) + 2] one. *Compare to undulipodia and centrioli*

Coevolution Process whereby distinct species reciprocally influence each other's future course of evolution

Computational evolution Field in computer science and artificial intelligence that develops computational models to investigate evolutionary problems

Cyanobacteria Chlorophyll pigment-containing and photosynthetic bacteria, previously known as blue-green algae, but algae are eukaryotes, while cyanobacteria are prokaryotes

Cytoplasm Cell liquid

Domains of life/3-domain classification According to Carl Woese, and based upon comparative molecular phylogenetics (in particular comparisons of sections of ribosomal RNA), life is classifiable into 3 major domains: Archaea, Bacteria and Eukaryota. This undoes the previous 5-kingdom classification

Eukaryota The third domain of life, consisting of protists, fungi, plants, and animals. Eukaryotes can be unicellular or multicellular organisms. Their distinctive feature is that their cells have nucleated genomes where the genes are packaged into separate chromosomes. Besides a nucleus, the cells of these organisms often also contain organelles, organ-like structures such as mitochondria and chloroplasts, peroxisomes, and Golgi that associate with specific metabolic functions

Flagellum/Flagella Bacterial motile extensions made up of flagellin protein

Fitness Reproductive success, measured by the number of offspring

Five-kingdom classification of life According to Whitaker and Margulis, and based upon the 3 symbiogenetic mergings proposed by the serial endosymbiotic theory, life can be classified into 5 kingdoms: prokaryotic Monera (that contain the Archaebacteria, and Eubacteria) and the eukaryotic Protoctist (alternatively known as Protists, Rothschild 1989), Fungi, Plant, and Animal kingdoms

Fungi An eukaryotic kingdom of life that evolved after archaea, bacteria, and protists, and distinct from animals and plants. They contain microorganisms such as yeast and molds, but also larger organisms such as mushrooms

Germ theory of disease Theories first introduced by scholars such as Pasteur and Koch that indentify microorganisms as causal agents of disease

Holobiont Term first introduced by Margulis and Fester (1991) to designate an organism and its symbiotically associating partners

Horizontal transmission Any type of exchange between distinct individuals that happens during their lifetime and outside of the germ line (in a non-Mendelian fashion)

Host The larger partner in a symbiotic association

Jumping genes Genes that can switch position in the genome they are part of, as well as travel to adjacent intracellular genomes (neighboring organelles for example), thereby causing deletions, insertions, and duplications in turn responsible for mutations, malfunctions, or the introduction of novel traits. Today known as transposons

Kinetosomes Basal body of undulipodia

Microbiome The complete ecological community of microorganisms that inhabit a species. *Compare to viriome*

Microtubules Polymers (strings) of tubulin proteins

Mitochondria Eukaryotic cell organelles that evolved from aerobe proteobacteria by symbiogenesis, functionally resembling power factories because they produce and store energy

Modern Synthesis The standard evolutionary paradigm that unites (aspects of) Darwinian selection theory with Mendelian hereditary laws, Boveri–Sutton's chromosome theory; Weismann's vertical hereditary descent theory; and de Vries' and others' mutation theory to explain the evolution of life. Alternatively known as neo-Darwinism

Monera Taxonomic unit previously known as the first Kingdom of life, subdivided into Archae- and Eubacteria

Nucleoid Prokaryotic genome, not bounded by a membrane, not packaged into separate chromosomes. *Compare to nucleus*

Nucleus Membrane-bounded cell organelle that contains DNA packaged into separate chromosomes, only present in eukaryotes

Pathogens Disease-causing agents such as bacteria, bacteriophages, viruses, prions, fungi, and other protozoan microorganisms

Phagocytosis The act of "eating" whereby a cell engulfs a solid particle that either becomes an organelle or vesicle

Phylogenetics The systematic study of the evolutionary relationship amongst species, phyla, and higher taxa

Plasmid Extrachromosomal, often circular DNA, often the seat of antibiotic resistance genes and crucial for bacterial conjugation

Prions Infectious pathogenic proteins

Prokaryotes All organisms that neither have a membrane-bounded nucleus nor organelles inside their cell. Instead, their genome floats freely inside the cytoplasm in a structure called the nucleoid

Speciation The origin of new species out of old ones, induced by evolutionary mechanisms including, among others, symbiogenesis, lateral gene transfer, hybridization, drift, virolution and natural selection; biotic factors including geographical barriers or species-mate recognition factors; and abiotic factors such as climate change

Spirochetes A phylum of gram-negative, anaerobe, double-membraned, corkscrew-shaped, mobile bacteria

Symbiont The smaller partner in a symbiotic association

Thermoplasma A genus of Archaea (prokaryotes), consisting of anaerobe and fermenting microorganisms

Undulipodium/undulipodia Motile extension of eukaryotic cells, visually resembling a tail. Undulipodia are typified by their [9(2) + 2] microtubullar pattern in their shaft (called the axoneme) and a [9(3) + 0] pattern in their basal body (called the kinetosome). They are similar and presumed evolutionary homologous to eukaryotic centrioli and cilia, and distinct from bacterial flagella. *Compare to cilia and centrioli*

Vector Any organism that functions as a medium for the distribution of pathogens or microorganisms

Viriome All viruses infectious for, and viral parts present in, a certain species. *Compare tomicrobiome*

Virus Infectious genetic agent

References

Anderson E (1949) Introgressive hybridization. John Wiley, New York

Anderson E, Stebbins GL Jr (1954) Hybridization as an evolutionary stimulus. Evolution 8:378–388

Albany M (1998) Conjunctive symbiosis. A Dictionary of Plant Sciences. http://www.encyclopedia.com

Archibald JM (2011) Origin of eukaryotic cells: 40 years on. Symbiosis 54:69–86

Archibald J (2014) One plus one equals one: symbiosis and the evolution of complex life. Oxford University Press, New York

Archibald JM, Richards TA (2010) Gene transfer: anything goes in plant mitochondria. BMC Biol 8:147

Arnold M (1997) Natural hybridization and evolution. Oxford University Press, New York

Arnold M (2004) Transfer and origin of adaptations through natural hybridization: were Anderson and Stebbins right? Plant Cell 16:562–570

Arnold M (2006) Evolution through genetic exchange. Oxford University Press, New York

Babcock EB, Stebbins GL (1938) The American species of crepis: their interrelationships and distribution as affected by polyploidy and apomixis. Carnegie Institution of Washington, Washington DC

Backhed F, Ley RE, Sonnenburg JL, Peterson DA, Gordon JI (2005) Host-bacterial mutualism in the human intestine. Science 307:1915 doi: 10.1126/science.1104816
Bapteste E (2014) The origins of microbial adaptations: how introgressive descent, egalitarian evolutionary transitions and expanded kin selection shape the network of life. Front Microbiol 3(5):83
Bapteste E, Susko E, Leigh J, MacLeod D, Charlebois RL, Doolittle WF (2005) Do orthologous gene phylogenies really support tree-thinking? BMC Evol Biol 5(1):33. doi:10.1186/1471-2148-5-33
Bastian FO, Sanders DE, Forbes WA et al (2007) *Spiroplasma* spp. from transmissible spongiform encephalopathy brains or ticks induce spongiform encephalopathy in ruminants. J Med Microbiol 56(9):1235–1242. doi:10.1099/jmm.0.47159-0
Blanchard JL, Lynch M (2000) Organellar genes: why do they end up in the nucleus? Trends Genet 16(7):315–320. doi:10.1016/S0168-9525(00)02053-9
Bonen L, Doolittle WF (1975) On the prokaryotic nature of red algal chloroplasts. PNAS 72(6):2310–2314
Bonen L, Doolittle WF (1976) Partial sequences of 16s rRNA and the phylogeny of blue-green algae and chloroplasts. Nature 261:669–673. doi:10.1038/261669a0
Bonen L, Cunningham RS, Gray MW, Doolittle WF (1977) Wheat embryo mitochondrial 18s ribosomal RNA: evidence for its prokaryotic nature. Nucleic Acids Res 4(3):663–671
Bravo JA, Julio-Pieper M, Forsythe P, Kunze W, Dinan TG, Bienenstock J, Cryan KF (2012) Communication between gastrointestinal bacteria and the nervous system. Curr Opin Pharmacol 12:6667–6672. doi:10.1016/j.coph.2012.09.010
Brucker RM, Bordenstein SR (2012) Speciation by symbiosis. Trends Ecol Evol 27(8):443–451
Buchner P (1921) Tier und Pflanze in intrazellularer Symbiose, Berlin
Buchner P (1939) Symbiose der Tiere mit pflanzlichen Mikroorganismen, Berlin
Campbell PD (2007–2015) The Germ theory calender. http://germtheorycalendar.com/
Carrapiço F (2006) The origins of life and the mechanisms of biological evolution. In: Proceedings of SPIE 6309: 63090O-1–63090O-5
Cowles HC (1915) Hereditary symbiosis. Bot. Gaz. 59(1):61–63
De Bary HA (1878) Über symbiose. Tageblatt 51 Versamml. Deutscher Naturforscher u. Aerzte, Cassel 121–126
De Duve C, De Barsy T, Poole B, Trouet A, Tulkens P, Van Hoof F (1974) Commentary: lysosomotropic agents. Biochem Pharmacol 23(18):2495–2531
Dole M (1965) The natural history of oxygen. J Gen Physiol 49(1) Suppl: 5–27. doi:10.1085/jgp.49.1.5
Doolittle FW (2000) Uprooting the Tree of life. SCIAM 282(2):72–77. doi:10.1038/scientificamerican0200-90
Douglas AE (2010) The symbiotic habit. Princeton University Press, New Jersey
Dunning Hotopp JC, Clark ME, Oliveira DCSG et al (2007) Widespread lateral gene transfer from intracellular bacteria to multicellular eukaryotes. Science 317(5845):1753–1756. doi:10.1126/science.1142490
Engels WR, Preston CR (1981) Identifying P factors in *Drosophila* by means of chromosome breakage hotspots. Cell 26(3):421–428
Finnegan DJ (1989) Eukaryotic transposable elements and genome evolution. Trends Genet 5:103–107
Flannery DT, Walter RM (2012) Archean tufted microbial mats and the great oxidation event: new insights into an ancient problem. Austr J Earth Scien 59(1):1–11. doi:10.1080/08120099.2011.607849
Flavell AJ (1981) Did retroviruses evolve from transposable elements? Nature 289:10–11
Foster JA, McVey Neufeld KA (2013) Gut–brain axis: how the microbiome influences anxiety and depression. Trends Neurosci 36(5):305–312. doi:10.1016/j.tins.2013.01.005
Frost LS, Leplae R, Summers AO, Toussaint A (2005) Mobile genetic elements: the agents of open source evolution. Nat Rev of Microbiology 3:722–732

Gaudillière JP, Löwly I (eds) (2001) Heredity and infection: the history of disease transmission. Cermes, Paris and Routledge, London
Gifford R, Tristem M (2003) The evolution, distribution and diversity of endogenous retroviruses. Virus Genes 26(3):291–315. doi:10.1023/A:1024455415443
Gogarten P (2000) Horizontal gene transfer: a new paradigm for biology. In: Esalen Center for Theory and Research Conference
Gontier N (2006) Evolutionary epistemology and the origin and evolution of language: taking symbiogenesis seriously. In Gontier N, Van Bendegem JP, Aerts D (eds) Evolutionary epistemology, language and culture. Springer, Dordrecht, pp 195–2260
Gontier N (2007) Universal symbiogenesis: an alternative to universal selectionist accounts of evolution. Symbiosis 44:167–181
Gontier N (forthcoming) Uniting micro- with macroevolution into an extended synthesis: reintegrating life's natural history into evolution studies. In: Serrelli E, Gontier N (eds) Macroevolution. Springer, Dordrecht, p 227
Gontier N (forthcoming) Symbiosis and symbiogenesis. In Kliman RM (ed) Encyclopedia of evolutionary biology. Elsevier, London
Haldane JBS (1949) Disease and evolution. Ric Sci Suppl A 19:68–76
Harrison RG (1990) Hybrid zones: windows on evolutionary process. Oxford Surv Evolution Biol 7:69–128
Holland HD (2006) The oxygenation of the atmosphere and oceans. Phil Trans R Soc B 361:903–915
Hurst G, Jiggins FM, Graf von Der Schulenburg JH, Bertrand D et al (1999) Male killing *Wolbachia* in two species of insects. Proc R Soc B 266(1420):735–740. doi:10.1098/rspb.1999.0698
Hussain M, Lu G, Torres S, Edmonds JH (2013) Effect of *Wolbachia* on replication of West Nile virus in a mosquito cell line and adult mosquitoes. J Virol. doi:10.1128/JVI.01837-12/
Kazazian HH Jr et al (1988) Haemophilia A resulting from de novo insertion of L1 sequences represents a novel mechanism for mutation in man. Nature 332(6160):164–166
Kingsbury N (2009) Hybrid: the history and science of plant breeding. University of Chicago Press, Chicago
Knerr I, Beinder E, Rascher W (2002) *Syncetin*, a novel human endogenous retroviral gene in human placenta: evidence for its dysregulation in preeclampsia and HELLP syndrome. Am J Obstet Gynecol 186(2):210–213. doi:10.1067/mob.2002.119636
Kozek WJ, Ramakrishna RU (2007) The discovery of *Wolbachia* in arthropods and nematodes—a historical perspective. Issues in Infectious Diseases 5:1–14. doi:10.1159/000104228
Lederberg J (1952) Cell genetics and hereditary symbiosis. Physiol Rev 32(4):403–430
Lederberg J (1998) Personal perspective: plasmid (1952–1997). Plasmid 39:1–9
Lederberg J (1999) JBS Haldane (1949) on infectious disease and evolution. Genetics 1(153):1–3
Lederberg J (2003) Infectious history. Science 288(5464):27
Lederberg J, Lederberg EM (1956) Infection and heredity. Symp Soc Growth Develop 14:101–124
Lederberg J, McCray AT (2001) Ome sweet 'omics—a genealogical treasury of words. Scientist 15:8
Ley R (2010) Obesity and the human microbiome. C Opinion Gastroent 26(1):5–11. doi:10.1097/MOG.0b013e328333d751
Ley RE, Peterson DA, Gordon JI (2006) Ecological and evolutionary forces shaping microbial diversity in the human intestine. Cell 124:837–848. doi:10.1016/j.cell.2006.02.017
Linnaeus C (1753 Species plantarum (2 volumes). Holmiae: Impensis Laurentii Salvii http://dx.doi.org/10.5962/bhl.title.669
Livingstone Bell PJ (2001) Viral eukaryogenesis: was the ancestor of the nucleus a complex DNA virus? J Mol Evol 53(3):251–256. doi:10.1007/s002390010215
López-Caamal II A, Tovar-Sánchez E (2014) Genetic, morphological, and chemical patterns of plant hybridization. Rev chil hist nat Santiago 87 doi:10.1186/s40693-014-0016-0

Lovelock J (1972) Gaia as seen through the atmosphere. Atmos Environ 6:579–580

Lovelock JE, Margulis L (1974) Atmospheric homeostasis by and for the biosphere: the Gaia hypothesis. Tellus Series A 26(1–2): 2–10 (Stockholm: International Meteorological Institute)

Löwer R, Löwer J, Kurth R (1996) The viruses in all of us: characteristics and biological significance of human endogenous retrovirus sequences. Proc Natl Acad Sci USA 93(11):5177–5184

Mallet J (2005) Hybridization as an invasion of the genome. Trends Ecol Evol 20:229–237

Mallet J (2007) Hybrid speciation. Nature 446:279–283

Marchi E et al (2013) Neanderthal and Denisovan retroviruses in modern humans. Curr Biol 23(22):R994–R995

Margulis L (1970) Origin of eukaryotic cells. Yale University Press, New Haven

Margulis L (1998) The symbiotic planet: a new look at evolution. Weidenfeld & Nicolson, London

Margulis L (ed) (1991) Symbiosis as a source of evolutionary innovation: speciation and morphogenesis. The MIT Press, Boston

Margulis L, Dolan MF (2001) Early life. Jones & Bartlett Publishers International, Sudbury

Margulis L, Fester R (1991) Symbiosis as a source of evolutionary innovation: speciation and morphogenesis. MIT Press, Boston

Margulis L, Dolan MF, Guerrero R (2000) The chimeric eukaryote: origin of the nucleus from the karyomastigont in amitochondriate protists. PNAS 97(13):6954–6959

Margulis L, Chapman M, Guerrero R, Hall J (2006) The last eukaryotic common ancestor (LECA): acquisition of cytoskeletal motility from aerotolerant spirochetes in the Proterozoic Eon. PNAS 103(35):13080–13085

Martin W, Herrmann RG (1998) Gene transfer from organelles to the nucleus: how much, what happens, and why? Plant Physiol 118(1):9–17

Maxmen A (2010) Virus-like particles speed bacterial evolution. Nature. doi:10.1038/news.2010.507

Mayr E (1942) Systematics and the origin of species, from the viewpoint of a zoologist. Harvard University Press, Cambridge

McClintock B (1950) The origin and behavior of mutable loci in maize. Proc Natl Acad Sci USA 36(6):344–355

McClintock B (1953) Induction of instability at selected loci in maize. Genetics 38(6):579–599

Melezhik VA (2006) Multiple causes of earth's earliest global glaciation. Terra Nova 18(2):130–137. doi:10.1111/j.1365-3121.2006.00672.x

Nelson PN, Hooley P, Molecular Immunology Research Group (2004) Human endogenous retroviruses: transposable elements with potential ? Clin Exp Immunol 138(1):1–9. doi:10.1111/j.1365-2249.2004.02592.x

Paracer S, Ahmadjian V (1986) Symbiosis: an introduction to biological associations. New England University Press, England

Pentecost A, Franke U (2010) Photosynthesis and calcification of the stromatolitic freshwater *cyanobacterium rivularia*. Eur J Phycol 45(4):345–353. doi:10.1080/09670262.2010.492914

Pietilä MK, Demina TA, Atanasova NS, Oksanen HM, Bamford DH (2014) Archaeal viruses and bacteriophages: comparisons and contrasts. Trends Microbiol 22(6):334–344. doi:10.1016/j.tim.2014.02.007

Pound R (1893) Symbiosis and mutualism. Am Nat 27:509–520

Prusiner SB (1982) Novel proteinaceous infectious particles cause Scrapie. Science 216(4542):136–144. doi:10.1126/science.6801762

Prusiner SB (1991) Molecular biology of prion diseases. Science 252(5012):1515–1522. doi:10.1126/science.1675487

Rieseberg LH (1995) The role of hybridization in evolution: old wine in new skins. Amer J Bot 82:944–953

Rieseberg LH (2001) Hybrid speciation. eLS, Wiley, New York. doi:10.1038/npg.els.0001753
Robert E, Kopp J, Kirschvink L, Hilburn IA, Nash C (2005) The Paleoproterozoic snowball earth: a climate disaster triggered by the evolution of oxygenic photosynthesis. Proc Natl Acad Sci USA 102(32):11131–11136. doi:10.1073/pnas.0504878102
Roberts HF (1929) Plant hybridization before Mendel. Princeton, New Jersey
Roossinck MJ (2012) Even viruses can be beneficial microbes. Microbiol Austr 33:111–112
Rothschild LJ (1989) Protozoa, protista, protoctista: what's in a name? JHist Biol 22(2):277–305
Rubin GM, Spradling AC (1982) Genetic transformation of Drosophila with transposable element vectors. Science 218(4570):348–353
Ryan FP (2005) Human endogenous retroviruses in health and disease: a symbiotic perspective. J R Soc Medicine 97(12):560–565
Ryan FP (2006) Genomic creativity and natural selection: a modern synthesis. B J Linnean Soc 88(4):655–672
Ryan FP (2009) Virolution. Harper Collins, London
Sagan L (1967) On the origin of mitosing cells. J Theor Biol 14:225–274
SanMiguel P, Tikhonov A, Jin YK et al (1996) Nested retrotransposons in the intergenic regions of the maize genome. Science 274(5288):765–768. doi:10.1126/science
Sapp J (1994) Evolution by association: a history of symbiosis. Oxford University Press, New York
Sapp J (2003) Genesis: The evolution of biology. Oxford University Press, New York
Sapp J (2009) The new foundations of evolution: on the tree of life. Oxford University Press, New York
Schneider A (1897) The phenomena of symbiosis. Minnesota Bot Stud 1:923–948
Shapiro J (1969) Mutations caused by the insertion of genetic material into the galactose operon of *Escherichia coli*. J Molec Biol 40:93–109
Shapiro JA (ed) (1983) Mobile genetic elements. Academic Press, London
Siefert JL (2009) Defining the Mobilome. In: Gogarten MB, Gogarten JP, Olendzenski LC (eds) Horizontal gene transfer: genomes in flux. Humana Press, Springer, Dordrecht
Singer MF (1982) SINEs and LINEs: highly repeated short and long interspersed sequences in mammalian genomes. Cell 28(3):433–444. doi:10.1016/0092-8674(82)90194-5
Smocovitis VB, Ayala FJ (2000) George Ledyard Stebbins. Biogr Mem Nat Acad Sci 85:290–313
Smocovitis VB (1997) G. Ledyard Stebbins, Jr. and the evolutionary synthesis (1924–1950). Am J Bot 84(12):1625–1637
Stanton TB (2007) Prophage-like gene transfer agents-novel mechanisms of gene exchange for *Methanococcus*, *Desulfovibrio*, *Brachyspira*, and *Rhodobacter* species. Anaerobe 13(2):43–49
Stebbins GL (1940) The significance of polyploid plants in plant evolution. Am Nat 74(750):54–66
Stebbins GL (1950) Variation and evolution in plants. Columbia University biological series 16, New York
Stebbins GL (1959) The role of hybridization in evolution. P Am Philos Soc 103:231–251
Sugimoto J, Schust DJ (2009) Review: human endogenous retroviruses and the placenta. Reprod Sci. 16(11):1023–1033. doi:10.1177/1933719109336620
Taylor AL (1963) Bacteriophage-induced mutation in *E. coli*. Proc Natl Acad Sci USA 50(6):1043–1051
Temin HM (1980) Origin of retroviruses from cellular moveable genetic elements. Cell 21:599–600
The NIH HMP Working Group (2009) The NIH human microbiome project. Genome Res 19(12):2317–2323
Turnbaugh PJ, Ley RE, Hamady M, Fraser-Liggett CM, Knight R, Gordon JI (2007) The human microbiome project. Nature 449:804–810. doi:10.1038/nature06244
Turnbaugh PJ, Hamady M, Yatsunenko T et al (2009) A core gut microbiome in obese and lean twins. Nature 457(7228):480–484. doi:10.1038/nature07540

van Beneden PJ (1873) Un mot sur la vie sociale des animaux inférieurs. Bull Acad R Belgique 2(36):779–796

van Beneden PJ (1875) Les comensaux et les parasites dans le règne animal. Biblio Sci Int, Paris

van den Hurk AF, Hall-Mendelin S, Pyke AT, Frentiu FD (2012) Impact of *Wolbachia* on infection with chikungunya and yellow fever viruses in the mosquito vector *Aedes aegypti*. PLoS Negl Trop Dis. doi:10.1371/journal.pntd.0001892

Venu I, Durisko Z, Xu J, Dukas R (2014) Social attraction mediated by fruit flies' microbiome. J Exp Biol 217:1346–1352. doi:10.1242/jeb.099648

Villarreal LP, Defilipps V (2000) A hypothesis for DNA viruses as the origin of eukaryotic replication proteins. J Virol 74(15):7079–7084

Villarreal LP, Witzany G (2010) Viruses are essential agents within the roots and stem of the tree of life. J Theor Biol:698–710

von Faber FC 1912 Das erbliche Zusammenleben von Bakterien und tropischen Pflanzen. Jahrbücher für wissensch Botanik 51:III

Wallin IE (1927) Symbionticism and the origin of species. Williams and Wilkins, Baltimore

Whittaker RH, Margulis L (1978) Protist classification and the kingdoms of organisms. Biosystems 10:3–18

Woese CR, Fox GE (1977) Phylogenetic structure of the prokaryotic domain: the primary kingdoms. PNAS 74(11):5088–5090. doi:10.1073/pnas.74.11.5088

Woese CR, Kandler O, Wheelis M (1990) Towards a natural system of organisms: proposal for the domains Archaea, Bacteria, and Eucarya. Proc Natl Acad Sci USA 87(12):4576–4579. doi:10.1073/pnas.87.12.4576

Woese CR, Magrum LJ, Fox GE (1978) Archaebacteria. J Mol Evol 11(3):245–251. doi:10.1007/BF01734485

Zinder ND (1953) Infective heredity in bacteria. Cold Spring Harbor Symp Quant Biol 18:261–269

Zirkle C (1934) More records of plant hybridization before Köelreuter. J Hered 25:3–18

Zirkle C (1935) The beginnings of plant hybridization. Morris Arboretum Monographs 1. University of Pennsylvania Press, Philadelphia

Symbiosis—Evolution's Co-Author

Douglas Zook

Abstract Symbiotic integration is a primary contributor to the centerpiece of evolution, genetic novelty. Acquisition of foreign organisms or parts thereof, and potential subsequent assimilation and often internalization of one or several different genomes into another different entity are the foundational expressions upon which natural selection acts, particularly in eukaryotic organisms. Thus, the entire landscape of life—from cells to biomes—is substantially an evolving collection of chimeric communities. Competition may be pronounced and successful in evolution in large part *because* the competing organisms do *not* function as, and indeed are not, individuals. Moreover, growing evidence indicates symbiosis to be on a flexible continuum of physiological expression, often with real plasticity in the organisms' integrating life cycles. Therefore, so-called "mutualism", "parasitism", and "commensalism" as symbiotic reference points and analyses may be outdated and perhaps of dubious use. For example, fundamental ecological principles show us that "parasitism" among two different organisms is often of significant advantage to not only the "parasite" but its "host." Symbiosis system examples are here reviewed and redefined on a more meaningful evolutionary context; namely, symbiosis is the acquisition of one organism(s) by another different organism(s), and through subsequent long-term integration, new structures and metabolism emerge.

Keywords Symbiogenesis · Holobiont · Symbiosis · Margulis · Mutualism · Microbiome

D. Zook (✉)
Global Ecology and Science Education,
Boston University, Two Silber Way, Boston, MA 02215, USA
e-mail: dzook@bu.edu

N. Gontier (ed.), *Reticulate Evolution*, Interdisciplinary Evolution Research 3,
DOI 10.1007/978-3-319-16345-1_2

1 Introduction

This journey reveals how symbiosis permeates the biosphere and its evolutionary history. It emphasizes new perspectives not only about what is in front of us every day and how it got there, but reinforces the revolution in science today—the emerging realization of individuals as ecosystems. Our travels, with stops at dozens of symbiotic examples, many only recently revealed, will thrust through old symbiosis definitions and offer a new workable one. It will dare to step away from the traditional mutualism–parasitism–commensalism gyre into new currents that reflect the fluid reality that is symbiosis. It will culminate with comments from an interview this author conducted a few years ago with longtime friend and sage for so much that we realize today in life science, Lynn Margulis. There is no more profound and revealing place to start than with the pervasive eukaryotes, the algae.

Algae dominate the biosphere. These autotrophic protists, the larger of which are commonly called "seaweeds," significantly impact every biome and nearly every ecosystem on earth. Most are microscopic and are in high densities in the colder regions of the world's oceans, which make up 71 % of the globe's surface. The algae (along with cyanobacteria) are the main fixers of carbon; the primary source of oxygen in the atmosphere; an essential food source for key marine and freshwater food webs; substantial biomineralizers, contributing much of the lithosphere's limestone; principle conduits for critical element flow; emitters of gases that serve as condensation nuclei in cloud formation; and serve as substrates, foundations, and "partners" for biodiverse communities such as mats, crusts, and films. They are the physiological glue of the biosphere, effectively keeping the earth's biosystems productive, efficient, and perpetual. And, their evolution, which extends back to nearly the dawn of eukaryotes two billion years ago, is the result of remarkable symbiotic infection and acquisition events. Indeed, the vast algal groups are among the most prominent evidence for symbiosis strongly sharing the biosphere stage with mutation and recombination as evolution's co-author—with natural selection as the essential and ultimately passive editor.

The first photosynthetic-centered symbiotic event is that which also led to the lineage that emerged as plants—the phagocytosis of a free-living cyanobacterium into a microscopic heterotrophic protist already equipped with other products of symbiosis, mitochondria and the nucleocytoplasm (Archibald 2011). In geologic time that amounts to a flash of lightning, one genome became embedded and functional within another, resulting in a novel now autotrophic organism. Referred to as a "primary symbiosis," this profound acquisition was the biological big bang that still expands outward today, producing phylogenies via little or no gradualism and with mutation as a more secondary influence.

2 Primary, Secondary, and Tertiary Symbiosis

This primary event of autotrophy acquisition resulted in three distinct lineages represented by the Chlorophyta (green algae), Rhodophyta (red algae), and Glaucocystophyta. The latter more obscure algae features a reduced cyanobacterium known as a "cyanelle" as its evolving photosynthetic organelle. This cyanelle includes the pervasive polymer and cell wall constituent "peptidoglycan," a revealing remnant of its prokaryotic, cyanobacterial acquisition ancestry. While the few representatives of this phylum are extant, this lineage, evolutionarily speaking, was a "dead-end" in that there is no evidence that any new forms branched from it. The same cannot be said of the other two primary symbiosis lineages (Delwiche 1999). At close to 470 mya, green algae from within either the Charophycean class (Lewis and McCourt 2004) or Zygnematales (Wodniok et al. 2011) transitioned from aquatic habitats to the land, eventually leading to the first plants. Thus, all green chlorophyll-containing eukaryotic photosynthesizers, such as plants, are the result of this first cyanobacterial acquisition, likely by a mitochondrion-containing amoeboid-like heterotrophic protist. However, remarkably, this critically significant event—termed a "primary symbiosis"—was only the start of a broad series of secondary symbiosis-generated lineages. Categorized as "secondary" symbiosis, members of what we now recognize as from the primary green and red lineages were phagocytized by another eukaryote. For example, certain chlorophyte algae were engulfed by a heterotrophic protist and emerged as Euglenophyta, while still others became Chlorarachniophyta (Palmer 2003). The latter group is made up of very few species, but has great evidential significance, for these microscopic, colonial forms reveal today a greatly reduced genome, essentially a remnant of the nucleus from the chlorophyte, which it phagocytized. Thus, chlorarachniophytes feature the original "host" heterotrophic eukaryote with its primary symbiosis-derived mitochondria and nucleocytoplasm, as well as a reduced "captured" alga with its now miniscule nuclear expression known as a "nucleomorph." The discovery of the nucleomorph indicated what had been merely suspected previously—that many diverse algal groups are actually well-integrated, multi-genomic consortia (Bhattacharya et al. 2003).

The evidence is further strengthened by the existence of a different nucleomorph that verifies yet another secondary symbiotic event leading to another lineage (Ludwig and Gibbs 1985; Moore and Archibald 2009). This remnant nucleus was that of a species of microscopic red alga, which was engulfed by a heterotrophic protist but not digested. As in the chlorarachniophytes, this red alga counterpart conferred relatively quickly natural selective advantages in the new consortium. This algal lineage, which emerged from the primary symbiont rhodophyte lineage, represents the phylum Cryptophyta. Cryptophytes are mostly freshwater and have two motility organelles ("undulopodia" or what is more traditionally called "eukaryotic flagella"), which it uses in conjunction with specialized ribbon devices known as "ejectisomes." These structures contract and expand and propel the microbe in various directions.

Other secondary symbioses led to other algae of incalculable importance to the biosphere. These include the glass-enclosed (SiO_2 encased) diatoms (Bacillariophyta), the limestone depositing coccolithophores (Haptophyta), and brown algae (Phaeophyta). While there are no remnant nuclei from an acquired symbiont in these and other algal phyla, evidence shows unequivocally secondary symbiosis in action. For example, membrane counting and analysis is a useful indicator. When a heterotrophic protist phagocytizes the alga, the alga becomes permanently surrounded by that host membrane with its characteristic lipids. Moreover, the plastid enclosed in the red or green algal symbiont also has at least one and often two (or more) membranes, one characteristic of the surrounding cytoplasm of the alga and another of the original cyanobacterial primary symbiosis. These membrane "layers" combined with ultrastructural and gene sequencing comparative data confirm the identity of the integrated genomes within nearly all other algal lineages and show their evolution as derived from secondary level symbiosis (Archibald 2009) (Fig. 1).

Remarkably, one of the most common algae in the oceans and particularly found in symbiosis with larger invertebrate "hosts" such as corals and anemones, dinoflagellates, are sometimes the result not only of the primary and secondary symbioses but a third symbiotic event. For example, species of haptophytes—itself the product of secondary symbiosis—have been phagocytized by yet another likely heterotrophic protist resulting in a tertiary autotrophic dinoflagellate (Inagaki et al. 2000). Such a dinoflagellate can be seen as the sum of up to a dozen genomes or genome remnants without of course counting bacterial gene transfer events over recent or deep time. Indeed, there is growing evidence of many other photoautotrophy-based tertiary symbioses among the protists (Vesteg et al. 2009).

3 Algal Phylogeny: Showcase for Genetic Novelty Through Symbiosis

The algae are deserving of focus from the outset, for there is no more profound example of symbiogenesis—the acquisition-centered impact of symbiosis on evolution. The autotrophic portion of an entire kingdom (or subkingdom) so central to biospheric systems, global biodiversity, and geological substrates is due to the process of symbiosis. In each lineage, genes foreign to an organism were tolerated and eventually incorporated, whole or in part, into the consortium. The heterotrophic protist host would have to undergo unimaginable mutational events to express eventually even a fraction of the consortium's traits. Mutation and recombination influence in the emerging eukaryotic algae without symbiosis is an oxymoron. It is plain to see that there would be no diverse phylogeny of algae as such. In this way, lasting symbiotic mergers through symbiogenesis are not only central to evolution and global ecology but foundational. New species, lineages, varieties can develop within and from the symbiotically constructed lineage, often in turn, leading to new symbiotic mergers. Algal taxonomy reveals life-forms much like bridging silk strands of an orb spider's web, a series of integrated connections that transform the concept of individual to one of a vast symbiotic community, or as the emerging symbiosis-based revolution in science now terms, the "holobiont," as first proposed by Margulis and Fester (1991, p. 2). The term later became more specifically associated with corals (Rowher et al. 2002) and more recently

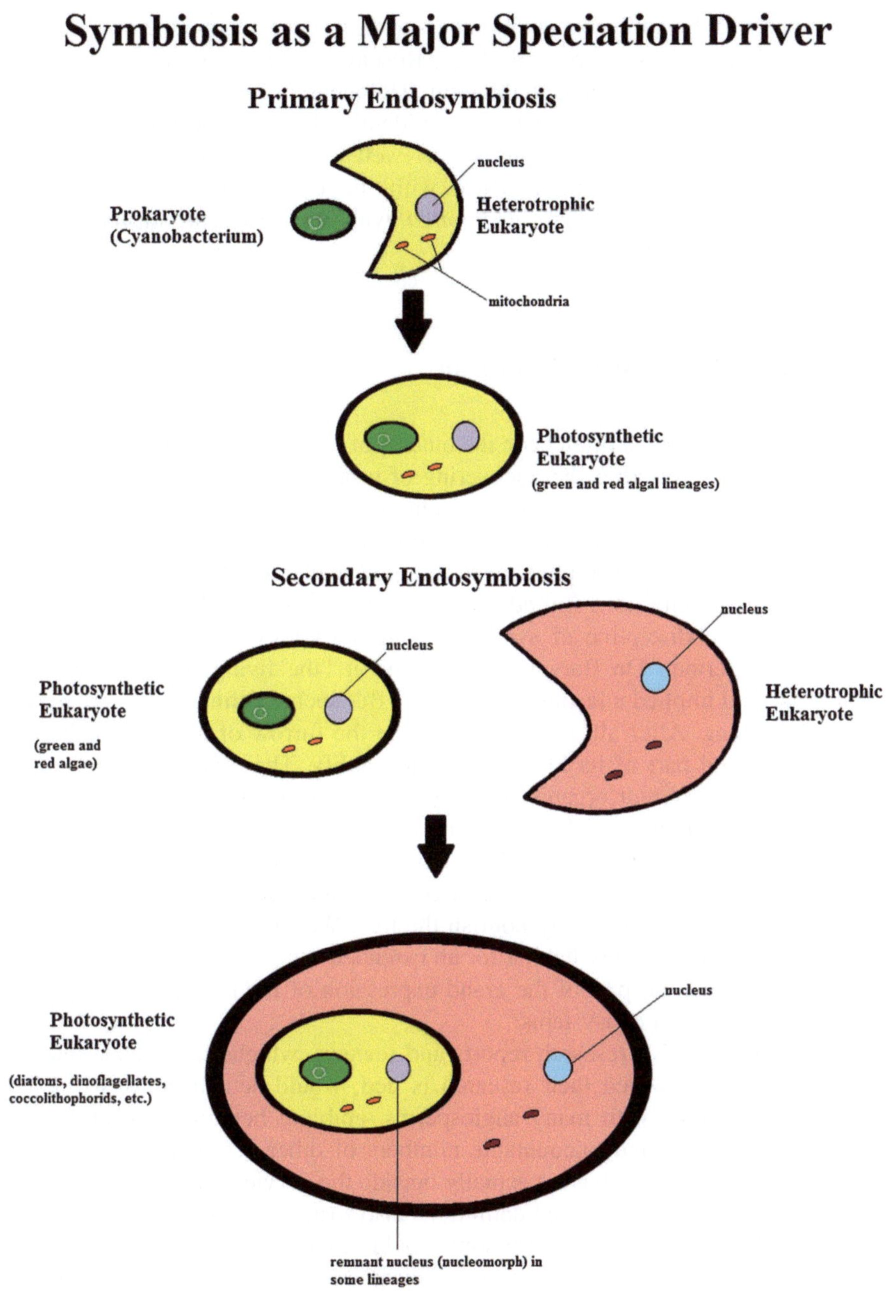

Fig. 1 Diagram by Olivia Hathaway

transformed to a dynamic concept in understanding the metagenomic unit of selection in evolution (Rosenberg and Zilber-Rosenberg 2011; Gilbert et al. 2012; McFall-Ngai et al. 2013). Indeed, the term "Hologenome" can be used to not only refer to the symbiont genomes but to those identified genes that were horizontally transferred from bacteria for example, as well as possible extra-chromosomal mainstays which may be in evidence with the extraordinary "symbiosis" of the sacoglossan mollusc *Elysia chlorotica* with its *Vaucheria litorea* plastids (Bhattacharya et al. 2013).

4 Outdated Symbiosis Definition

Ironically, symbiosis has often been the outcast of biology, at best seen as an interesting curiosity. Even after the uncovering of prokaryotic DNA (characterized as within a circular chromosome and not containing key histone proteins) in mitochondria in the 1960s (Nass and Nass 1963; Nass 1969), there was antagonism toward any evidence that might suggest the powerful role of symbiosis in evolution. Its original definition, still advocated by many, may not have helped in fostering a clear subdiscipline of symbiosis within Biology. In the late nineteenth century, the German De Bary (1879) labeled it "the living together of unlike organisms" and implied a lasting relationship. But such a definition has proved too all-encompassing. After all, an insect living in the furrow of a particular bark of a tree for a good part of its life cycle could qualify. The tree and the insect are certainly vastly different organisms. They are living together and even in physical closeness for an extended period, with one nestled within the other. If natural selective advantages are considered, we could perhaps find that the bits of waste material from the insect, which get carried down the tracks of the tree bark, end up in the rhizosphere and partly nourish the tree. We can further surmise that the insect in turn gains a secure habitat for an extended time and so on. Of course this is not a symbiosis, but part of the grand expression of fundamental ecology. It is the ubiquitous stuff of ecosystems.

Moreover, extensive research reports and reviews, which clearly state that this is the definition to which their research is tied, would actually have to include systems—pollinators with many angiosperms, epibiont heterotrophic protists on marine macro-algae, and uncountable numbers of other ecological relations—in their data and discussion that are actually outside the purview of symbiosis.

The vagueness of the original definitions also fostered a sense of new categorizations, such as mutualism (that for some could fit the above simplistic insect–tree bark example), parasitism, and commensalism. Much of the symbiosis research over the past one hundred years and right to the present is seemingly intoxicated with having to place symbiosis in one of these boxes. Strangely, it can even guide research, wherein one of the ultimate purposes of many symbiotic studies is to determine the degree of mutualism or "shared benefit" or whether one form is more parasitic or simply there for the ride without any significant contribution. This has led to subcategories of "cheaters" and "freeloaders," and other terms that seem to not be cognizant that *all* organisms appear to seize on opportunities to enhance their life cycle and balanced or altruistic fits are seldom in play.

Does one really need to use these terms to define or even connote symbiosis? To what degree are these terms actually meaningful and strongly reflective of the biological and evolutionary reality? Do these terms potentially move us away from ecological thinking and replace it with anthropogenic, human chauvinistic thinking? To what degree does such an obsession with these terms skew how we should be investigating and interpreting our findings? In other words, are we subjecting ourselves to research processes that are far less than open ended but rather designed to see how they fit into some prescribed, small set of categories, slots that may reflect more human analysis than nature's reality? I posit that the continued reinforcement of the original definitions and the dogmatic emphasis on the three categories with analysis of the degree of "benefit" or "antagonism" is neither representative nor particularly useful in the now mainstream discipline of symbiosis. The data collected can be outstanding and revelatory but the language and context is often more convenient, habitual, and simply scientifically inappropriate. It is difficult to find a symbiosis research paper that does not become focused on "benefits" and "costs," as well as the mutualism, parasitism, and commensalism.

Moreover, there appears to be little recognition that entire studies and chapters of books within the overarching discipline of ecology discuss "mutualism" and in so doing are referring to both the widespread behavior of pollination and the association of fungi (mycorrhizal) with plant root cells. The latter a symbiosis, the former, in most cases, is not. Mutualisms are very common ecological expressions and for clarity sake alone should not be used to analyze and judge symbiotic systems. To do so only risks greater confusion and again makes symbiosis appear to be synonymous with ecology when it is a central reasonably identifiable discipline within ecology. Further, one can argue with a reasonable degree of validity that most associations of any kind are "mutualistic." Pathogenic organisms that cause death are essential to the continuance of that species (the "victim"). Commensals die and the decaying biochemistry from it becomes part of the ongoing nutrient supply. Parasites ultimately can strengthen the resilience of the species in that natural selection can often favor new varieties more fit for the threatening environmental conditions.

Symbiosis analysis also implies for some a denial of the centrality of competition in ecology and evolution. Rather, the reality can be seen in the context that some competitors are often more fit *because* they have symbiotic "partners" and alliances. Combined with the fact that many eukaryotic organisms (holobionts), as well as bacteria, are naturally selected for efficiency, energy-consuming competition may be less of an evolutionary driver and often more a life strategy that is embedded with frequent caution signs.

5 Symbiosis Redefined More Concretely and as a Better Reality Fit

Those working in symbiosis research know that there are very clear components that make the discipline more concrete. Thus, I found it refreshing when the outstanding symbiosis researcher Angela Douglas in her book *Symbiotic Interactions* (1994, now out of print) indicated some new, clearer criteria, albeit less emphasized in her latest book, *The Symbiotic Habit* (2010). I have adapted some of those ideas into a definition that over the years have helped my students truly identify symbiotic systems less ambiguously, more accurately, and in a more appropriate evolutionary context. Symbiosis is *the acquisition of an organism(s) by another unlike organism(s), and through subsequent long-term integration, new structures and metabolism(s) emerge.*

This definition makes the focal point of symbiosis the specific physical and metabolic outcomes of the symbiosis. For example, the prototype symbiosis can arguably still be seen as the lichen (Sapp 1994). In most lichen symbioses, we have an alga and a fungus, two fundamentally, phylogenetically distinct life-forms. If I have an alga isolated species with its own morphology and indeed its own genome(s) and I have a compatible fungus separately with its own morphology and genome(s), these organisms are fundamentally its own discreet "selves." But, given genetically programmed signaling and recognition factors, if I axenically bring them physically together in the laboratory for growth on an appropriate nutrient medium and mineral substrate, as was done many years ago by pioneering lichenologist Ahmadjian (1993)—or even in its natural setting, the morphology and indeed the ontogeny change dramatically. Both original forms become substantially unrecognizable as a growing entity. So much so that we are forced to give the new multi-genomic morphotype—this grand "holobiont"—that one can see with the naked eye, a name, the "thallus" (Fig. 2).

Fig. 2 *Cladonia cristatella* ("British soldier") lichen on right and Cladina sp. on left at Parker River Wildlife Sanctuary, Newburyport, MA. Photograph by D. Zook

This construct by both organisms by way of signaling, recognition, environmental conditions, assimilation, and integration is the principal defining characteristic of symbiosis.

Likewise, alga and fungus have vastly different physiological features, such that their metabolic properties are very dissimilar. Fungi cannot photosynthesize. Algae generally do not feed by direct uptake from surface materials. But together in symbiosis, they each bring different degrees of new metabolic expression to such an extent that the fungi have become photosynthesizers through the de facto integration of a "foreign" eukaryote with chloroplasts embedded within the new consortium, resulting in this now lichen holobiont, i.e., the integrated multi-genome entity. Thus, the second defining characteristic is the relatively rapid emergence of new metabolite expressions (e.g., photoautotophy) essential to the holobiont—a physiology(s) and biochemistry that was not there previously in either of the free-living proto-symbionts (Fig. 3).

A third critical component of symbiosis is the process of acquisition. Entire sets of genes with a genome that is "foreign" to one symbiotic partner become sorted, rejected, accepted, and integrated to various degrees in the new multi-genomic holobiont. Acquisition is defined by "coming to control or possess something." Thus, a genome's important functions and components of the partner organism are acquired in that they then belong to a new emerging entity. The acquisition can often best be considered reciprocal in the sense that both once-independent entities are acquiring significant degrees of the gene expression of the other, if not in some cases the whole genome. Often, the acquisition becomes both intimately syntrophic and synergistic.

The proposed symbiosis definition differentiates clearly between relatively short-term, ephemeral, and non-integrated relationships involving different organisms and those that are long term, persistent, and highly integrated. For example, many ecologists consistently cite mutualism for both a coral-dinoflagellate reef-building relationship and a honeybee pollinating a flower. The latter is short term, ephemeral, and relatively unpredictable and is simply one of thousands of

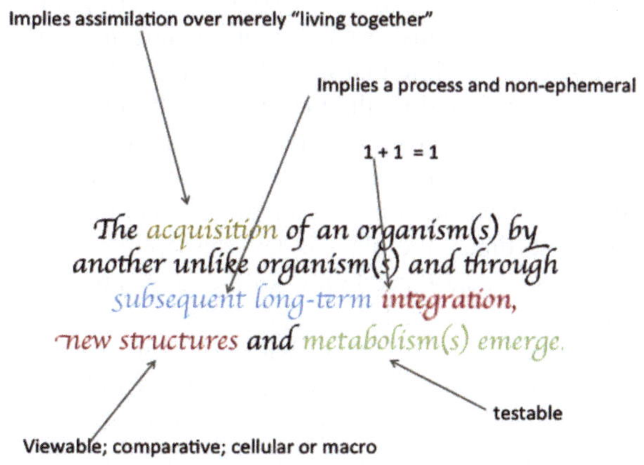

Fig. 3 A new definition of symbiosis as proposed by D. Zook

ecological expressions. The former is long term, very persistent, and well integrated, what with specific dinoflagellates changing their morphology and encysting within coral-created membranes in the anthozoan's cells. Within this analysis, the honeybee's relationship to flowers is ecologically mutualistic and not a symbiosis or holobiont with respect to each other, but the bee with its endosymbiotic bacteria is a symbiosis itself (Martinson et al. 2011) and a holobiont. Yet, frequently, we read that both systems represent "mutualists." This only serves to confuse, oversimplify, and relegates symbiosis to a kind of taboo "cooperation" word that should be avoided. The new symbiosis definition does not include ambiguous and confusing "mutualist" and various traditional anthropogenic terms, but rather emphasizes viewable and measurable outcomes that emerge from an acquisition-centered reality. In this context, these novel structures—the thallus of lichens, the rumen or ruminants, the reefs of dinoflagellate–corals, the trophosome of Riftia tube worms, the arbuscles of fungi within root cells, the intercellular Hartig net of ectomycorrhizae, the paunch of termites, the bacteriome of many insects, the light organ of the bobtail squid–Vibrio, the trichome-lined cavity of *Azolla*, the syconium of *Ficus*, the subterranean nests of attine ants, the symbiosome membranes around many intracellular symbionts, and the nodules of Rhizobium with legumes—are all defining central characteristics of symbiosis and reinforce this new definition.

6 Ecosystem Thinking Replaces Compartmentalizing

This labeling of systems as "mutualist" or "parasitic" can also be misleading, given the nature of symbiotic systems. Over both the diurnal and full life cycle of the holobiont, any of the given integrated genomes can, often through disrupted signaling and alternative feedbacks, be more dominant or subservient than the other. Prominent examples reside in the endophyte symbioses, such as those involved in Epichloë (fescue) plants and *Claviceps* fungi (Schardl 2001). In these holobionts, switching via enzyme triggering (Tanaka et al. 2008) to a sexual cycle in the fungus fosters a more pathogenic expression such that hyphal growth becomes so prolific that it chokes out floral development of the grass. This switch may be promoted by the fact that metabolites of the grass, which typically restrict the fungal partner growth, become relegated instead to the energy needs for grass reproductive structure and function. Yet, a different selective advantage expression dominates during asexual fungal periods much of the year in the same holobiont. At these times, the plant is less susceptible to drought conditions and herbivory due to secondary metabolites of the fungus residing intercellularly in the grass leaf and often in the seed (Eaton et al. 2011).

There is also evidence that very limited genetic changes can move a symbiont from necessary holobiont entity to assimilated food source. This is the case in the amoeba *Dictyostelium discoideum* which harbors among many bacteria, two strains of *Pseudomonas fluorescens*. The protective anti-fungal more "mutualistic" strain is converted by a single point mutation at the activator gene to the edible and thus rapidly ingested strain (Stallforth et al. 2013). Recently, Wooldridge (2010) challenged the long-standing compartmentalizing of hermatypic coral animals with dinoflagellate symbionts by emphasizing a kind of animal winnowing mechanism that results in the most effective photosynthate-transferring varieties, as well as suggesting significant fitness cost for the algal symbionts. Such characteristics among many suggest less of a "mutualism" than a kind of "controlled parasitism," a term originally proposed for lichens many years ago (Ahmadjian and Jacobs 1981) and again which could apply at various parts of the holobiont life cycle to many symbioses.

We are starting to see a refreshing perspective emerge that the guide for symbiosis enquiry and interpretation needs to be on open-minded exploration of physiology, ecology, ontogeny, and cell communication. There is no prescribed screenplay but an ongoing series of images that tell a story within an ecosystem context. For example, the review of microbial symbiont transmission by Bright and Bulgheresi (2010) states at the outset, "The key question is how the symbiont is transferred to the host progeny, *regardless of the type of symbiosis" (italics mine).* They go on to emphasize in their astute and comprehensive analysis "how the conversation between partners… is initiated." That said, they also show a contradiction in that they are focusing on those organisms that maintain "protracted physical contact and involve most of the host population," caveats that are not clear within the de Bary definition to which on the next line they pledge allegiance.

While some symbiosis researchers are facing this compartmentalizing stigma head on by offering such terms as "context-dependent symbioses" (Daskin and Alford 2012), this could be considered a malapropism as it is hard to realize the life history of any holobiont as being independent of context. Arguably, there can be gradations of "context dependency." The "poster child" of symbiosis dating back nearly 150 years has always been the lichen. These algal–fungal and occasionally cyanobacteria-inclusive extracellular consortia have evolved an impressive array of biochemical and physiological features that allow them to secure strongly their niches across nearly all terrestrial biome conditions and even on many aquatic substrates. No feature is more valuable to their fitness than their production of a laboratory full of secondary compounds, commonly labeled as lichen acids. These not only break down substrates to their mineral or particulate constituents and thus enhance element cycling, but they are strongly anti-herbivorous and often antimicrobial. Even those lichens that do not have such acids, such as those that have cyanobacteria as a main phycobiont, may compensate trait loss for such biochemistry through the presence of antibiotic-producing actinobacteria (Zook 1983). Moreover, there is growing evidence that the lichen holobiont may involve and perhaps require a consistent community of bacteria (Cardinale et al. 2006; Grube et al. 2009). Yet, even in this stalwart symbiosis, there is a fluidity and plasticity that defies simple categorization. The algal and fungal symbionts within the holobiont are tantamount to mammalian organs, certainly subject to break down but coordinated through positive and negative feedbacks, often involving pH changes that allow for functional and even thriving life cycles. At any given time and dependent on environmental conditions such as alternations of wetting and drying, either the algal or the fungal genome can be controlling or dominant. The emerging realization of lichen symbiont fluidity can be seen in many Cladoniaceae lichens which have even been placed outside of key co-speciation possibilities and instead adapting to the environment needs of the holobiont through frequent algal symbiont switching (Piercy-Normore and Depriest 2001). Various fungal symbionts such as *Colletotrichum* spp. function across the spectrum of mutualistic to parasitic and so-called commensal depending on with which plant species they are associating, as well as environmental conditions (Redman et al. 2001).

The emergence of holobiont thinking, recently represented also within the hologenome theory (Rosenberg et al. 2009), emphasizes that the unit of selection is the multiplicity of genomes and genome constituents in what has usually been called an "individual," but in reality is with all eukaryotes a symbiotic community (Gilbert et al. 2012). This outlook reflects a new and necessary ecological and environmental framework, which in turn reveals the fluidity inherent in context-dependent nature. A key extension of this fluidity is that many symbioses now appear to be highly variable, flexible, and adaptable, consistently utilizing associated "foreign" genomes or genomic remnants to fulfill essential metabolic expressions.

7 Symbiogenesis Rooted in Lamarck, Darwin, and Kozo-Polyansky

Acquisition has actually been central to evolution thinking since the early nineteenth century. For example, Lamarck developed the first organized view of evolution, in which he proposed that characteristics can be acquired by organisms and then be inherited into the next generation (1809, reprint 2011). While this idea was subsequently derided, his views became a running thread through Charles Darwin's *Origin of Species* (1859, reprint 2011). A variation of this thinking that Lamarck would not have been able to realize at that time is a reality today in the concept of acquisition-based evolution which we term "symbiogenesis," originally proposed by Russian biologist Kozo-Polyansky (1924), later resurrected by Margulis (1990) and in a new translation and interpretation by Margulis and Fet (2010). Kozo-Polyansky originated the term and summarized its meaning as the origin of evolutionary novelty by the merger of different organisms into one. Ironically, Darwin himself had some sense of this, "We cannot fathom the marvelous complexity of an organic being; but on the hypothesis here advanced this complexity is much increased. Each living creature must be looked at as a microcosm—a little universe, formed of a host and a self-propagating organism, inconceivably minute and as numerous as the stars in heaven." (Darwin 1858, p. 453).

Today, symbiogenesis connotes the emergence of this acquired multi-genomic entity (holobiont) over evolutionary time perpetuated by natural selection.

8 Holobiont Selection Allows for More Rapid Adaptation and Greater Fitness

Two insect symbiotic systems dramatically illustrate this intricate coordination and fluidity of various symbionts as a community, the holobiont.

The tsetse fly, made up of over 30 species in the genus *Glossina,* are large, biting flies which are prolific in North African arid and desert regions. They produce four generations each year and are hematophagous. The tsetse holobiont consists not only of itself and its mitochondria but an interlocking array of microbes. Two are obligate gammaproteobacteria of *Wigglesworthia* spp. A third is *Sodalis glossinidius,* a more recent symbiont as evidenced by its ability to be extracted and cultured (Snyder and Rio 2013). All three are vertically transmitted through the maternal milk glands (Balmond et al. 2013), which carry specific proteins and lipids to the uterus for the viviparous symbiosis-accommodating offspring development (Attardo et al. 2008; Ma and Denlinger 1974).

Wigglesworthia are mostly intracellular, being located in specialized cells known as "bacteriocytes." The collection of bacteriocytes make up a defined region of the insect, the "bacteriome." Such a new structure, a proposed defining characteristic of symbiosis, is commonly found in various insect holobionts (Baumann 2005). Because the fly lacks B vitamins in its blood diet, selection has favored these bacteria which provide not only vitamins but stabilizes the fly immunological development and digestion, and influences the degree of trypanosome infection (Snyder and Rio 2013). Verifications of this symbiont dependency have been shown by providing *Wigglesworthia* cell extracts to aposymbiotic, immune-weak mother flies. Such a treatment restores immune vitality (Weiss et al. 2012). While the role of the *Sodalis* bacterial symbiont, also vertically transmitted via the milk glands, in the holobiont community remains unclear, it is undergoing considerable genome reduction, which indicates likely integration through gene elimination and possible transfers to the other holobiont symbionts. Perhaps indicative of the tight community nature of this holobiont, the demise of *Wigglesworthia* causes a corresponding loss of *Sodalis* (Snyder and Rio 2013).

A fourth (facultative) symbiont in the tsetse fly is a *Wolbachia* species within the bacterial family Rickettsiaceae. *Wolbachia* is the most common bacteria affecting the reproductive system of animals known. It is most commonly found in arthropods and confers dominance of females through various male-reducing and male-eliminating strategies (Werren et al. 2008). In the tsetse fly, it induces cytoplasmic incompatability, which ultimately means that females that are uninfected by *Wolbachia* cannot mate with males which are infected. Because *Wolbachia* can only be transmitted by females, this promotes *Wolbachia* reproduction and viability (Werren 1997).

The salivary gland hypertrophy virus (SGHV) of the Hytrosaviridae family can be considered as another tsetse holobiont genome, albeit a facultative virus. This viral infection of the tsetse fly may confer gonad abnormalities and reduce reproductive success (Sang et al. 1999). Thus, this genome within this holobiont community mirrors large ecosystems conventionally studied, in that population regulation through disease and death is an ongoing necessity for optimal fitness and viability of the whole.

Because the holobiont approach tends to minimize the necessity for anthropogenic and often misleading terms such as "host," we can of course consider the sixth major symbiont in this holobiont to be the tsetse fly genome itself. For example, Wang and Aksoy (2012) founded that a fly peptidoglycan recognition protein PGRP-LB, similar to that found in Drosophila, prevents immune deficiency signaling stimulation and thus is closely associated with *Wigglesworthia* infection and maintenance. It is produced by adults and also transferred via milk glands to offspring after the latter's initial blood ingestion (Wang and Aksoy 2012).

Findings with the *Planococcus citri* (mealy bug) from the Pseudococcidae family reinforce and even expand the symbiosis-centered holobiont community concept. These cosmopolitan scaly insects are only female in the adult stage and commonly feed on plant sap. Males do not feed and live only until fertilization of the female. Recently, mealybugs have been found to contain not just a bacterial symbiont, but the smallest known bacterial genome at 139 kb, considerably less than both free-living bacteria and other symbiotic bacteria. Husnik et al. (2013) surmised that such gene reduction may be similar to organelle development as in the endosymbiotic origins of mitochondria and plastids in eukaryotes. However, quite rare for prokaryotes, they found that the bacterium *Tremblaya* had acquired a 538 kB genome bacterium, *Moranella*, now completely within its cytoplasm. A considerably larger genome at 538 kb than the near-organelle level of *Tremblaya*, this bacterium was found to code for many of the essential metabolites needed by its bacterium in which it is situated. Moreover, key enzymes and proteins for mealybug function were not merely the result of genes coded within these holobiont bacteria, but were substantially due to lateral transfers of genes from three diverse bacterial lineages over recent evolutionary time. In essence, this tripartite mealybug symbiosis is a holobiont mosaic that may be a model for many holobiont systems across the phyla of life. It is noteworthy that this symbiotic story indicates that pathways other than transfer of symbiont genes to a "host" nucleus, as in the case with many organelles, may be at play among holobionts, given that little evidence was found that the reduction of the genomes among the symbiotic bacteria was due to gene movement to the mealybug nuclei (Lopez-Madrigal et al. 2011).

Studies which reveal the complex holobionts of tsetse flies and mealybugs actually evolved from the many years of research on the aphids–*Buchnera* bacteria symbiosis—except now we know that this is not a pair-wise holobiont, and indeed, it may not be obligate or even appropriate to pigeonhole as a so-called mutualist. Pea aphid holobiont includes associated facultative bacteria not located in the aphid's bacteriocytes. Koga et al. (2003) showed in a landmark study that aposymbiotic aphids infected with only y-proteobacteria secondary symbionts appeared to compensate for much of the *Buchnera* contributions in that the aphid was able to reproduce successfully through several generations. These non-*Buchnera* symbiotic aphids were smaller, and their fecundity was less, but nevertheless, they were fully functional. Interestingly, these substitute secondary symbionts were found to not only be in the usual intercellular regions but intracellularly within the primary bacteriocytes usually occupied by *Buchnera.* The question arises, however, as to how these secondary symbionts—so effective at compensation in laboratory experiments—confer advantages to the holobiont in nature. The answer may lie in the heat stress to which *Buchnera* is susceptible. The secondary symbionts were found to positively impact aphid reproduction under usually detrimental high heat conditions (Montilor et al. 2002). Generally, when both *Buchnera* and the secondary symbionts exist with the pea aphid, the secondary symbionts convey periodic negative effects.

These data indicate that facultative symbionts, once thought to be unimportant or solely detrimental, can under certain environmental conditions compensate for *Buchnera* weakness or loss. To think that a vertically transmitted obligate symbiosis likely "locked in" for over 100 million years (Moran et al. 1993) evolved a compensation factor involving facultative bacterial genomes on standby reinforces not only the community mosaic of symbiotic holobionts but the fluidity and resilience that argue against static categorization of symbiosis.

Bark beetles are another prime example of holobiont community dynamics. These prolific insects dwell in tree phloem somewhat devoid of nutrients and among regions where there are plant-produced anti-herbivory toxins. They thus depend on an array of microbial symbionts—an "expanded genetic repertoire" as leading insect symbiologist Six (2013) calls them. Several bark beetle species colonizing conifers feature novel symbiotic structures called "mycangia," which house obligate associated fungi that provide nutritional selective advantages (Six 2012). Some beetles carry additional fungi, which tap into sapwood and transport it to the phloem, where it is available for the larvae which gain significant amounts of nitrogen, a particularly limiting nutrient in these substrates (Bleiker and Six 2008). A few associated fungi produce sterols that are necessary for the hormones that stimulate reproductive metabolism. The determining factor in the degree of integration for many of the fungal–beetle associations is often temperature. Sudden or unexpected temperature changes can alter fungal populations, a particular concern with increasing anthropogenic climate change threats. Yeasts are also prevalent among bark beetles as well as other insects, with indications that some may even be involved in converting tree chemical compounds to pheromones, but much of their functional importance remains unclear (Six 2013).

A holobiont community would seemingly not be complete without the implications of bacterial genomes. While gut microbes are in low diversity in bark beetles likely due to the more sugar centered as opposed to cellulose diet within the phloem, the nitrogen-fixing bacterium *Rahnella aquatilis* is consistently found in all stages of the beetle life cycle (Six 2013). Tree defense compounds and toxins may be degraded by bacterial symbionts within some beetle species (Boone et al. 2013).

Tropical rain forest biomes are particularly dominated by symbiotic systems (Zook 2010), with one of the most revealing being the attine ant holobiont with its the cascade of adapting players in a symbiosis that likely dates back 65 million years (Mueller et al. 2001). This extraordinary holobiont features the leaf-cutter ant in association with a fungus from the Lepiotaceae family which it cultivates for food but which is consistently threatened by growth of the ascomycete (order Hypocreales) fungus, *Escovopsis*. Symbiotic actinobacteria of the genus *Pseudocardia* populate the ant's surface and convey antibiotic protection often targeted to the specific variety of the invading fungus (Poulsen et al. 2010). Black yeast species in turn tend to limit the actinobacteria growth not through resistance to antibiotics but more through outcompeting the bacteria for food (Little and Currie 2008). While the gut microbiota of leaf-cutter ants is still unknown, a wide variety of ant species are known to harbor specific bacterial symbionts which mediate diet and digestion. Bacteria species of Burkholderiales, Pseudomonadales, Rhizobiales, and others are consistently a part of ant holobiont communities. Russell et al. (2009) concluded that bacteria have facilitated convergent evolution of herbivory across many ant groups and suggested that "symbiosis has been a major force in ant evolution."

9 Symbiogenic Foundation of Earth Biomes

Much of the earth's biosphere is a geosymbiotic construct, indeed often microbiogenic. The topography of terrestrial and marine regions on earth results from the remnants of symbiotic processes. The coral-dinoflagellate holobiont builds rocky substrates, the calcium carbonate reef, which then becomes one of the most biodiverse ecosystems on earth. The process evolves around free-living *Symbiodinium* algal varieties encysted within specialized membranes—symbiosomes—of coral polyp cells transferring as much as 95 % of its photosynthate, usually as glycerol (Stat et al. 2006), to its surrounding animal partner, albeit the degree and timing of transfer dependent on the dinoflagellate clade representative and environmental conditions (Cantin et al. 2009). Without this infection and subsequent multi-genomic integration, there is not the energy or the metabolites to express a reef. The resulting alkaline excretion allows the polyp cells to return to a more acidic, functioning pH (Goreau et al. 1979), albeit the primary selection for such a hermatypic symbiosis may be that coral larvae have a definitive substrate upon which to affix as well as habitats that support organisms which the coral tentacles can capture as sustenance in their multi-trophic lifestyles.

Oceanic reef regions represent only about 0.1 % of the area of the oceans' surface area with approximately 90 % of that total being in the Indo-Pacific. While oceanic reefs—including many of the result of calcareous sponges likely with symbionts—were more prolific in more ancient eras, these water "oases" were always in relatively small patches given the reality that nutrients in tropical waters, which are basically devoid of upwelling, are in short supply. Reef biomass is highly correlated with the diversity of organisms, which depend on the reef structure, not only as habitat but often as a location for pelagic forms to lay eggs before returning to more open waters. In the Great Barrier Reef off the northern coastline of Australia, 30 species of cetaceans live in or visit; 40 species of seabirds, 5000 species of bivalves, 6 breeding species of sea turtles, and 1500 fish live amidst the coral reef architecture http://www.reef.crc.org.au/discover/plantsanimals/

facts_plantanimal.htm. Examined from a symbiosis perspective, these biodiversity numbers increase exponentially when we realize that most of these megaorganisms are themselves holobionts made up numerous microbial symbionts. Moreover, the reef itself supports varieties of free-living microbes, most of which have yet to be discovered, let alone researched.

Through orogeny and terrestrial subsidence, these reefs become part of the lithospheric crust and pedosphere. The reef can then be seen as a limestone-dominated mountain, mountain ridge, peak, mountain chain, rolling hills, plains, or karstic caves. More than 25 % of the surface area of the People's Republic of China is limestone. This includes massive cave regions as well as extraordinary mountain regions in Guangxi region. Even much of the Gobi desert features remnant limestone fine "sand," the result of biogenic rocks ground down by ancient glacial retreat. Biogenic and microbiogenic limestone geology is prolific around the planet, including in North America where one of the largest limestone quarries exists in the state of Michigan.

But, many of the limestone zones are derived from yet another vast holobiont diversity. The most common eukaryote on earth could be *Emiliania huxleyi* and its varieties. This haptophyte alga produces intricate calcium carbonate "tests" known as coccoliths, as it floats within the photic zones in mostly northern temperate seas (Shutler et al. 2010; Holligan et al. 1983). As these massive blooms of algae die, most of the limestone tests gradually reach the benthic regions and accumulate tens of meters thick over tens of thousands of years. Much of the uppermost lithospheric crust of Europe—not merely well-known outcroppings such as in Dover, UK—is remnant coccoliths as well as some foraminifera tests (Huxley 1868). As with all haptophytes, the coccolithophorids are the result of a secondary symbiosis involving a heterotrophic protist phagocytizing a microbial red alga (Archibald 2009), which in turn had of course internalized a free-living cyanobacterium originally.

Thus, limestone-based geology common around the globe and critical to global ecology is a crucial extension and visible reminder of the dominance of symbiogenesis not only in macroevolution but in the emergence and maintenance of the biosphere. The origins and life cycles of karst-depositing hermatypic corals and hapotophytic algae have an impact far beyond its own singular body or colonial structure. The boundaries of holobionts are therefore fluid as well, for they involve the expressions of readily viewable geology, geomorphology, biogeography, and ecosystem dynamics. Symbiosis, as manifested through holobiont communities, is a central component of global ecology.

10 Anthropogenic Threats to Holobiont Global Ecology

If, as the evidence shows, the very foundations of how biomes and its ecosystems emerged and are maintained are substantially symbiosis-reliant, then, one can imagine identifying many symbionts as "keystone" species, i.e., usually inconspicuous, smaller organisms that have a disproportionately significant impact on the greater biodiversity (Zook 2002). An example is the *Ficus* (fig tree) symbiosis with highly specific fig wasps of the superfamily Chalcidoidea. The flower of this prolific tropical rain forest tree is an enclosed receptacle with often hundreds of florets inside. This evolving fruit is called a "syconium" and can only be entered by specific pollinating female wasps through an ostiole. Using its ovipositor, the wasp lays its eggs deep within the stamens, and offspring later fly out carrying fig pollen. *Ficus* trees are critically important to the biodiversity of the rain forests in that a single tree can mast (produce fruits) up to four times each year, providing abundant food for organisms from throughout the phyla (Janzen 1979). The fig wasp is clearly a keystone species. Indeed, conservation policy directed at preserving fig trees will "automatically" help to conserve a wide range of other species, nearly all of which are likely holobiont symbiotic communities themselves. The *Symbiodinium* spp of hermatypic corals are another classic example of keystone species, and how identifying and conserving such symbionts may be essential in the process of not only understanding the symbiotic system and its environment but establishing policies and initiating actions to maintain biodiversity.

While the demise of coral-dinofagellate reefs due to bleaching out of the dinoflagellate algae within the coral cells is the most prominent example of anthropogenic climate change effects on symbiotic systems, emerging research indicates other potentially problematic holobiont changes with significant ecosystem implications. Kiers et al. (2010) in a review paper pointed out that in the last forty years, fertilizer use by humans has increased 700 %, which in turn resurrects the long prevalent concern that such excess over an extended period can translate to demise for some mycorrhizal–plant symbioses, as well as *Rhizobium*–Fabaceae nitrogen fixers. Nutrient-rich sites commonly show replacement of strong mycorrhizal strains with weaker, less advantageous (to the plant) strains (Johnson 1993). Wang and Qui (2006) pointed out that some plants in Brassicaceae that typically thrive in high nutrient soils have lost their ability to form symbioses with mycorrhizae. Kiers et al. (2010) warned of a worrisome picture for the near future with symbiotic systems. They emphasize the likelihood of partner switching as "mutualistic" relationships are threatened and even indicate the actual replacement of a symbiont by antagonistic species. However, while the warnings ring true, the overall analysis cites symbiotic and other ecological systems with minimal consideration of bacteria impacts, now well recognized as critically important in holobiont metabolism, viability, and ontogeny.

Problems in this analysis are compounded by the traditional ecology usage of "mutualists." Mycorrhizae fungi with specific plants are lumped into the same group as bee generalists in ephemeral relationships as pollinators. Indeed, the entire article avoids the terms "symbiosis" or "holobiont." This is all the more confounding when in the same paper, the authors readily admit to fluidity in "mutualisms" (which presumably include some symbioses) pointing out how at ecological and evolutionary timescales the partners shift on a bidirectional continuum from beneficial to antagonistic. Key questions of environmental impacts on partnered organisms are on target, but lost in the questionable uniting of ubiquitous ecological relationships with actual symbioses, as discussed earlier in this chapter.

There are some growing indications that lethal diseases affecting both bats and amphibians worldwide may be related to climate-related temperature changes affecting microbial populations associated with the animals (Daskin and Alford 2012). In bats, there are grounds to speculate that the lethal affect of the fungus *Pseudogymnoascus destructans* (formerly *Geomyces destructans)*, known as white nose syndrome (WNS), may have become pronounced due in part to changes in the bat microbiota. If any of the six species of bats extensively affected are shown to be a holobiont with interacting multiple genomes such as most mammals, some climate change, or environmental effects that helped to foster the fungi could be ameliorated by the fluidity inherent in many bacterial-influenced symbioses. Studies such as that of Daniel et al. (2013) have identified key members of the gut microbiota in the shortnosed fruit bat (*Cynopterus brachyotis*), albeit the authors characterize their work as a search for pathogens. A good start in Chiroptera-microbiota into enquiry of the microbiota of Chiroptera is represented by Phillips et al. (2012) who used comparative metagenomic analysis to not only identify the likely endemic gut microbiota but to indicate how such populations vary dependent on geography, stage of the bat life cycle, and diet.

The destructive agent for amphibians worldwide appears to be fungus *Batrachochytrium dendrobatitdis* (*Bd*) (Kilpatrick et al. 2010). The prevalence and severity of the disease with amphibians have been at higher elevations in the tropics. It is possible that the effects of possible symbiotic bacteria in the animals may have reduced impact on immunity against the fungi in the new temperature regimes influenced by current climate (Daskin and Alford 2012). This view is credible in light of recent work (Myers et al. 2012) that shows antimicrobial peptides (AMP) of the frog *Rana muscosa* secreted onto its skin may work synergistically with metabolites from endemic frog bacteria to confer resistance to the lethal chytridiomycosis. More specifically, *Plethodon cinereus* and skin bacterium *Pseudomonas fluorescens* may be a holobiont in that the bacterium limits the amount of AMP necessary from the frog.

Such findings further promote the concept of bioaugmentation in the face of environmental degradation and climate change. For example, probiotics using anti-Bd bacteria on amphibian skin in vitro reduced the harmful infection (Harris et al. 2009). Administration of specific bacteria to augment immunity in the amphibians could be a necessary conservation measure. Such human intervention is not without risks, for probiotic use could reach other organisms in and beyond the food web or certainly beyond the holobiont. Myers et al. (2012) suggested using an ecological ethics framework such as that of Minteer and Collins (2008) to consider and balance such risks and promote appropriate decision-making that is more conservation helpful than harmful.

Amphibian dependency on its microbiota is perhaps not so surprising given the historic findings of Kerney and colleagues (Kerney 2011; Kerney et al. 2011) and Graham et al. (2013). The eggs deposited as gelatinous masses in shallow waters by North American spotted salamander *Ambystoma maculatum* are later penetrated by a green alga, *Oophila amblystomatis*. Since first discovered decades ago (Gilbert 1942; Hammon 1962; Goff and Stein 1978), it was presumed that this association was an epibiotic ecological association in that perhaps oxygen emitted from the algae through photosynthesis provided an appropriate environment for egg development in an ecosystem context. However, the recent work shows a deeper story that fits in well with the growing holobiont perspectives. The algae actually enter the developing embryo capsules near the blastopore and settle within the cells and tissues of the salamander embryo. Moreover, while oxygen can be a selective advantage for the animal in the holobiont, the alga actually translocates photosynthate to the salamander embryo as well as inhibiting invasive bacterial growth. Comparative studies with non-infected spotted salamanders confirm that the infecting algal symbiont is essential for optimal growth and viability of the salamander. These discoveries open the door for important follow-up enquiries such as whether the algae foster antibiotic production through associated bacteria; how the holobiont, in particular the chlorophyte alga, populations are regulated; and, of course, the obvious developmental biology and immunology questions of how this infection evolved and emerged as obligate. Moreover, this work is especially noteworthy in modern science, for they represent first definitive evidence of algae in symbiosis with a vertebrate; the latter previously considered a completely foreign domain for photosynthesizers.

Much like the bacteria and algae, with respect to the amphibian sustainability, certain mycorrhizal fungi could be a partial solution to both human-caused and natural environmental threats of a quite different nature: The human-created toxic waste sites scattered around the world. Mining for metals and minerals may allow for a supply of consumables deemed important in our societies, but the extraction process results in massive tons of hazardous waste products. For example, in Poland, as in many countries of the world, metals such as zinc, lead, and silver have been extracted for industrial purposes. Entire natural areas have been transformed and degraded. In some cases, excavation and removal of minerals and the corresponding waste has gone on since the twelfth century, but more intensively since the industrial revolution start in the mid-nineteenth century.

Case in point is the once active Trzebionka Mining Works within a major karst belt in southern Poland. Each year over many decades around two million tons of ore had been extracted. The ground-down waste rock and soils from the process were deposited as a 60 m-high heap covering about 64 ha (158 acres). Now, with the site essentially abandoned, there is little effort to water down the dry barren hazardous waste hill, albeit doing so would only be a very short-term measure. Therefore, some of the waste area is completely devoid of nearly all plant life including what was once there as part of a temperate zone forest biome. The dominant elements in these tailings (ore mining waste) are not organic matter but tons of crushed rock resulting in essentially zinc, cadmium, and lead "sand," all at levels far beyond what is tolerable for most life. These toxic-laced particles, usually about 0.3 mm in diameter, easily blow off from the tailing heap surface in even light winds. Rain and melting snows on the tailings tend to run off into nearby greener zones and can potentially percolate to regions where water is used for gardens, farms, or drinking. The area is surrounded by fragmented forest zones, a highway, and some farming and village communities. Phytostabilization of the tailing heap is the only viable practical way to ensure reduction of contamination into neighboring villages and ecosystems. Until recently, even this possibility was far-fetched as it was unimaginable that any plant with its roots could grow and take hold on such a low nutrient and toxic substrate. However, within the emerging subdiscipline of "applied symbiosis," the possibility of remediation is now realized through utilizing selected plants that show some evidence of tolerating extremely harsh soil conditions in association with mycorrhizae (Turnau et al. 2012) (Fig. 4).

Mycorrhizae in association with these "extreme" plants not only can act as root extensions and reach limited phosphorus and water, but its mycelium (extensive hyphal network in the soil) can accumulate and store massive amounts of toxic metals. For example, one arbuscular mycorrhizae type can accumulate 10–20 times more cadmium than the plant roots to which it is associated. Identifying and collecting those plants that grow sporadically at the site, its perimeter, or nearby downslopes have resulted in identifying a growing inventory of those plant–mycorrhizal holobionts which may have the best chance at populating the tailings and then continuing to grow and reproduce into distance future generations. Thus, the field of phytoremediation in once-mined regions where toxic metal waste remains situated substantially depends on the capabilities of the mycorrhizal symbiont in symbiotic association with specific plants (Turnau et al. 2012).

All the examples posed and the many not mentioned usually involve a holobiont community interacting with another holobiont community. Nowhere is this more evident than with the spruce beetle and its microbiota involved in mycorrhiza-supported spruce tree substrates. In Alaska and the adjacent Yukon region in the 1990s, consistently warmer than normal temperatures during summers promoted an extra beetle reproductive cycle, such that eggs were annually doubled in what would usually be over a two-year span (Raffa et al. 2013). With some beetle outbreaks, it is not only the increased reproduction as a result of increased temperature, but also the spread into new regions. For example, the mountain pine beetle expansion in western Canada has expanded over the past 40 years into more

Fig. 4 Abandoned heavy metal mining site in southern Poland where bioremediation via the use of specific mycorrhizal plants is being investigated. Photograph by D. Zook

northerly latitudes and higher altitudes with a 1 °C increase. Because bark beetle bacterial symbionts are known to detoxify tree defense chemicals (Adams et al. 2013; Boone et al. 2013), evolving research is focusing on some manipulation of the bacterial community to alleviate the growing invasive strength of the beetles, especially with increasingly alarming data on anthropogenic climate change.

11 Symbiosis as an Ancient Strategy in Evolution

While symbiosis is front and center in the emerging crises involving anthropogenic-caused climate change and related issues, the evidence indicates that as a prevalent system in the biosphere, symbiosis is both ancient and resilient.

For example, it is likely that there were major selection pressures for the endosymbiotic evolution of the eukaryotic cell two billion years ago. We now know that mitochondria resulted from a free-living facultatively aerobic bacteria being assimilated into a chimeric archaea–eubacterium "host." It is likely that this critically important symbiogensis occurred in part due to the environmental pressures of relatively toxic oxygen levels emerging in a substantially anaerobic world. As has so often been expressed, the serial endosymbiotic theory (SET) for the origin of eukaryotic cells resurrected, restructured, and promulgated by the late Lynn Margulis (Sagan 1967) shows clearly the powerful role acquisition-oriented behavior exemplified by symbiogenesis plays in shaping evolution. For nearly a half century, the energy transforming centers of eukaryotes, mitochondria, and plastids have been the *sine qua non* of symbiosis significance in evolution. Yet, it has always appeared as a kind of strange omission or bias that this endosymbiotic basis of so many life-forms and their metabolism—foreign, greatly reduced, but assimilated genomes resulting from symbiotic acquisition—was and remains relegated in textbooks from high school and upward to a page or two or a special sidebar box. With the holobiont-centered revolution in science real and prominent today, this is finally likely to join the newer prolific discoveries as an exemplar of the new evolution paradigm.

The deep time symbiogenesis story is not only about the essential eukaryotic cell components, for there is significant micro- and plant–fossil evidence that symbiosis was an entrenched lifestyle for a variety of organisms through ancient time. One can even think of the dominant microbiogenic features dating back to nearly 3.5 billion years ago (Schopf and Kudryavtsev 2012) and forward through the Paleozoic, the stromatolites, as a kind of ubiquitous precursor to symbiosis on a grand scale. After all, these lithified structures due to binding and trapping of sediment in usually shallow salty water were the creations of a community of bacteria led by specific polysaccharide-excreting cyanobacteria. Moreover, we can be assured that this prokaryotic layering through photoautrophic growth and post-metabolic mineral deposition consistently included substantial gene transfers, such that any given individual bacterium in the community was likely housing genes from neighbors and the past. These stromatolitic structures when still living entities feature a blue-green color on the rock surface indicating their continued colonization by cyanobacteria and continued growth. Much like the latter biogenic geomorphology represented by limestone generated from secondary symbiont coccolithophorid algal and coral-dinoflagellate holobionts as discussed earlier, stromatolite communities were a dominant biospheric feature with great global ecological importance. These prokaryotic communities became greatly reduced by the Cambrian Period (541–489 mya) as ocean regions became less shallow and less salty and the emergence of a wide variety of algal and cyanobacterial-feeding animals appeared (Schopf 1999).

Dating back to at last 600 million years are the oldest unicellular ancestors of Animalia, the choanoflagellates. Pre-dating sponges, an extant choanoflagellate protist *Salpingoeca rosetta,* has been found to respond to sulfonolipid signaling from associated bacteria that initiates colony formation (Alegado et al. 2012). This is the seed of a fascinating possibility—that multi-cellularity may have arisen through a choanoflagellate–bacterial symbiosis (McFall-Ngai et al. 2013). Sponges actually have choanocytes or "collared cells," much like the choanoflagellate protists. Moreover, nearly all marine sponges are considered now to be symbiotic with wide varieties of bacteria and algae prevalent (Thacker and Freeman 2012). It is striking that these earliest animal forms that remain a highly successful phylum today may be among the most dense and diverse holobiont communities.

Evidence indicates that well before bryophytic and vascular plants, fungi and photoautotrophs were evolving as likely symbioses. The primary terrestrial lifeform most widely associated with symbiosis, lichens, appears now to have had its origins more than 600 million years ago, with the report by Yuan et al. (2005) of hyphae and coccoid cyanobacteria or algae in likely biogenic phosphorite-rich sedimentary rock at Weng'an S. China. In a landmark study, Lutzoni et al. (2001) examined the small and large subunits of nuclear rRNA genes for 52 species from 24 orders of ascomycete fungi that associate as lichens in order to infer the occasions of lichenization and losses of lichenization, as well as to get indications of lichens in more accurate phylogenic placement. The work not only showed lichen symbiosis as more ancient that originally surmised being Late pre-Cambrian in origin, well before the first plants, but that major ascomycete fungal lineages are actually derived from lichen-forming ancestors.

Moreover, electron micrograph examination of fossilized lichens from the lower Devonian (approximately 400–385 mya) indicates, similar to extant lichens, actinobacteria in the medulla layer beneath the photobiont as well bacterial colonies on its surface (Honegger et al. 2013). Reports from the same specialists (Honegger et al. 2009) clearly show well-stratified lichens featuring both cyanobacteria and algae in approximately 415 my strata, while other findings at the Rhynie chert reveal a likely ancient lichen, *Winfrenatia reticulate,* with what are probable filamentous and coccoid cyanobacteria (Karatygin et al. 2009).

One current symbiosis stands out as both very unique and yet with likely deep linkages to ancient terrestrial ecosystems. *Geosiphon pyriforme–Nostoc punctiforme* is one of the few known symbioses involving a fungus and a cyanobacterium. This holobiont grows on soil surfaces and features unicellular bladders about

2 mm long and 0.5 mm in diameter, which house the recognized *Nostoc* filaments. The cyanobacteria are in symbiosomes derived from the fungal plasma membrane. Hyphae are prolific between the symbiosomes (symbiosis-created membranes), and the bladders are substantially chitinous. The *Nostoc* grows and divides within the bladders and produces the non-photosynthesizing specialized spheres (heterocysts) on the filament for nitrogen fixation. It is also photosynthetic in both its sessile colony and its motile hormogonia stages. In fact, there is some evidence it has a higher photosynthetic capacity when associated with the fungus than when isolated (Bilger et al. 2004). The fungus appears to be a likely ancestor of arbuscular mycorrhizal (AM) fungi. Schüssler et al. (2001) showed through SSU rRNA sequencing that both AM fungi and the *Geosiphon* holobiont are a monophyletic group so distinctly separated from other fungi that it constitutes its own new phylum Glomeromycota. This *Geosiphon* symbiosis can be seen as a modern-day remnant of ancient forms that led to mycorrhizal fungi, which in turn later associated with eukaryotic algae en route to initial land plant formation or as an extant more direct AM precursor from which its variations developed into fungal–plant associations. New findings through phylogenetic analysis reveal that six species of liverworts from the earliest diverging clade of land plants, two hornworts and a fern among others associate with Endogone-like fungi (Mucuromycotina) and pre-date the Glomeromycota ancestry back to the mid-Ordovician (475 my) (Bidartondo et al. 2011).

Whether the new endogonaceae family of fungal mycorrhizal data supersede by age or given that the fungi of both the *Geosiphon* and those involved in AM fungi are so similar in features and of the same clade—in either case it is likely that all plant-based terrestrial and even estuarial biomes are and have been foundationally dependent on symbiogenesis at all stages of their evolutionary history.

The initial hypothesis to explain the emergence of plants from a charophycean algal lineage via early mycorrhizal fungi during the Late Ordovician or Early Devonian dates back several decades (Pirozynski and Malloch 1975) has gained further acceptance in more recent years (Turmel et al. 2007; Selosse and Le Tacon 1998). TEM evidence from the fossilized axial prevascular plant *Aglaophyton major* recovered from Early Devonian (419–400 my) strata of the famed Rhynie chert in Aberdeenshire, Scotland, repeatedly shows mycorrhizae fungal infection (Taylor et al. 1995, 2005). Remarkably, other TEM fossil evidence from the same plant and region shows extensive filamentous cyanobacteria colonizing the intercellular spaces of the outer cortex as well as penetrating parenchyma cells within the plant root zone of arbuscular mycorrhizal infection. Often the filaments are seen coiled within the plant cells. Electron micrographs also indicate that entry into the plant is commonly through stomata (Krings et al. 2007a, b). Surface plant openings are often a means of entry in today's plant–cyanobacterial symbioses. For example, in the extant ancient plants *Gunnera* and a variety of cycads, cyanobacterial symbionts enter via surface openings, spread intercellularly, and some become embedded intracellular deeper into the plant structure. Named after a Swedish botanist of the eighteenth century, the herbaceous flowering plant *Gunnera* often features very large leaves of up to 2 m long, and its symbiosis with the cyanobacteria *Nostoc punctiforme* is characterized by prominent glands at the base of its long petioles through which the cyanobacterial symbiont enters and colonizes. The *Nostoc* fixes nitrogen and is vertically transmitted (directly transferred via in the holobiont germ line rather than horizontally, i.e., being acquired each generation from the environment). This unique plant has been dated back to nearly 100 million years through its distinctive fossilized pollen (Jarzen 1980). These findings in specimens from the lower Ordovician through the Cretaceous lend further credence to the view that symbiosis, even apart from eukaryotic cell origins, is ancient and likely had high selective advantages for organisms, including for transitions to very new environments, adaptation to climate changes, and procuring better access to sustaining resources. Moreover, if the unit of selection is the holobiont as is now being widely considered, natural selection would favor those forms that were able to adapt most quickly, that is, without the extremely slow and usually lethal process of point mutation change.

This speed of symbiogenesis is most readily revealing in the pioneering work of Kwang Jeon. Jeon discovered that one of the amoeba cultures he had been growing in his laboratory become infected with colonies of a still unidentified *Legionella*-like bacterium that could not be separately cultured. These gram-negative rods had the effect of killing off most of the amoebae. However, several amoebae appeared to tolerate the bacterium (Jeon and Lorch 1967). Their numbers peaked regularly at 42,000 per amoeba cell, each sequestered as groups within amoeba-generated membranes or "symbiosomes." Within 18 months or approximately 200 amoeba cell generations, the two genomes became obligately dependent on each other. Indeed, the new symbiosis based on the bacterial infection could no longer coexist with the original amoeba and became restricted to narrow temperature regimes and conditions (Jeon 1995). The emerging amoeba–bacteria holobiont was essentially a new species in the geological time equivalency of a blink of an eye, became the centerpiece of important evolutionary and symbiosis investigations, and continued to thrive through thousands of generations for years after. While the laboratory and its nutrient-filled petri dishes represent an artificially created environment, rather than in nature per se, this series of longitudinal studies extending from 1965 to the present day are nevertheless suggestive of how quickly acquisition of genomes can occur, be viable, and result in potentially new taxa. Increased rates of evolution are also indicated in metagenomic enquiries, including with lichens, wherein Lutzoni and Pagel (1997) showed much higher rates of nucleotide substitutions in nuclear ribosomal DNA in the symbiotic lichenized state and with liverworts associated with fungi than with non-symbiotic associated fungi.

12 The Human Microbiome: A Centerpiece of Symbiogenesis

As 2013 closed out, there were about 1,200 refereed, published articles in journals that appear when the keywords "human microbiome" are inserted. The majority of the titles are mainly in the past six years but date back about ten years. Prior to that time, there were perhaps a half dozen. Nothing has spurred the renewed recognition of the centrality of symbiosis and bacterial gene movement in our biosphere than this "new" discovery of a biome literally under and including our nose. We as humans and all the mammalian kin and indeed all those that emerged from blastula developmental architecture have joined the rain forests and coral-dinoflagellate reefs as key centers of biodiversity. Due to our proclivity to know as much about ourselves as we can—some would say due to our egocentricity and correspondingly minimal humility in the face of nature—we have poured time, monies, and resounding inquisitiveness into finding out who is inhabiting us and why. Only of course to find out that the us is not *Homo sapiens*, the individual member of a species, but rather *Homo sapiens* the mobile ecosystem comprised of millions of life-forms, indeed more microbes in and on one human body than that human being's total number of cells or to realize that the microbiota of one human body has nearly 100 times more genes that its associated animal "self" (Nelson et al. 2010). While a first reaction might well have been there is more of them than me, we know that we are on the verge of discovering that each one of us was never "I" but always "we" (Gilbert et al. 2012). Could there be alien microbial life in the solar system has now been replaced with what is the function and meaning of the "alien" life in us, the human holobiont community?

The human microbiome, inclusive of interacting bacteria and the less studied viral populations, can be functionally envisioned as a classic wheel model in that the hub of governance and stability is the intestinal organs—the six meter coiled small intestine and the slightly shorter but much wider large intestine. This extraordinary gut system is akin to the hermatypic corals' calcium carbonate reef, as its folds and crevices maximize volume, and house a remarkable diversity of microbial life. Indeed, extensive genomic studies by Eckburg et al. (2005) led them to conclude, "Bacterial diversity within the human colon and feces is greater than previously described, and most of it is novel." The spokes of this hub are the specific array of often bidirectional and biochemical signals to and from the gut microbiota to and from the respective organ systems; namely, immune, circulatory, digestive, reproductive, neuroendocrine, musculoskeletal, and so on, while the wheel rim are these systems to which the spokes are spatially, chemically attached. One could say the outer tire represents the direct contact of this mammalian holobiont with the greater surrounding ecosystems through which the "wheel" traverses. But, what are the evidences for such a scenario and to what degree are such interactions "symbiotic"?

Work in the field of gnotobiology (artificially raised "germ-free" animals) allows one to see whether there are functional deficiencies or defects as compared to those populations raised in a normal microbe-colonized environment. In such studies, gut microbiota were found to be essential for intestinal immune maturation, warding off infections by inducing increased "T" cells (called such for they mature in the thymus). Moreover, the bacterial inducers must be the "correct" recognized ones (Chung et al. 2012). This has implications for medical treatments as well as suggesting that interactions between the human cell and bacterial genomes are likely well-coordinated and not happenstance. In a remarkably thorough review citing scores of studies, Nicholson et al. (2012) emphasized how the human cells and bacteria are involved in an ongoing "cross talk" through signaling pathways within the immune system and beyond. They point out, "These immune-mediated signaling processes, together with direct chemical interactions between the microbe and the host, act upon multiple organs such as the gut, liver, muscle, and brain…. Multiple bacterial genomes can sequentially modulate metabolic reactions resulting in a combined metabolic process by the microbiome and host genome."

The growing evidence of widespread microbiota controls on the human immune system mirrors the findings with other animal holobionts which often incorporate and maintain a bacterium that confers protection against common infectious agents. For example, *Drosophila neotestacea* is susceptible to infections which lead to sterility by various nematodes. However, those Drosophila that included *Spiroplasma* bacteria as part of its holobiont community were more tolerant of such nematodes and did not become sterile. The bacteria were found to inhibit the actual size and therefore potential reproductive output of the adult female worms (Jaenike et al. 2010). In one of many examples involving *Streptomyces* bacteria, *Philanthus* (a beewolf wasp) larval nests can be overcome by infecting bacteria and fungi, not unlike the threats to the subterranean nests of the leaf-cutter ant. However, most such beewolf wasps are now able to be more protective of offspring through the development of glands at the base of the mother's antenna which house the antibiotic-producing *Streptomyces* spp. The mother wasp actually actively secretes the liquid containing the actinobacteria onto the developing offspring as they spin their cocoons (Seipke et al. 2011). In another example, this one involving vertebrates, the colorful bird revered in Egyptian history and other venues, known as the European Hoopoe (*Upupa epops*) accesses secretions from its uropygial gland through preening. These secretions contain specific volatile chemicals produced by bacterial symbionts, such as *Enterococcus*, which reduce potentially deleterious high numbers of diverse bacteria in this avian holobiont (Martin-Vivaldi et al. 2010).

The human microbiome also shows some affinity with other animals when evaluating the digestive tract. Bacteria as well as some protists and fungi in many vertebrates, birds, reptiles, and amphibians convert food materials to absorbable nutrients and ferment carbohydrates into short-chain fatty acids which foster energy and ultimately allows for more efficient absorption of salts and water. In some herbivores, the larger gut capacity through the presence of a foregut allows for additional fermentation by a microbial community that synthesizes proteins and B vitamins (Stevens and Hume 1998). In the human digestive tract, while the diversity of microbes changes radically with different food intake, Wu et al. (2011) founded specific characteristic bacteria or "enterotypes" associated with long-term diets that dominated the gut microbiome and were not easily altered. *Bacteroides* spp. predominated in diets with high animal fat and protein, while *Prevotella* spp. was the enterotype for high carbohydrates diets. This reinforces the view that food intake is a significant contributor to the human microbiome and once acclimated are not in the short-term susceptible to major change. Pepper and Rosenfeld

(2012) emphasized the need more than ever to see animal bodies as ecosystems and suggest that the consistently more static enterotypes combined with often shifting larger microbial populations reported in the human microbiome may be an example of "multi-stability," in that the bacteria–human holobiont has evolved to stay relatively stable under wide-ranging conditions. This can be compared to migrating species or seasonal eutrophication in larger ecosystems wherein the stability of the latter is not fundamentally altered. The diet-microbiome linkage has ramifications for sickness and obesity study (Ley et al. 2006) in that the pathway of chosen external foods to core gut bacteria to then degree of body size and eventually "good health" may be significantly intertwined. Moreover, the linkage of the mammalian microbiome to global ecology resonates profoundly with research reports from Dominguez-Bello and her team (Clemente et al. 2015). Their extended microbiome studies on the isolated Yanomami indigenous peoples of Venezuela show levels of microbial diversity far in excess to what has been measured in the microbiome of modern western civilization cultures. The results imply that modern day eating habits and related behaviors may strongly limit microbiome potential and ultimately human health. This research opens an exciting and potentially a revealing pathway to understanding the evolution of the microbiome within the mammalian holobiont.

At first, suggestions that the brain may be subject to microbiome influence seems far-fetched, even science fiction, until we simply realize that the brain like all other body organs depends on intake of nutrients conveyed by the bloodstream from the intestines. And, if nutrient supplies, catalysts, processes, degraders, recyclers, and signalers are substantially microbial, the connection becomes profoundly logical. Indeed, Nicholson et al. (2012) pointed out and McFall-Ngai et al. (2013) reinforced that as much as one-third of the metabolites that are distributed through our blood circulatory system to our body organs are of gut microbial origin.

Neuroscience, microbiology, and ecology have begun a prolonged and essential meeting at the human microbiome. The growing number of research papers on this aspect is a testament to this. Particularly noteworthy are the detailed studies such as by Heijtz and his team in Stockholm, Sweden (2010). They found in repeated testing with mice that germ-free mice and normal microbiota (specific-pathogen-free) mice differed significantly in motor control and anxiety behaviors. However, if germ-free mice were exposed to normal gut microbiota very early in life, they display behaviors and motor control similar to the mice with normal microbiota. Human microbes particularly target, they discovered, two key synaptic proteins, PSD-95 and synaptophysin. Intriguing and profound linkages usually involving complex chemical signaling of the gut to the brain and vice versa are being consistently reported (Wang 2002; Forsythe et al. 2010).

The unfolding of the human microbiome energized by the Human Genome Project certainly puts ecology front and center as the science of what we formerly would call the individual. Still more revelatory is that under our symbiotic definitions, including the new one proposed in this chapter, symbiosis can be seen as both prevalent and governing in the functioning of all megafauna and megaflora.

13 Summary

The impacts of symbiosis and symbiogenesis on evolution, ecology, and earth science include the following:

1. *Establishes essential novelty upon which natural selection "acts" through the acquisition of nonself genomes which have a vastly different phylogeny. The emerging holobiont is then further acted upon by natural selection, resulting in a new organism and often the start of expansive lineages. The primary unit of selection is the multi-genomic holobiont.*
2. *Reveals that the integration of genomes from vastly different lineages often fosters new geodynamic substrates—reefs, calcium carbonate/marble deposits, caves, and Ficus-enriched forest canopies—that become physical substrates and habitats for the emergence of novel "communities" and expanding lineages.*
3. *Biome and ecosystem foundations extend deep into the fossil record. Symbiosis was likely ubiquitous in the biosphere from the late Proterozoic through the Phanerozoic to the present. Symbionts can thus often be seen as foundational and serve as "keystone" expressions for both the specific holobiont within the larger ecosystem in a macroevolution perspective and for holistic systems development from an earth history and homeostasis view.*
4. *Renders the concept "individuals" among eukaryotes as mythical. The "self" is incomplete and non-functional without the integration of foreign genomes and frequent gene transfers from "foreign" bacteria and viruses. The reality in the biosphere is that all eukaryotes are actually metagenomic entities functioning as an integrated community, the holobiont. Prokaryotes are often significant symbionts in and on eukaryotic holobionts, albeit the prokaryotic cell itself is a holobiont more from consistent gene transfers than whole genome assimilation.*

14 Epilogue: The Insightful Proponent of Symbiogenesis and the Concept of the Holobiont, Lynn Margulis

The distinguished researcher Margaret McFall-Ngai and her colleagues conducted revealing and often elegant work with the dynamic *Euprymna–Vibrio* bioluminescence research over many years and thereby helped pave the way for the new symbiosis-centered paradigm for life on earth. Her review of this new perspective published with many accomplished symbiosis research colleagues (2013) as well as the brilliant treatise of Gilbert et al. (2012) are already seen as historic contributory bridges to the holobiont perspective and symbiogenesis. In the former paper, McFall-Ngai et al. remarked, "For much of her professional career, Lynn Margulis (1938–2011), a controversial visionary in biology, predicted that we would come to recognize the impact of the microbial world on the form and function of the entire biosphere, from its molecular structure to its ecosystems. The weight of evidence supporting this view has finally reached a tipping point…."

In this context, I share a recorded, previously unreleased interview excerpt that I conducted with the late Lynn Margulis, who was a friend and frequent mentor for over three decades and whose course she passed on to me when she left Boston University and which I taught with the appropriate major updates for twenty-five years. Designed for those just beginning to explore the importance of symbiosis, Lynn informally reviews in this excerpt a few Darwinian basics and discusses some key differences of symbiogenesis and neo-Darwinism. Rather than excerpting words from her extensive publications, which are deserving of the reader's more prolonged attention and study, I share here this brief portion of the interview, focusing particularly on the centrality of symbiogenesis in evolution. (Margulis 2009, interview by Douglas Zook, video recorded by Michael Lee and video/audio edited by Divya Madhavan):

> What do virtually all eukaryotes, even diatoms, do that no prokaryotes ever do? Eukaryotes can take up new genomes which may ultimately be symbionts essentially a genome at a swallow. And, that is the crucial point. Eukaryotes have steroid-containing membranes. They open the membranes and they take things in, and they can of course digest those organisms, but they do not have to… If the digester is resisting and under conditions where the digester and the potentially digested then live together for an extended period, you tend to have these associations. So with eukaryosis, there is this ability to open membranes, close them with a foreign genome enclosed, and both survive! That is something you don't see in bacteria. Now, we know that bacteria have invented just about all the main metabolic processes for life….nitrogen fixation, methanogenesis, sulphide reduction, sulfur oxidation, and of course chemo- and photoautrophy. We could go on and on… But the getting together is pretty weird in bacteria. While they form tight communities, their relations are substantially external. They are practitioners of syntrophy, where one produces one product and the other uses it. Are you not amazed with fertilization in eukaryotes?! In fertilization, you open a membrane and something comes in exclusively and closes it again. That's what is going on in symbiogenesis… We have phagocytosis, exocytosis, endocytosis…. We have all these fancy words, but we don't have the intellectual understanding yet that these are all words for basically the same kind of common central process in evolution. And its prevalence in evolution shows us that symbiogenesis becomes the rule of speciation, innovation, higher taxa formation, once you have a eukaryotic world which is always superimposed on a prokaryotic world. The prokaryotic world of course remains and thrives, but members can also be assimilated into the eukaryotic structure.
>
> Ernst Mayr said it well when he pointed out that when you are concerned with evolution, you cannot simply be an evolutionist. It is a multi-component theme. There are many processes involved. What are they? Darwinian evolution has these main components. The tendency of all populations of organisms is to grow exponentially, beyond what the resources available can support. An example is the fungi *Alternaria fusarium* which make 100-150,000 spores per minute for six months. Of course they are growing on a tree. Humans have the potential to have 20-21 children per couple. The bacteria that we can see and count…a single bacterium doubles to two, four, eight, sixteen, thirty two, and so on. A single bacterium can generate the weight of the earth unchecked in less than a week. The potential to grow is everywhere, and that potential is never remotely close to being reached…. And it can be studied in orchids with their tiny seeds, plants that grow vegetatively like the philodendrons here where we sit…every organism can theoretically have a number associated with it, which we call its biotic potential, that is the number of organisms produced per unit time or translated to the number of organisms produced per generation. This is characteristic of all life, always. The fact that the biotic potential is not reached…that we don't have a bacterial planet that is only saturated with bacteria, that is what we call natural selection. Natural selection is the elimination of organisms, the what is left over – because they always have "checks," as Darwin would call it. Checks

are, among many, lack of food, lack of water, lack of space, disease. Those are among the agents of selection. We have wonderful examples of protective coloration where the animal is in a proper environment and it is completely hidden. Those organisms will not be selected against relative to that same animal just a few meters away that is exposed. Natural selection is the fact of biotic potential, which is measurable and is not reached.

So natural selection within evolution maintains what is already there with respect to that environment….It is all about the ones who have made it through to reproductive age and had offspring which then had offspring. And, of course the vast majority of all species to have been recorded on the earth are extinct and the vast majority of all offspring do not move on to produce more offspring indefinitely. In the human species, it acts mostly at the level of two billion sperm per ejaculation and often not even one gets through to fertilize! So, there is a huge example… So with every organism you can show that there is the potential to grow new offspring, and it is not reached.

Now what is the essential difference between the symbiogenic view of evolution and the standard neo-Darwinian view of evolution? Darwin was quite different than neo-Darwinian, indeed he was more Lamarckian in many ways. Well. you and I were taught that the source of all variation, the differences from parents, are the accumulation of random mutations. I remember being told that there was direct evidence that all offspring are not exactly like their parents, and there are lots of reasons for that. And, as Darwin said, we are only interested in the variation that is important to us, and by that he meant the inherited variation. So we are looking at inherited variation – color of our eyes, your blood type, skin, hair qualities and so on with respect to people. There's this variation in traits that are of real interest to evolutionary processes because they have 100 % heritability potential for example. These high heritability traits can be measured. From generation to generation the probability of laying 12 eggs during a week in a season or something like that. This can be inherited. Now here are variations from parent to offspring whether they are non-sexually produced from one parent or whether there are two parents, the source of the inherited variation as told to me and in every book is random mutations. And, when there is enough random mutations accumulated, you have new species. So the main unit of variability is said to be mutational changes in base pairs of DNA, and there's of course recombination and immigration and emigration in natural populations. These are listed as the sources of inherited variation. This is where I part company, not with Darwinism but with neo-Darwinism.

Take a Drosophila and induce random mutations. You will get a sick or dead Drosophila. You don't get a new species. It is nearly always deleterious. I have looked for years for examples of how mutations produce a new species in any literature. Even the best examples from neo-Darwinists involve the acquisition of mycoplasms or other bacteria. The main way that inherited variation is positive, that is it gives you new changes that Lamarck did not understand, is not of inherited characteristics but of entire genomes, bacterial genomes or fungal genomes. There's lots of different examples of course. The random mutations hone, modify, modulate and yes this is important. But when you acquire and integrate a whole genome, you gain the key component in evolution - variability - which often results in speciation. For example, you get a slug that gropes around eating in translucent environments and it is taking in chloroplasts and that animal turns green relative to its non-chloroplast relatives, and in one step, much like punctuated equilibrium, you get a new species. My favorite example is actually the Convoluta symbiogensis examples. Convoluta convoluta is a little flatworm, and it eats and digests all sorts of algae on the western European shorelines but does not retain them. But *Convoluta roscofensis* is a new species from that non-symbiotic lineage. It is green because it took in but did not digest certain Platymonas algae. Every member of the population is green and has phototactic responses. They are all photosynthetic except the eggs, which hatch out, feed and digests other microbes and eventually assimilates the alga it is programmed to recognize. *Convoluta paradoxa* on the other hand is brown, solitary, grows in a different way, is found in a different habitat and has different symbiotic algae, diatoms. There, through these three we can see genome acquisition, variation and hence speciation through symbiogenesis (Fig. 5).

Fig. 5 Author of this chapter, Biologist Douglas Zook with Lynn Margulis at Boston University, 2009 in an image from previously unreleased video interview. Photograph by Michael Lee

Glossary

Following are selected terminology defined by the author that may be of use to some readers.

Arbuscular Branching tree like hyphae of mycorrhizal fungi within, but not entirely enclosed, plant root cells

Actinobacteria Filamentous bacteria commonly found in soils and featuring an array of antibiotic chemistry

Archaeans Microscopic organisms that thrive in "extreme" temperature or saline conditions. They have many biochemical and genetic features that are closer to eukaryotes than prokaryotes

Ascomycete Small craterlike features on the surface of fungi and lichens, from which spores are emitted

Bacteriocytes Specialized intracellular regions of many insects that house symbiotic bacteria which are transmitted via the insect egg and often grouping during the life cycle to form functional organs known as bacteriomes

Bioaugmentation Any intervention by humans that seeks to promote the viability and fitness of a holobiont (organism) living in non-anthropogenic nature

Chimera In the context of a holobiont, it is a collection of different genomes interacting as one entity

Coccoliths The plates of calcium carbonate (limestone) surrounding holobionts known as coccolithophores. These algae in the group haptophta build these structures as part of their outer covering

Endemic A species that is characteristic of a biogeographical region over a significant period of geologic time

Extant In the context of biology and evolution, organisms or conditions from more ancient geological time that have persisted to the present

Endophytes Bacteria or fungi that live symbiotically in between or within plant cells

Epibiotic An organism lives on the surface of another different organism. It may or may not be symbiotic

Facultative An organism that functions with clear options such as being to live in either aerobic or anaerobic conditions

Gnotobiology The study of organisms living in an artificially created environment, namely in conditions where no other living organisms are present

Heamatophagous The ability of certain animals to penetrate body parts of other organisms and feed on blood

Hermatypic Coral–dino holobionts that build exoskeletons known commonly as reefs, as opposed to many corals which do not extrude limestone and thus known as a hermatypic

Holobiont Any living entity (all eukaryotes and rarely some prokaryotes) made of two or more different symbionts—minimally a so-called host species and different symbiont species

Horizontal transmission The passing of a symbiont to following generations through one symbiont acquiring the other symbiont from the environment

Karst Geological formations usually created by the dissolution of carbonate rocks such as limestone

Lithosphere The outermost section of the solid earth, frequently referred to as "crust" but encompassing as well somewhat deeper layers, such as the upper region of the mantle. Much of the lithosphere can be considered part of the region where life can be found, known as the biosphere

Metagenomic The collection of genomes from different organisms as collected directly from the natural environment as opposed to laboratory cultures

Microbiogenic Geological structures and features which are the result of living microbial processes and depositions

Nucleomorph A genetic fraction or remnant of a previously complete nucleus from an alga and now embedded in a new alga with its own nucleus

Pedosphere The outermost layer of the solid earth composed of the soil and rock eroding regions

Peptidoglycan A chemical compound made up of sugars and amino acids that forms a mesh-like cell layer known as the bacterial cell wall. It is the defining characteristic of eubacteria, for it is not found in the microbial domain, Archaea

Phagocytosis The process whereby a cell, usually a eukaryotic one, or an organism envelopes and then internalizes materials or other organisms from the surrounding environs

Rhizosphere The soil regions among the roots of plants, including the organisms and all their interactions

Rumen The specialized first section of the alimentary canal of many hooved animals, wherein fermenting, cellulose-producing microbes are housed

Stromatolites Lithified structures built by the trapping, binding, processing, and then deposition of sediment by cyanobacteria. They are prominent in the fossil record and serve as evidence that our oxygenated atmosphere was substantially the result of cyanobacterial metabolism

Syconium The section of the Ficus (fig) tree that becomes a fruit, but initially is a completely enclosed structure with numerous internal flowers. Only its holobiont specific partners, certain fig wasps can gain entry and promote the necessary pollination

Symbiosome A specialized membrane usually substantially formed by the "host" member of a holobiont which completely encloses the entering or captured symbiont

Syntrophy One species lives off the products of another organism

Thallus The living structure built by the algal–fungal lichen symbiosis. It bears little or no resemblance to the morphology of either the fungus or the alga. Some lichens have a cyanobacterial holobiont partner which also contributes to its development

Trophosome A specialized symbiosis-created food-processing organ which houses sulfur oxidizing and other bacteria, in deep sea vent tube worms

Vertical transmission The persistence from generation to generation of a symbiont(s) through direct transfer via the "host," often through incorporation within or attachment to an egg

Viviparous Animals which produce live young emerging from the body as opposed to the deposition externally of eggs

References

Adams AS, Aylward FO, Adams SM, Erbilgin N, Aukema BH, Currie CR, Suen G, Raffa K (2013) Mountain pine beetles colonizing historical and naïve host trees are associated with a bacterial community highly enriched in genes contributing to terpene metabolism. Appl Environ Microbiol 79(11):3468–3475

Ahmadjian V (1993) The lichen symbiosis. Wiley, New York

Ahmadjian V, Jacobs JB (1981) Relationship between fungus and alga in the lichen *Cladonia cristatella* Tuck. Nature 289:169–172

Alegado RA, Brown LW, Cao S, Dermenjian RK, Zuzow R, Fairclough SR, Clardy J, King N (2012) A bacterial sulfonolipid triggers multicellular development in the closest living relatives of animals. eLife. doi:10.7554/eLife.00013

Archibald JM (2009) The puzzle of plastid evolution. Curr Biol 19:R81–R88. doi:10.1016/j.cub.2008.11.067

Archibald JM (2011) Origin of eukaryotic cells: 40 years on. Symbiosis 54:69–86. doi:10.1007/sI3199-011-0129-z

Attardo GM, Attardo M, Lohs C, Heddi A, Alam UH, Yildirim S, Aksoy S (2008) Analysis of milk gland structure and function in *Glossina morsitans*: milk protein production, symbiont populations and fecundity. J Insect Physiol 54(8):1236–1242

Balmond S, Lohs C, Aksoy S, Heddi A (2013) Tissue distribution routes for the tsetse fly endosymbionts. J Invertebr Pathol 112:S116–S122

Baumann P (2005) Biology of bactericyte-associated endosymbiosis of plant sap-sucking insects. Annu Rev Microbiol 59:155–189

Bhattacharya D, Yoon HS, Hackett JD (2003) Photosynthetic eukaryotes unite: endosymbiosis connects the dots. BioEssays 26:50–60

Bhattacharya D, Pelletreau KN, Price DC, Sarver KE, Rumpho ME (2013) Genome analysis of *Elysia chlorotica* egg DNA provides no evidence for horizontal gene transfer into the germ line of this kleptoplastic mollusc. Mol Biol Evol 30(8):1843–1852. doi:10.1093/molbev/mst084

Bidartondo MI, Read DJ, Trappe JM, Merck J, Ligrone R, Duckett JG (2011) The dawn of symbiosis between plants and fungi. Biol Lett 7:574–577

Bilger W, Budel B, Mollenhauer R, Mollenhauer D (2004) Photosynthetic activity of two developmental stages of a *Nostoc* strain isolated from *Geosiphon pyriforme*. J Phycol 30(2):225–230

Bleiker K, Six D (2008) Competition and coexistence in a multi-partner mutualism: interactions between two fungal symbionts of the mountain pine beetle in beetle-attacked trees. Microb Ecol 57:191–202

Boone C, Keefover-Ring K, Mapes AC, Adams AS, Bohlmann J, Raffa RF (2013) Bacteria associated with a tree-killing insect reduce concentrations of plant defense compounds. J Chem Ecol 39(7 special issue):1003–1006

Bright M, Bulgheresi S (2010) A complex journey: transmission of microbial symbionts. Nat Rev Microb 8(3):218–230

Cantin NE, van Oppen MJH, Willis BL, Mieong JC, Negri AP (2009) Juvenile corals can acquire more carbon from high-performance algal symbionts. Coral Reefs 28:405–414

Cardinale M, Puglia AM, Grube M (2006) Molecular analysis of lichen-associated bacterial communities. FEMS Microbiol Ecol 57:484–495

Chung H, Pamp SJ, Hill JA, Surana NK, Edelman SM, Troy EB, Reading NC, Villablanca EJ, Wang S, Mora JR, Umesaki Y, Mathis D, Benoist C, Relman DA, Kasper DL (2012) Gut immune maturation depends on colonization with a host-specific microbiota. Cell 149:1578–1593

Clemente JC, Pehrsson EC, Blaser M, Kuldip S, Zhan G, Wang B, Magris M, Hidalgo G, Contreras M, Noya-Alarcón Ó, Lander O, McDonald J, Cox M, Walter J, Oh PL, Ruiz JF, Rodriguez S, Shen N, Song SJ, Metcalf J, Knight R, Dantas G, Dominguez-Bello MG (2015) The microbiome of uncontacted Amerindians. Science Advances 1(3). doi:10.1126/sciadv.1500183

Daniel DS, Ng YK, El Chua, Arumugam Y, Wey LW, Kumaran JV (2013) Isolation and identification of gastrointestinal microbiota from the short-nosed fruit bat *Cynopterus brachyotis brachyotis*. Microbiol Res 168(8):485–496

Darwin C (1858) The variation of animals and plants under domestication, vol 2. Orgame Judd, New York

Darwin C (1859) On the origin of species. Empire Press, New York (reprint 2011)

Daskin JH, Alford RA (2012) Context-dependent symbioses and their potential roles in wildlife diseases. Proc R Soc B. doi:10.1098/rspb.2011.2276

De Bary A (1879) Die Erscheinung der Symbiose. In: Vortrag auf der Versammlung der Naturforscher und Arzte zu Cassel. KJ Trübner, Strassburg

Delwiche CF (1999) Tracing the thread of plastid diversity through the tapestry of life. Am Nat 54:S164–S177

Douglas AE (1994) Symbiotic interactions. Oxford University Press, Oxford (out of print)

Douglas AE (2010) The symbiotic habit. Princeton University Press, Princeton

Eaton CJ, Cox MP, Scott B (2011) What triggers grass endophytes to switch from mutualism to pathogenism? Plant Sci 180:190–195

Eckburg PB, Bik EM, Bernstein CN, Purdom E, Dethlefsen L, Sargent M, Gill SR, Nelson KE, Relman DA (2005) Diversity of the human intestinal microbial flora. Science 308:1635–1638

Forsythe P, Nobuyuki S, Dinan T, Taylor VH, Bienenstock J (2010) Mood and gut feelings. Brain Behav Immun 24:9–16

Gilbert PW (1942) Observations on the eggs of *Ambystoma maculatum* with special reference to the green algae found within the egg envelopes. Ecol 23:215–227

Gilbert SF, Sapp J, Tauber AI (2012) A symbiotic view of life: we have never been individuals. Q Rev Biol 87(4):325–341. doi:10.1086/668166

Goff LJ, Stein JR (1978) Ammonia: basis for algal symbiosis in salamander egg masses. Life Sci 22:1463–1468

Goreau TF, Goreau NI, Goreau TJ (1979) Corals and coral reefs. Sci Am (August):124–136

Graham ER, Fay SA, Davey A, Sanders RW (2013) Intracapsular algae provide fixed carbon to developing embryos of the salamander *Ambystoma maculatum*. J Exp Biol 216(3):452–459
Grube M, Cardinale M, de Castro Jr JV, Müller H, Berg G (2009) Species-specific structural and functional diversity of bacterial communities in lichen symbioses. ISME J 3:1105–1115
Hammen CS (1962) Carbon dioxode assimilation in the symbiosis of the salamander *Ambystoma maculatum* and the alga *Oophila amblystomatis*. Life Sci 10:527–532
Harris RN, Lauer A, Simon MA, Banning JL, Alford RA (2009) Addition of antifungal skin bacteria to salamanders ameliorates the effects of chyridiomycosis. Dis Aquat Org 83(1):11–16
Holligan PM, Vollier M, Harbour DS, Camus P, Champagnephilippe M (1983) Satellite and ship studies of coccolithophore production along a continental shelf edge. Nature 304:339–342
Honegger R, Edwards D, Axe L (2009) The earliest records of internally stratified cyanobacterial and algal lichens from the Lower Devonian of the Welsh Borderland. New Phytol 197(1):264–275
Honegger R, Axe L, Edwards D (2013) Bacterial epibionts and endolichenic actinobacteria and fungi in the Lower Devonian lichen *Chlorolichenomycites salopensis*. Fungal Biol 117(7–8):512–518
Husnik F, Naruo N, Koga R, Ross L, Duncan RP, Fujie M, Tanaka M, Satoh N, Bachtrog D, Wilson ACC, von Dohlen CD, Fukatsu T, McCutcheon JP (2013) Horizontal gene transfer from diverse bacteria to an insect genome enables tripartite nested mealybug symbiosis. Cell 153:1567–1578
Huxley T (1868) On a piece of chalk. MacMillan's Mag (reprint available: Kessinger Publishing, LLC 2010)
Inagaki Y, Dacks JB, Doolittle WF, Watanabe KI, Ohama T (2000) Evolutionary relationship between dinoflagellates bearing obligate diatom endosymbionts: insight into tertiary endosymbiosis. Int J Syst Evol Microbiol 50:275–281
Jaenike J, Unckless R, Cockburn SN, Boelio LM, Perlman SJ (2010) Adaptation via symbiosis: recent spread of a Drosophila defensive symbiont. Science 329:212–215
Janzen DH (1979) How to be a fig. Ann Rev Ecol Syst 10:13–51
Jarzen DM (1980) The occurrence of *Gunnera* pollen in the fossil record. Biotropica 12:117–123
Jeon KW (1995) The large, free-living amoebae: wonderful cells for biological studies. J Eukaryot Microbiol 42:1–7
Jeon KW, Lorch IJ (1967) Unusual intra-cellular bacterial infection in large, free-living amoebae. Exp Cell Res 48:236–240
Johnson NC (1993) Can fertilization of soil select less mutualistic mycorrhizae? Bull Ecol Soc Am 3(4):749–757
Karatygin IV, Snigirevskaya NS, Vikulin SV (2009) The most ancient terrestrial lichen *Winfrenatia retuculata*: a new find and new interpretation. Paleontol J 43(1):107–114
Kerney R (2011) Symbiosis between salamander embryos and green algae. Symbiosis 54(3):107–117
Kerney T, Kim E, Hangarter R, Heiss AA, Bishop CD, Hall BK (2011) Intracellular invasion of green algae in a salamander host. Proc Nat Acad Sci USA 108(16):6497–6502
Kiers T, Palmer TM, Ives AB, Bruno JF, Bronstein JL (2010) Mutualisms in a changing world: An evolutionary perspective. Ecol Lett 13:1459–1474
Kilpatrick AM, Briggs CJ, Daszak P (2010) The ecology and impact of chytridiomycosis: an emerging disease of amphibians. Trends Ecol Evol 25(2):109–118
Koga R, Tsuchida T, Fukatsu T (2003) Changing partners in an obligate symbiosis: a facultative endosymbiont can compensate for loss of the essential endosymbiont *Buchnera* in an aphid. Proc R Soc Lond B 270:2543–2550
Kozo-Polyansky BM (1924) Novyi printsip biologii: ocherk teorii simbiogeneza (The new principle of biology: an essay on the theory of symbiogenesis). Puchina, Moscow-Leningrad
Kozo-Polyansky BM (2010) Symbiogenesis: a new principle of evolution (trans: Fet V, Fet V, Margulis L (eds)). Harvard University Press, Cambridge

Krings M, Kerp H, Hass H, Taylor TN, Dotzler N (2007a) A filamentous cyanobacterium showing structured colonial growth from the Early Devonian Rhynie chert. Rev Palaeobot Palynol 146:265–276

Krings M, Taylor TN, Hass H, Kerp H, Dotzler N, Hermsen EJ (2007b) Fungal endophytes in a 400-million-yr-old land plant: infection pathways, spatial distribution, and host responses. New Phytol 174:648–657

Lamarck JB (1809) Philosophie zoologique: an exposition with regard to the natural history of animals. Cambridge University Press, Cambridge (reprint 2011)

Lewis LA, McCourt RM (2004) Green algae and the origin of land plants. Am J Bot 91:1535–1556

Ley RE, Turnbaugh PJ, Klein S, Gordon JI (2006) Human guts associated with obesity. Nature 444:1022–1023

Little AEF, Currie CR (2008) Black yeast symbionts compromise the efficiency of antibiotic defense in fungus-growing ants. Ecology 89(5):1216–1222

Lopez-Madrigal MJ, Latorre A, Porcar M, Moya A, Gil R (2011) Complete genome sequence of "*Candidatus Tremblaya princeps*" strain PCVAL, an intriguing translational machine below the living-cell status. J Bacteriol 193:5587–5588

Ludwig M, Gibbs SP (1985) DNA is present in the nucleomorph of cryptomonads—further evidence that the chloroplast evolved from a eukaryotic endosymbiont. Protoplasma 127(1–2):9–20

Lutzoni F, Pagel M (1997) Accelerated evolution as a consequence of transitions to mutualism. Proc Nat Acad Sci USA 94(21):11422–11427

Lutzoni F, Pagel M, Reeb V (2001) Major fungal lineages are derived from lichen symbiotic ancestors. Nature 411:937–940

Ma WC, Denlinger DL (1974) Secretory discharge and microflora of milk gland in tsetse flies. Nature 247:301–303

Margulis L (1990) Words as battle cries—symbiogenesis and the new field of endocytobiology. Bioscience 40(9):673–677

Margulis L (2009) Recorded interview at Boston University conducted by Douglas Zook, videorecorder Michael Lee, audio/video editor Divya Mahdavan

Margulis L, Fet V (2010) Symbiogenesis: a new principle of evolution. The rediscovery of Boris Mikhaylovich Kozo-Polyansky (1890–1957). Paleontological J 44(12):1525–1539

Margulis L, Fester R (1991) Symbiosis as a source of evolutionary innovation: speciation and morphogenesis. MIT Press, Boston

Martinson VG, Danforth BN, Minckley RL, Rueppell O, Tingek S, Moran N (2011) A simple and distinctive microbiota associated with honey bees and bumble bees. Mol Ecol 20(3):619–628

Martin-Vivaldi M, Aránzazu P, Peralta-Sanchez JM, Sánchez L, Ananou S, Ruiz-Rodriguez M, Soler JJ (2010) Antimicrobial chemicals in hoopoe preen secretions are produced by symbiotic bacteria. Proc R Soc B 277:123–130

McFall-Ngai M, Hadfield MG, Bosch TCG, Carey HV, Domazet-Loso T, Douglas AE, Dubilier N, Eberl G, Fukami T, Gilbert SF, Hentschel U, King N, Kjelleberg S, Knoll AH, Kemer N, Mazmanian SK, Metcalf JL, Nealson K, Pierce NE, Rawls JF, Reid A, Ruby EG, Rumpho M, Sanders JG, Tautz R, Wernegreen JJ (2013) Animals in a bacterial world, a new imperative for the life sciences. Proc Nat Acad Sci USA 110(9):3229–3236. doi:10.1073/pnas.1218525110

Minteer BA, Collins JP (2008) From environmental to ecological ethics: toward a practical ethics for ecologists and conservationists. Sci Eng Ethics 14(4):483–501

Montilor CB, Maxmen A, Purcell AH (2002) Facultative bacterial endosymbionts benefit pea aphids *Acyrthosiphon pisum* under heat stress. Ecol Entom 27:189–195

Moore CE, Archibald JM (2009) Nucleomorph genomes. Ann Rev Gen 43:251–264

Moran N, Munson MA, Baumann P, Ishikawa H (1993) A molecular clock in endosymbiotic bacteria is calibrated using the insect hosts. Proc R Soc Lond B 253:167–171

Mueller UG, Currie CR, Schultz TR, Adams RM, Malloch D (2001) The origin of the attine ant-fungus mutualism. Q Rev Biol 76(2):169–197

Myers JM, Ramsey JM, Blackman JP, Alison L, Blackman A, Nichols AE, Minbiole KPC, Harris RN (2012) Synergistic inhibition of the lethal fungal pathogen *Batrachochytrium dendrobatidis*: the combined effect of symbiotic bacterial metabolites and antimicrobial peptides of the frog *Rana muscosa*. J Chem Ecol 38(8):958–965

Nass MMK (1969) MitochondriaL DNA: advances, problems, and goals. Science 165:25–35

Nass MM, Nass S (1963) Intra-mitochondrial fibers with DNA characteristics: fixation and electron staining reaction. J Cell Biol 19(3):593–611

Nelson KE, Weinstock GM, Highlander SK, Worley KC, Creasy HH, Wortman JR, Busch DB, Makedonka M, Sodergren E, Chinwalla AT, Feldgarden M, Gevers D, Haas BJ, Madupu R, Ward DV (2010) A catalog of reference genomes from the human microbiome. Science 328:994–999

Nicholson JK, Holmes E, Kinross J, Burcelin R, Gibson G, Jia W, Pettersson S (2012) Host-gut microbiota metabolic reactions. Science 336:1262–1267

Palmer JD (2003) The symbiotic birth and spread of plastids: how many times and whodunit? J Phycol 39:4–11

Pepper JW, Rosenfeld S (2012) The merging medical ecology of the human microbiome. Trends Ecol Evol 27(7):381–384

Phillips CD, Phelan G, Dowd SE, McDonough MM, Ferguson AW, Hanson JD, Siles L, Ordonez-Garcia N, San Francisco M, Baker RJ (2012) Microbiome analysis among bats describes influences of host phylogeny, life history, physiology and geography. Mol Ecol 21(11):2617–2627

Piercy-Normore MD, Depriest PT (2001) Algal switching among lichen symbioses. Am J Bot 88(8):1490–1498

Pirozynski KA, Malloch DW (1975) Origin of land plants—matter of mycotropism. Biosystems 6(3):153–164

Poulsen M, Cafaro M, Erhardt D, Little AEF, Gerardo NM, Tebbets B, Klein BS, Currie CR (2010) Variation in *Pseudonocardia* antibiotic defense helps govern parasite-induced morbidity in *Acromyrex* leafcutting ants. Environ Microbiol Rep 2(4):534–540

Raffa K, Piwell EN, Townsend PA (2013) Temperature-driven range expansion of an irruptive insect heightened by weakly co-evolved plant defenses. Proc Nat Acad Sci USA 110(6):2193–2198

Redman RS, Dunigan DD, Rodriguez RJ (2001) Fungal symbiosis from mutualism to parasitism: who controls the outcome, host or invader? New Phytol 151(3):705–716

Rosenberg E, Zilber-Rosenberg I (2011) Symbiosis and development: the hologenome concept. Birth Defects Res (Part c) 93:56–66

Rosenberg E, Sharon G, Zilber-Rosenberg I (2009) The hologene theory of evolution: a fusion of neo-Darwinism and Lamarckism. Environ Microbiol 11:2959–2962

Rowher F, Seguritan V, Azam F, Knowlton N (2002) Diversity and distribution of coral-associated bacteria. Mar Ecol Prog Ser 243:1–10

Russell JA, Moreau CS, Goldman-Huertas B, Fuliwara M, Lohman DJ, Pierce NE (2009) Bacterial gut symbionts are tightly linked with the evolution of herbivory in ants. Proc Nat Acad Sci USA 106(50):21236–21241

Sagan L (Margulis) (1967) On the origin of mitosing cells. J Theor Biol 14(3):255–274

Sang RC, Jura WG, Otieno LH, Mwangi RW, Pgaja P (1999) The effects of a tsetse fly virus infection on the functions of the male accessory reproductive gland in the host fly *Glossina pallidipes* (Diptera:Glossinidae). Curr Microbiol 38:349–354

Sapp J (1994) Evolution by association: a history of symbiosis. Oxford University Press, Oxford

Schardl CL (2001) *Epichloë festucae* and related mutualistic symbionts of grasses. Fungal Genet Biol 33:69–82

Schopf JW (1999) Cradle of life: the discovery of earth's earliest fossils. Princeton University Press, Princeton, pp 183–200

Schopf JW, Kudryavlsev AB (2012) Biogenicity of Earth's earliest fossils: a resolution of the controversy. Gondwana Res 22(3–4):761–777

Schüssler A, Schwarzott D, Walker C (2001) A new fungal phylum, the Glomeromycota: phylogeny and evolution. Mycol Res 105:1413–1421

Seipke RF, Kaltenpoth M, Hutchings MJ (2011) Streptomyces as symbionts: an emerging and widespread theme? FEMS Microbiol Rev 36:862–876

Selosse M-A, Le Tacon F (1998) The land flora: a phototroph-fungus partnership? Trends Ecol Evol 13(1):15–29

Shutler JD, Grant MG, Miller PI, Rushton E, Anderson K (2010) Coccolithophore bloom detection in the northeast Atlantic using SeaWiFS: algorithm description, application and sensitivity analysis. Remote Sens Environ 114(5):1008–1016

Six D (2012) Ecological and evolutionary determinants of bark beetle symbioses. Insects 3:339–366

Six D (2013) The bark beetle holobiont: why microbes matter. J Chem Ecol 39:989–1002

Snyder AK, Rio RVM (2013) Interwoven biology of the tsetse holobiont. J Bacteriol 195(19):4322–4330

Stallforth P, Brock DA, Cantley AM, Tian X, Queller D, Strassman JE, Clardy J (2013) A bacterial symbiont is converted from an inedible producer of beneficial molecules into food by a single mutation in the gacA gene. Proc Nat Acad Sci USA 110(36):14528–14533

Stat M, Carter D, Hoegh-Guldberg O (2006) The evolutionary history of *Symbiodinium* and scleractinan hosts—symbiosis, diversity, and the effect of climate change. Persp Plant Ecol Evol Syst 8(1):23–43

Stevens CE, Hume ID (1998) Contributions of microbes in vertebrate gastrointestinal tract to production and conservation of nutrients. Physiol Rev 78(2):393–427

Tanaka A, Takemoto D, Hyon GS, Park P, Scott B (2008) NoxA activation by the small GTPase RacA is required to maintain a mutualistic symbiotic association between *Epichloë festucae* and perennial ryegrass. Mol Microbiol 68:1165–1178

Taylor TN, Krings M (2005) Fossil microorganisms and land plants: associations and interactions. Symbiosis 40:119–135

Taylor TN, Remy W, Hass H, Kerp H (1995) Fossil arbuscular mycorrhizae from the Early Devonian. Mycologia 87:560–573

Thacker RW, Freeman CJ (2012) Sponge-microbe symbiosis: recent advances and new directions. Adv Mar Biol 62:67–111

Turmel M, Pombert JF, Charlebois P, Otis C, Lemieux C (2007) The green algal ancestry of land plants as revealed by the chloroplast genome. J Plant Sci 168(5):679–689

Turnau K, Gawronski S, Ryszka P, Zook D (2012) Mycorrhizal-based phytostabilization of Zn–Pb tailings: lessons from the Trzebionka mining works (southern Poland). In: Kothe E, Varma A (eds) Bio-geo interactions in metal-contaminated soils. Soil Biology 31:327–348. Springer, Berlin

Vesteg M, Vacula R, Krajcovic J (2009) On the origin of chloroplast, import mechanisms of chloroplast-targeted proteins, and loss of photosynthetic ability—a review. Folio Microbiol 54(4):303–321

Wang X (2002) Evidences for vagus nerve in maintenance of immune balance and transmission of immune information from gut to brain in STM-infected rats. World J Gastroenterol 8:540–545

Wang J, Aksoy S (2012) PGRP-LB is a maternally transmitted immune milk protein that influences symbiosis and parasitism in tsetse's offspring. Proc Natl Acad Sci USA 109:10552–10557

Wang B, Qui YL (2006) Phylogenic distribution and evolution of mycorrhizas in land plants. Mycorrhiza 16(5):299–363

Weiss BI, Maltz M, Aksoy S (2012) Obligate symbiosis activate immune system development in the tsetse fly. J Immunol 188:3195–3403

Werren JH (1997) Biology of Wolbachia. Ann Rev Entomol 42:587–609

Werren JH, Baldo I, Clark ME (2008) Master manipulators of invertebrate biology. Nat Rev Microbiol 6:741–751
Wooldridge SA (2010) Is the coral-algae symbiosis really "mutually beneficial" for the partners? BioEssays 32(12):615–625
Wodniok S, Brinkmann H, Gloeckner G, Heidel AJ, Philippe H, Melkonian M, Becker B (2011) Origin of land plants: do conjugating green algae hold the key? BMC Evol Biol 11(104). doi:10.1186/1471-2148-11-104
Wu GD, Chen J, Hoffman C, Bittinger K, Chen Y-Y, Keilbaugh SA, Bewtra M, Knbights D, Walters WA, Knight R, Sinha R, Gilroy E, Gupta K, Baldassano R, Nessel L, Hongzhe L, Bushman FD, Lewis JD (2011) Linking long-term dietary patterns with gut microbial enterotypes. Science 334:105–108
Yuan XL, Xao SH, Taylor TN (2005) Lichen-like symbiosis 600 million years ago. Science 308(s724):1017–1020
Zook D (1983) A study of the role of bacteria in lichens. MA thesis, Clark University, Worcester, MA, USA
Zook D (2002) Prioritizing symbiosis to sustain biodiversity: are symbionts keystone species? In: Seckbach J (ed) Symbiosis: mechanisms and model systems. Kluwer, Dordrecht, pp 3–12
Zook D (2010) Tropical rainforests as dynamic symbiospheres of life. Symbiosis 51(1):27–36

Can We Understand Evolution Without Symbiogenesis?

Francisco Carrapiço

> *...symbiosis is more than a mere casual and isolated biological phenomenon: it is in reality the most fundamental and universal order or law of life.*
>
> Hermann Reinheimer (1915)

Abstract This work is a contribution to the literature and knowledge on evolution that takes into account the biological data obtained on symbiosis and symbiogenesis. Evolution is traditionally considered a gradual process essentially consisting of natural selection, conducted on minimal phenotypical variations that are the result of mutations and genetic recombinations to form new species. However, the biological world presents and involves symbiotic associations between different organisms to form consortia, a new structural life dimension and a symbiont-induced speciation. The acknowledgment of this reality implies a new understanding of the natural world, in which symbiogenesis plays an important role as an evolutive mechanism. Within this understanding, symbiosis is the key to the acquisition of new genomes and new metabolic capacities, driving living forms' evolution and the establishment of biodiversity and complexity on Earth. This chapter provides information on some of the key figures and their major works on symbiosis and symbiogenesis and reinforces the importance of these concepts in our understanding of the natural world and the role they play in the establishing of the evolutionary complexity of living systems. In this context, the concept of the symbiogenic superorganism is also discussed.

Keywords Evolution · Symbiogenesis · Symbiosis · Symbiogenic superorganism · New paradigm

---

F. Carrapiço (✉)
Centre for Ecology Evolution and Environmental Change (CE3C);
Centre for Philosophy of Science, Department of Plant Biology, Faculty of science,
University of Lisbon, Lisbon, Portugal
e-mail: fcarrapico@fc.ul.pt

N. Gontier (ed.), *Reticulate Evolution*, Interdisciplinary Evolution Research 3,
DOI 10.1007/978-3-319-16345-1_3

1 Introduction

The idea of evolution applied to the biological world was used for the first time in the eighteenth century (1779) by the Swiss naturalist and philosopher Charles Bonnet (Bowler 1975), who developed this concept in the context of egg fertilization by spermatozoon. In his work, the author considers that the egg contains the embryo preformed with all the parts of the future organism present, the sperm cell being the trigger for such development. The unfolding of the pre-existent embryo was called "evolution" (Rieppel 2011). However, the use of this term in a more modern sense began to emerge when new data were obtained from different expeditions around the world carried out by French and English naturalists in the eighteenth and nineteenth centuries. These data, which included geological and biological information on different continents, undermined the official version of the Earth's formation and its age, as well as the universal tenet of the creation of species, questioning the validity of the biblical version and building a new tree of life on a dynamic planet (Mayr 2001; Kutschera 2011).

The first modern scientific ideas on evolution were presented in 1809 by Jean Baptiste Pierre Antoine de Monet, Chevalier de Lamarck, in his book *Philosophie Zoologique, ou Exposition des Considérations Relatives à l'Histoire Naturelle des Animaux*. The latter envisioned evolution as a progression from less to more complex organisms, where the notion of progression was represented by a straight line (Lamarck 1809). The shift from the belief in a static approach to a dynamic understanding of the evolution of the natural world was brought about by the publication of Alfred Wallace's works and especially, in 1859, by Charles Darwin's book *On the Origin of Species by means of Natural Selection or the Preservation of Favored Races in the Struggle for Life* (Darwin 1859). Influenced by the works of Thomas Malthus and Charles Lyell, Darwin built a theory that contributed to change radically the idea of constancy of species, which allowed for the development of the theory of common descent and the challenging of the natural theology principles that had ruled natural science for centuries (Kutschera 2011).

At the beginning of the twentieth century, new scientific data were published by several authors, among them the German biologist, Theodor Boveri, and the American biologists, Thomas Hunt Morgan and Hermann Joseph Muller (Reif et al. 2000), contributing to a new understanding of the evolution concept. Among these data, the discovery of the nature and role of the chromosome in heredity—which lays at the core of the chromosome theory of inheritance—was of primordial importance. Further research, namely in mathematical and field studies population genetics, developmental biology, biogeography, and paleontology, contributed to a better understanding of evolution and the formation of evolutionary synthesis, which constituted the core of the synthetic theory of evolution. This theory was based on five evolutionary factors: mutation, recombination, selection, isolation, and drift (Reif et al. 2000). In 1942, Julian Huxley published *Evolution: The Modern Synthesis*, opening a new chapter on the understanding of evolution, merging the Darwinist ideas with new concepts in genetics and evolutionary biology, developed previously by authors such as John B.S. Haldane and Theodosius Dobzhansky (Huxley 1942). The same year, Ernst Mayr published *Systematics and the Origins of Species, from the Viewpoint of a Zoologist* (Mayr 1942), an important work on modern evolutionary synthesis. It was the beginning of the neo-Darwinist period, which is still considered the mainstream approach to evolution studies.

Bearing this background in mind, this work is a contribution to the literature and knowledge on evolution, which takes into account the biological data obtained in the last few years. This chapter tries to find new answers to old questions, which neo-Darwinism in its "ivory tower" has not been able to cope with, having driven evolutionary science to a dead end regarding some topics in the field, reinforcing the importance of symbiogenesis to understand the natural world and in the establishment of evolutive complexity of living systems.

2 Roots and Paths of Symbiogenesis

Evolution is traditionally considered as a gradual process essentially consisting of natural selection conducted on minimal phenotypical variations, which are the result of random mutations and genetic recombinations to form new species. However, "Mutation accumulation does not lead to new species or even to new organs or new tissues," and "99.9 % of the mutations are deleterious" (Margulis and Sagan 2002).

In contrast, the biological world presents and involves symbiotic associations between different organisms to form consortia, a new structural life dimension and a symbiont-induced speciation. This reality implies a new understanding of the natural world, in which symbiogenesis plays an important role as an evolutive mechanism, with symbiosis as the key to the acquisition of new genomes and new metabolic capacities, which drives living forms' evolution and the establishment of biodiversity on Earth (Margulis and Sagan 2002). So, we can say that "Symbiosis is simply the living together of organisms that are different from each other" (Margulis and Sagan 2002) and symbiogenesis can be seen as the "origin of evolutionary novelty via symbiosis" (Margulis 1990). Even one of the well-known neo-Darwinists of our time, Richard Dawkins, in his book *The Selfish Gene* (Dawkins 1976, p. 182), introduced the idea that "Each one of our genes is a symbiotic unit" and "We are gigantic colonies of symbiotic genes." Nevertheless, he refused to admit that symbiosis and co-operation can have a crucial role in nature and reinforced the importance of gene selfishness in the evolutive process.

It was only with Peter Corning's work, *The Co-operative Gene…* (Corning 1996), that these ideas moved to a new level of understanding, highlighting co-operation and saying that "Synergy is a multi-leveled phenomenon that can take many different forms," and "has played a significant causal role in the evolution of complexity."

In a certain way, "Co-operation represents an often advantageous survival strategy" and in "a complex organism or superorganism [it] represents a collective survival enterprise" (Corning 1996, p. 205). It was also this author who in his book *Holistic Darwinism* clarifies the relation between symbiosis and synergy, saying

> That symbiosis refers to relationships of various kinds between biological entities and the functional processes that arise from those relationships. Synergy, on the other hand, refers to the interdependent functional effects (the bioeconomical pay offs) of symbiosis, among other cooperative phenomena. In short, all symbioses produce synergistic effects, but many forms of synergy are not the result of symbiosis (Corning 2005, p. 82).

As Yves Sciama states in his article "Penser coopération plutôt que competition (Think cooperation instead of competition)," it is important to consider as the main project for twenty-first century biology, "*Repenser le vivant à partir de la notion de symbiose* (Rethinking the living from within the notion of symbiosis)" (Sciama 2013).

Despite these open-minded ideas related to a more co-operative and synergistic approach to the evolutive process, symbiosis and symbiogenesis have been considered by the majority of the scientific community as "stepdaughters or stepsons" of evolutionary theory (Pereira et al. 2012), or in the case of symbioses, as biological jokes (Selosse 2000). This reveals a limited understanding of evolutive process and does not correspond to the reality of the facts nor to the structure of the web of life on our planet. The symbiogenic view also enables a coherent conceptual and epistemological rupture with some evolutionary ideas of the past, indicating and building a new approach to understanding the development and evolution of life. To comprehend this new approach and paradigm to the evolution process and diversity of life on our planet, we must go back in time and begin our narrative when the first modern scientific ideas on evolution appeared, namely after *The Origin of Species* by Darwin in 1859. On the topic of origins, let us start at the beginning…

The year of 1867 is better known for the publication of the first volume of *Das Kapital* by Karl Marx, but it was also in that year that Simon Schwendener, a Swiss botanist, proposed at the Swiss Natural History Society annual meeting, held in Rheinfelden, an interesting dual hypothesis. In order to explain the nature of lichens, this hypothesis indicated that they are an association of two organisms, a fungus and an alga, behaving as "master and slave" (Honegger 2000). The idea that an organism could be formed by two or more genetically separate organisms living together and working as one unit was regarded as so unusual at the time that it was largely rejected by the scientific community. The dual hypothesis was a revolutionary concept for the biology of the nineteenth century, as well as a rupture in the traditional concept of an organism. The proposal, however, was not easily accepted, as can be seen from the example of William Nylander's book *Les Lichens des Environs de Paris*, published almost 20 years after Schwendener's statement. In his book, Nylander states that "*On sait bien aujourd'hui que la formule 'les lichens sont des champignons vivant en symbiose avec des algues' est une assertion de pure fantaisie ou une calomnie* (Today, we know very well that the formula 'lichens are fungi living in symbiosis with algae' is an assertion resting on pure fantasy, or a calumny)" (Nylander 1896).

Another example of this situation was the living experience of Beatrix Potter who worked with lichens at the end of the nineteenth century and who was not allowed to continue her scientific work because she supported Schwendener's ideas, vand also because she was a woman. The traditional English scientific community was not supportive of her work (Sapp 1994; Taylor et al. 1995). However, as society lost a scientist, it gained a children's story writer. *Peter Rabbit and his Friends* probably did more for the establishment of an environmentally friendly behavior for new generations than many of her co-fellows who rejected her as a scientist.

The next important step was the introduction of the symbiosis concept by the German naturalist Heinrich Anton De Bary in 1878, which was based on studies of the nature of lichens and the role of algae and fungi in this association. He also used the example of the aquatic fern *Azolla* to develop this concept, referring to the permanent presence of the cyanobacterium *Anabaena azollae* in the leaf cavity and in the sexual structures of this plant. He further expanded on this presence by explaining that at no stage of its life cycle is the fern free from cyanobacterium and that the latter is in no way harmful to *Azolla* (Carrapiço 2010a). This concept was presented in a communication entitled "Ueber Symbiose" (About Symbiosis), at the Congress of Naturalists and German Doctors in Cassel (De Bary 1878), and was defined as "the living together of unlike named organisms," which is at present the best definition for this phenomenon (Carrapiço 2010a). However, it is important to note that this concept follows two previously introduced concepts. The first was mutualism, which was put forward by Pierre-Joseph Van Bénéden in 1875, and constituted an application of Pierre-Joseph Proudhon's social ideas to the animal kingdom (Van Bénéden 1875; Boucher 1985). The second concept was symbiotismus, which was introduced by Albert Bernhard Frank, in 1877, in a publication on the biology of lichens (Frank 1877). This author, who is better known for the study and introduction of the term "mycorrhiza" in 1885, defined symbiotismus in a similar way to De Bary's symbiosis in 1878. In 1879, De Bary published a new article related to this subject entitled "Die Erscheinung der Symbiose" (The Phenomenon of Symbiosis). In both works, De Bary considers the association *Azolla–Anabaena* to be a classic example of van Bénéden's mutualistic cases applied to the plant kingdom. Even though this association was previously studied by Eduard Strasburger in 1873, De Bary noted, as already mentioned, that no stage of the life cycle of the fern was free of the cyanobacteria and that they did no harm *Azolla*.

In 1895, the Danish botanist Eugenius Warming published *Plantesamfund* (*Ecology of Plants*) and considers the *Azolla–Anabaena* association an example of mutualism and an exception to normal behavior in plant communities: "In plant community egoism reigns supreme" (Sapp 1994).

In 1902, Petr Kropotkin published *Mutual Aid. A Factor of Evolution*. This work was written while in exile in England, and argues that, despite the Darwinian concept of the survival of the fittest, co-operation rather than conflict is the main factor in the evolution of species. Kropotkin, better known as a leader of the anarchist movement, developed his ideas of the natural world based on the experience he lived during a five-year expedition in Siberia (1862–1867). He was also influenced by the work of the Russian zoologist Karl Kessler, who in 1879 presented a paper entitled "On the Law of Mutual Aid" at the Society of Naturalists of St. Petersburg (Kropotkin 1902; Todes 1989).

However, the main core of the symbiogenic ideas was developed by the Russian biologist Constantin Merezhkowsky during his stay as professor at Kazan University (1902–1914) where he conducted research on symbiotic associations, namely on lichens. His research, however, goes well beyond these organisms. Between his stay in Kazan and his death in Geneva in 1921, this author published several papers on the origin of chloroplasts and the role of symbiosis in evolution (Sapp et al. 2002). In particular, in 1905 he published the article "Uber Natur und Ursprung der Chromatophoren im Pflanzenreich" (On the Nature and Origin of Chromatophores in the Plant Kingdom) where, for the first time, coherent scientific arguments showed that plastids arose from free-living cyanobacteria (Martin and Kowallik 1999; Merezhkowsky 1905). In 1909, he published "The Theory of two Plasms as Foundation of Symbiogenesis, New Doctrine on the Origin of Organisms" in Russian (Merezhkowsky 1909). The German version was published one year later. As a professor at Kazan University, Constantin Merezhkowsky developed this work introducing the concept of symbiogenesis as "The origin of organisms by the combination or by the association of two or several beings which enter into symbiosis." In this paper, he introduced not only new concepts on symbiogenesis and evolution, but he also developed some important ideas about the origin of life, namely related to the role of extremophiles in that scenario. A new classification of the living world was proposed using associations between organisms (Fig. 1; Merezhkowsky 1909).

In 1920, several months before committing suicide in Geneva, Constantin Merezhkowsky published the article "La Plante Considerée comme un Complexe Symbiotique" (The Plant Considered as a Symbiotic Complex) where the author developed his previous ideas on the symbiotic origin of chloroplasts and nucleus. In opposition to contemporary views of the time (Guilliermond 1918), Merezhkowsky defended that chloroplasts did not evolve from mitochondria or protoplasm, but from free-living cyanobacteria, as he had presented in 1905 (Merezhkowsky 1920).

It should be mentioned that another Russian botanist, Andrey Famintsyn, contemporary of Merezhkowsky and also working in the symbiotic field, published in 1907 *On the Role of Symbiosis in the Evolution of Organisms*, where he developed the idea that symbiosis has an important evolutionary, or even adaptive, meaning (Khakhina 1992; Sapp 1994; Sapp et al. 2002; Corning 2005). He states that the increasing complexity of the organization and function of organisms during the process of evolution may occur not only through the differentiation of simpler, early forms, but also on the basis of symbiotic unification of independent organisms into a living unit of a higher order (Khakhina 1992). In his point of view, the idea that symbiosis could be involved in evolution was important to understand the origin of life on Earth and its development (Khakhina 1992).

The same year that Constantin Merezhkowsky published his last work, Hermann Reinheimer published a book entitled *Symbiosis. A Socio-Physiological Study of Evolution* (Reinheimer 1920). The author points out the importance of specific interrelations in the development of organisms as a whole, giving us a holistic perspective on organismal evolution:

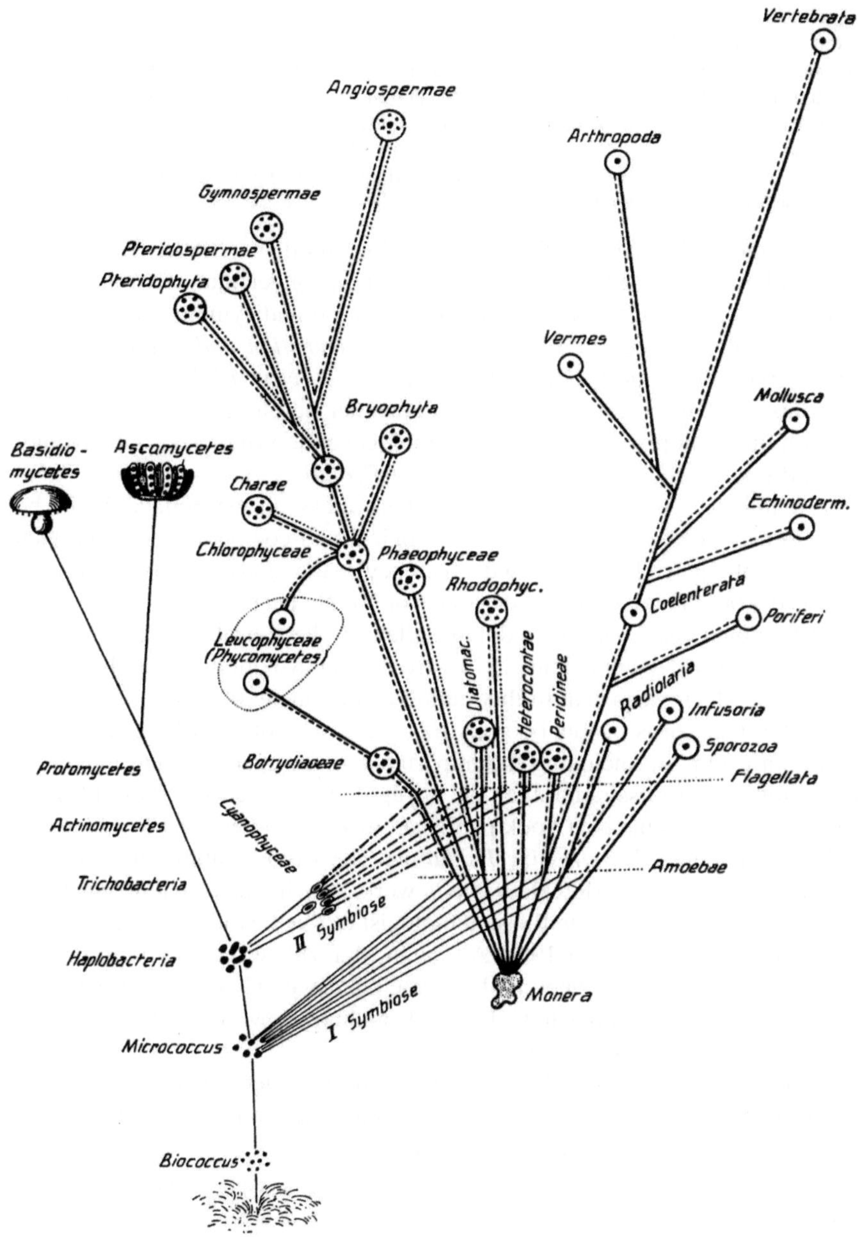

Fig. 1 The tree of life proposed by Constantin Merezhkowsky in 1909. In this, the organization of the living world is presented using, for the first time, associations between prokaryotic and eukaryotic organisms

> The main conclusion which I wish to enforce is that the normal relations between organisms, more particularly those having regard to food, involve, quite indispensably, a stupendous amount of systematic biological reciprocity, so that upon all organisms, be they high or low in the scale of life, there devolve definitive duties and obligations, on pain of degeneration or destruction, viz., to contribute in their several ways to the welfare of the organic family as a whole. (…) I regard the totality of organisms as a kind of world-society, the various species and families of plants and animals being the individuals of which this worldsociety is made-up.

This author, who lived in Surbiton (London) until the 1950s, is not particularly well known among biologists, which is strange given that he wrote 19 books during his life, several of them related to evolution and symbiogenic topics. His first book was published in 1909 with the title *Nutrition and Evolution.* A year later, he published *Survival and Reproduction. A New Biological Outlook.* It was in his three following books that Reinheimer developed his ideas about co-operation in a more coherent way: symbiogenesis, symbiosis, and evolution. The third book, published in 1913 and titled *Evolution by Co-operation. A Study in Bio-Economics*, is a good example of these ideas. In the preface of the book he mentions: "To the study of the physiological and combined economic factors productive of 'general stability and efficiency'—the study of biological eugenics—freed from the misleading side-issues of 'single peculiarities,' I have devoted myself for some years, and in so doing, I claim to be contributing to and furthering Darwin's work" (Reinheimer 1913). He was an evolutionist, but also believed in eugenics, which was usual in that period among many of Darwin's supporters. One of the main ideas of this book is the significance of what he calls "bio-economics" in evolution, including the importance of co-operation and mutuality in the evolutionary process rather than the "struggle for existence." These ideas were further developed in his next book published in 1915 and entitled *Symbiogenesis; the Universal Law of Progressive Evolution.* The word symbiogenesis was used without any reference to Merezhkowsky's work, which means that he either eventually omitted the work of the Russian biologist or that he did not have knowledge of his works, namely that of 1909. Although Merezhkowsky published this work in German in 1910 and Reinheimer knew the language, its diffusion was very limited and probably did not reach the United Kingdom. However, it is interesting and intriguing to notice the use of the same term.

To understand the nature of the content of Reinheimer's book and the way he perceived biology, we transcribe parts of the introduction that are relevant for the nature of our work. On page XIII he mentions:

> The first chapter is particularly devoted to the subject of symbiosis, which is generally defined as a physiological partnership between individuals of different species, but which is of far more universal meaning and occurrence than is suggested by this definition. The term must be particularly applied also to the wider bio-economic form of co-operation which underlies evolution and unites all organisms in one vast web of life.

On pages XIV and XV he defines symbiogenesis as

> By symbiogenesis I mean the production and increase of values throughout organic life by means of a symbiotic principle of co-operation or reciprocity between different organs

of the individual, but evolved and complex body, as well as between different organisms in a species or different species, genera, orders, etc., even in the last and most fundamental way between plant and animal in the web of life. By the term symbiosis I refer to that obvious phenomenon of co-operation of parts and organisms as they occur, while by symbiogenesis I mean the principle underlying such symbiosis and indeed all instances of mutuality in the progressive transmutation of biological values generally. Symbiosis, further, may be domestic when it is between the organs of one organism and between the members of a family; biological when it refers to physically separate partners, even when widely separated and unconscious of partnership.

And on pages XVI and XVII he states:

The grand importance of symbiosis consists in the fact that it evolves and safeguards those very modes of reciprocal differentiation which we must recognise as the universal means of the creation and elaboration of physiological and psychological values, including those which perhaps may be more especially regarded as genetic in character and influence. In other words, symbiosis is more than a mere casual and isolated biological phenomenon: it is in reality the most fundamental and universal order or law of life. So much so is this the case that I claim the great principle underlying all Creative Life, all Progressive Evolution to be that of "Symbiogenesis"; i.e., the mutual production and symbiotic utilisation of biological values by the united and correlated efforts of organisms of all descriptions. It is a well-known saying of Aristotle that the City exists for the sake of its good citizens, and I would apply it to the biological society, which also exists for its "good" citizens—those organisms, namely, which by symbiotic endeavour at once earn the right of biological citizenship and contribute to the welfare, permanence and progress of their "society."

At last, a sentence that summarizes his ideas related to symbiosis: "... Biologically speaking, I should say: 'La symbiose fait la force.'" This was said when he argues that "l'union fait la force" (Reinheimer 1915).

A final note about this author and his background. As we previously mentioned, Reinheimer is almost unknown among the authors working on symbiontology, especially taking into consideration the number of works he published related to this area of science. In many of these works, he used expressions that were ahead of his time, such as "web of life", "bio-economics", and "antibiotics". To understand how some of his works were not well accepted by established biologists, we transcribe a sentence included in a review of his 1915 book, which was authored by the American biologist William L. Tower, from the University of Chicago, and published in *The American Journal of Sociology*: "... in the whole book nothing to commend it, nor any possible escape from characterizing it as the least logical, worst constructed, most inaccurate and irrational book upon evolution that has appeared in a long time" (Tower 1916). Reinheimer was born in Germany (Hesse), but he was naturalized as a British citizen in 1901. In 1911, the England Census reported that he was 38 years old, single, and worked as a self-employed stockbroker. He lived in London (Surbiton) until the 1950s and subscribed to an alternative view of society, with the majority of his books being published by editors associated with anarchism, metaphysics, theosophy, and vegetarianism. A good example of alternative editors is the publisher Charles William Daniel, an anarchist and pacifist who founded his own company in 1902 for editing books on such topics. Another example is John M. Watkins, a publisher involved in the subjects of mysticism and metaphysics. Although Reinheimer refers to his occupation as stockbroker, his knowledge of natural sciences and namely of evolution suggest that he had a biological background, despite there being no indication that he had any affiliation with academia in England.

Several other authors were related to the development of symbiogenic ideas in biology during the first decades of the twentieth century. Among them, we must refer to the French biologist Paul Portier who published *Les Symbiotes* in 1918. In this work, Portier developed the idea that all organisms are constituted of an association of different beings. In the particular case of mitochondria, he argues that those cell organelles were symbiotic bacteria, which the author calls "symbiotes" (Portier 1918; Sapp 1994). He also refers to the positive role of these prokaryotic organisms in the human body at a time when germ theory was the mandatory rule in biology and medicine. These ideas shocked the French scientific community that reacted negatively. The following year, Auguste Lumière published a critical response in the book *Le Mythe des Symbiotes* (Lumière 1919).

In the United States, Ivan Wallin, working at the University of Colorado, developed similar ideas to Portier's concepts, and in 1923 and 1927 published two important works on the subject. The first, titled *The Mitochondria Problem*, emphasized the symbiotic origin of these organelles against the cytoplasmic point of view. In the second work, titled *Symbionticism and the Origin of Species*, the author defends the importance of symbiotic mechanisms in evolution, with emphasis on the symbiotic origin of mitochondria. Wallin also underlines the importance of microsymbiosis in this process, pointing out the idea "That bacteria, which are popularly associated with disease, may represent the fundamental causative factor in the origin of species" (Wallin 1923, 1927; Sapp 1994). He considers symbioticism as a mechanism of speciation, suggesting that the primary source of genetic novelty for speciation was the periodic repeated fusion of bacterial endosymbionts with host cells (Taylor 1979). Although he claims that it was possible to cultivate mitochondria outside of the cell, like Portier did in 1918, these data were incorrect as they resulted from culture contamination. It was only after his death, in 1969, that evidence began accumulating that his theory was partially correct concerning the bacterial origin of mitochondria, and the prokaryotes' role in evolution. *Symbionticism and the Origin of Species* was published in 1927, the year in which Hermann J. Muller published the paper "Artificial Transmutation of the Gene" in *Science*. This article opened the way to the explanation for species formation under the neo-Darwinian theory, showing that X-rays could dramatically increase the frequency of gene mutations in *Drosophyla*, and overshadowed Wallin's explanation of bacteria as a factor of speciation (Muller 1927; Wallin 1927; Sapp 1994; Brucker and Bordenstein 2012).

Another author, who must be referred to, is the Russian biologist Boris Kozo-Polyansky, who published an important book in 1924 entitled *A New Principle of Biology: An Essay on the Theory of Symbiogenesis*. This book gave symbiosis a determinant role in evolution, building the bridge between symbiogenesis and the Darwinian theory, and introducing the idea of the organism as a consortium (Kozo-Polyansky 2010). This concept was initially presented in 1873 by the German botanist Johannes Reinke, to refer to the relationship between the fungi and algae in lichens (Reinke 1873; Sapp et al. 2002). According to Kozo-Polyansky, the theory of symbiogenesis was a theory of selection relying on the phenomenon of symbiosis (Khakhina 1992).

All these ideas had criss-crossed in an elegant and outstanding way in the 1967 work of Lynn Margulis published in the *Journal of Theoretical Biology* under the title "On the Origin of Mitosing Cells" (Sagan 1967). In this paper, a theory of the origin of eukaryotic cells was presented, explaining the transition bridge between the prokaryotic and the eukaryotic levels of biological organization. Mitochondria, basal bodies of the flagella and chloroplasts, are considered to have derived from free-living prokaryotes, and eukaryotic cells are seen as the result of the evolution of ancient symbioses. All this pioneering work formed the basis of serial endosymbiotic theory, and it constituted the beginning of both remarkable work and contributions to the rehabilitation and development of symbiogenic ideas applied not only to the cellular world, but also to the construction of a new biology for the twenty-first century. Furthermore, it represented a clear and sustained rupture with the traditional neo-Darwinian understanding of biological evolution. Beginning with eukaryotic cell formation, symbiogenesis appears to be the main evolutionary mechanism in the establishment and maintenance of different ecosystems, as well as the foundation for biodiversity on Earth, based on rather sudden evolutionary novelties, and not in conventional gradualism or mutagenic processes (Carrapiço 2010b).

Among the numerous works published by Lynn Margulis, we would like to refer to two important works that changed the way biology is seen and understood nowadays. The first, published in 1970, is *Origin of Eukaryotic Cells*, considered a landmark in the understanding of the origins of eukaryotic cells. In the well-expressed words of John M. Archibald in a recent commemorative review published on the 40th anniversary of its publication, "This influential book brought the exciting and weighty problems of cellular evolution to the scientific mainstream, simultaneously breaking new ground and 're-discovering' the decadesold ideas of German and Russian biologists" (Archibald 2011). The other book is *Acquiring Genomes. A Theory of the Origin of Species*, in which Margulis and her co-author Dorion Sagan provide a solid critique of neo-Darwinism and identify the acquisition of new genomes involving symbiogenic processes as the main driving force in evolution, not random mutations (Margulis and Sagan 2002) (Fig. 2). These ideas include new research themes in order to develop the understanding of the evolutionary process and the complexification of life, namely the existence of horizontal DNA transfer between organisms and the mechanisms to explain it. These new paradigms in biology and in the evolution of biodiversity include bacteria and virus–host symbiosis and their composite dynamics in the establishment of the symbiogenic web of life (Sapp 2003; Carrapiço 2010b; Villarreal and Ryan 2011).

At the same time that the 1967 Margulis' article was published, an oft-forgotten short paper by the Norwegian microbiologist Jostein Goksoyr appeared in *Nature*, providing a similar endosymbiotic theory for the origin of eukaryotic cells (Goksoyr 1967). In this paper, the author suggested that the evolutionary development of the eukaryotic photosynthetic cell was based in prokaryotic forms. He

Fig. 2 The author of this chapter with Lynn Margulis at the Gulbenkian Foundation in Lisbon in 2009

also suggested that this evolution could have been of a polyphyletic nature, as stated in the conclusion of his work:

> A further logical conclusion is that the eucaryotic cell which developed would take its genetic material mainly from the procaryotic forms making up the coenocytic system. Such coenocytic systems may develop a number of times, from different procaryotic forms. Present-day eucaryotic organisms do not necessarily, therefore, have to be developed from one original species. This might even explain some of the rather puzzling parallels that exist between groups of procaryotic and eucaryotic organisms.

Before concluding this part of the text, we would like to refer to the work of the Canadian biologist F.J.R. (Max) Taylor, a renowned expert on dinoflagellates, who has published several papers on cell evolution and endosymbiosis theory (Taylor 1974, 1976, 1979). He was also one of the first researchers to understand the significance and importance of symbiotic bacteria in the origin of chloroplasts and mitochondria in eukaryotic cells and independently to develop similar ideas to Margulis' serial endosymbiosis theory, as well as the role of symbiosis in evolution. His ideas were ahead of his time as we can see in the 1979 work *Symbioticism Revisited: A Discussion of the Evolutionary Impact of Intracellular Symbioses*:

> From the evolutionary standpoint, a symbiotic event represents the union of two or more previously divergent genomes into a new coevolutionary unit. The subsequent fate of this unit will depend on both the survival effectiveness of the new unit interacting with external selective forces, and also the continued integrative and competitive interactions between the two symbionts.
>
> In terms of genetic novelty symbiosis represents a quantum leap of a magnitude far greater than that arising from intrinsic sources such as mutation, hybridization or ploidy changes. The component species can exist independently, but the structure formed by the union of the two may be equal or more successful than the individual species. Integrative factors are therefore crucial in intracellular symbioses (Taylor 1979).

Although we have referred mainly to the symbiogenic studies applied to the biological field, symbiogenesis can be related to other scientific fields beyond biology and evolution, such as in social studies. One pertinent example is the work of Nathalie Gontier from 2007, which states that "Besides the obvious application of the universal scheme in micro-evolutionary symbiosis studies and the origin of eukaryotic beings, it will be argued that universal symbiogenesis can also include the study of viruses and their hosts, hybridization, and even extra-biological phenomena such as culture and language" (Gontier 2007). We believe that economics, medical sciences, and education may also potentially benefit from this theory's application.

3 The "Big One" and the Concept of the Symbiogenic Superorganism

The concept of superorganic evolution was first introduced into the scientific literature by Herbert Spencer in 1876, in the first volume of *The Principles of Sociology* (Spencer 1876). Although the term "superorganism" was not used explicitly, the work implied the existence of a new approach to the classical concept of organism, with consequences at both the biological and social levels. In 1911, the American entomologist William Morton Wheeler, in his paper, "The Ant-Colony as an Organism," compared ant society to an organism when observing the biology and social behavior of these insects in colonies. However, it was only in 1928 that he concluded in his book *The Social Insects, Their Origin and Evolution* that the "insect colony or society may be regarded as a super-organism and hence as a living whole bent on preserving its moving equilibrium and integrity." In this case, the entire colony acts in unison as an independent "creature," feeding itself, expelling its wastes, defending itself, and looking out for its future (Wheeler 1911, 1928). The idea of the superorganism was applied to different levels of biological organization and was subsequently developed by other authors, such as Wilson (1975), Wilson and Sober (1989), Sapp (2003), Corning (2005), Carrapiço (2006a, 2010a, b), and Holldobler and Wilson (2009). Based on these ideas, we have introduced the concept of the *symbiogenic superorganism* (Carrapiço 2012b), applied to new entities or consortia formed by the integration of individual organisms, that possess characteristics that go beyond the sum of the individual properties of each element of the association, resulting in the development of new attributes and capacities as an integrated whole. In this process, these new entities also agglutinate and dynamize synergies not present in the individual organisms. This symbiogenic process also involves genetic sharing at the level of the organisms constituting the consortium, forcing the genomes to be incurred by synchronization and harmonization processes. These processes are aimed at establishing a proper functioning for the new organism as a whole. It indicates that the association depends not only on the intrinsic symbiont–host's properties, but also on the internal and external system environmental conditions. By way of example, a single organism formed by the association of two composite organisms could be demonstrated by way of mathematical formula. The result, however, would not be

Fig. 3 Sporophyte of *Azolla filiculoides* showing overlapping scale-like bilobed leaves and numerous microsporocarps (*yellow small spheres*)

$1 + 1 = 2$, but $1 + 1 =$ a larger 1, characterized by the following principles: (a) the new organism is formed by different species of organisms that work towards a common goal; (b) this new entity is a polygenomic one, in which the different genomes operate together in a complementary and synergistic way for the whole; (c) the parts and units of this entity modify themselves qualitatively, compared to the same units when isolated; and (d) the final outcome is not the mere qualitative and/or quantitative sum of the units that constitute the consortium, but acquire new collective synergies and characteristics. In reality, this phenomenon is widespread in nature and allows a coherent reconceptualization of the traditional epistemological concepts of the past, helping to form a new evolutionary approach to the web of life as well as a contribution to a new idea for the organism concept.

These ideas can be included in the concepts of holobiont (the host with its symbionts as a whole) and hologenome (the sum of the genetic information of the host and its microbiota), developed by several authors (Zilber-Rosenberg and Rosenberg 2008; Guerrero et al. 2013). These principles are similar to the symbiome concept introduced in 2003 by Jan Sapp (Sapp 2003; Carrapiço 2006b). The symbiome concept reinforces the principle that eukaryotic organisms are not genetically unique entities, and the concept of individual must be seen as a complex biological ecosystem, composed of multiple interdependent parts living symbiotically. It is at the symbiome level, composed of an integrated multigenomic genetic pool, that natural selection acts (Carrapiço 2006b). In a recent book, John Archibald explores and elaborates these related topics in an elegant way (Archibald 2014).

Some examples of these kinds of consortia are lichens, termites, and their symbionts, the symbiotic system *Azolla–Anabaena*–bacteria (Carrapiço 2006a, 2010a, b), and in many animal bodies, including humans, with their microbiota community (Sapp 2003). All of these relationships can be considered as constituting symbiogenic superorganisms.

In the case of *Azolla* (Fig. 3), the superorganism is constituted of the association of two types of prokaryote organisms (cyanobacterium and bacteria) living symbiotically inside the leaf cavity of the fern (host). This implies and involves the development and acquisition of new metabolic and organic capabilities and also genome sharing by the partners in syntony with the host, to establish a new level of organization, extending beyond the capability of each individual forming the association. One good example of this can be found at the pathway of the biological nitrogen fixation present in this symbiotic system and shared by the different elements of the consortium. Another is at the level of sexual reproduction of the fern, involving cooperative and synchronous efforts, taking into consideration that the cyanobacterium and the bacteria are also involved and incorporated in this process (Carrapiço 2010a). Due to these latter characteristics, this association can be considered both as an example of a hereditary symbiosis and a synergistic complex biological system, with the symbionts always present in the fern's life cycle, suggesting a phylogenetic parallel co-evolution of the associated partners with the fern.

4 The Symbiogenic Theory of Evolution

The biological world presents and involves symbiotic associations between different organisms to form consortia, a new structural life dimension and a symbiont-induced speciation. This implies a new understanding of the natural world, in which symbiogenesis plays an important role as an evolutive mechanism, with symbiosis being the key for the acquisition of new genomes and new metabolic capacities, which drives living forms' evolution and the establishment of biodiversity on Earth. One good example of the importance of symbiosis in evolution can be found in plant transition from aquatic to terrestrial environments. In a recent work, Lipnicki (2015) states that symbiosis played a very important role in the crucial stages of the transition of life onto land, namely through lichenization and mycorrhization. In this sense, explanations of evolutionary changes must include an integrated synergistic co-operation between organisms, in which symbiosis acts, not as an exception, but as the main rule in nature, based on rather sudden evolutionary novelty and the increased complexity of living systems (Carrapiço 2010b; Corning 2005, 2014; Corning and Szathmáry 2015; Reid 2007). These ideas constitute the development of novel concepts for a better understanding of life on our planet and beyond, including the foundation of a new biological theoretical framework that can integrate and explain the dynamical organismal interactions and synergistic relationships present on Earth and in other planets. In this sense, we would like to share in this work a set of principles that could be integrated into a new approach to the evolutive process, helping to build a symbiogenic theory of evolution (Carrapiço 2006a, 2010b, 2012a, b). This theory includes Darwinian principles, but does not limit itself to the latter in its attempt to promote and explain the development, organization, and evolution of the biological world in a symbiogenic and synergistic sense. To integrate these ideas in the scientific literature, we need to develop a new approach to the analysis of evolution based on six themes: (1) Darwinian principles, (2) symbiosis concept, (3) symbiogenesis as an evolutive mechanism, (4) serial endosymbiotic theory,

(5) horizontal gene transfer and other genetic recombinations, and (6) epigenetic changes. These tenets should be considered as a contribution to a new epistemological perception of the natural world and also to the understanding of the true complexity, organismal interactions, and relationships present in the different ecosystems on Earth.

5 Conclusion

Life is evolution, a dynamic continuum existing unbroken since its emergence. Nevertheless, we must go beyond the traditional approaches to the understanding of evolution based on competition and gradualism, and integrate symbiogenic, synergistic, and co-operative principles as potential sources of evolutive novelty and quick transition. In symbiotic relationships, the central aspect is the creation of evolutive novelty (metabolic, anatomical, and organismal), which also involves the sharing of genomes among the organisms constituting the consortium, forcing these genomes to be incurred by synchronization and harmonization aimed at the proper functioning of the new organism as a whole. All these data should be incorporated into a new field of biological science, symbiogenic developmental biology, or informally, *symbio-devo*, merging symbiogenic evolution with developmental biology. These ideas imply the development of novel concepts for a better understanding of life and the emergence of complexity in nature, including the foundation of a new biological theoretical framework that can integrate and explain the dynamical organismal interactions and synergistic relationships present on Earth. This reality can be embodied and built in a symbiogenic theory of evolution. The development of such a theory could contribute towards a new epistemological approach to symbiotic phenomena in evolution specifically, and indeed biology in general, presenting new perspectives that allow for a better understanding of the web of life on our planet and beyond.

6 Main Milestones in Symbiogenic Studies Until 2003

1840 Pierre-Joseph Proudhon (1809–1865) develops the idea of *mutualism* applied to the social and political arena in the book *Qu'est-ce que la propriété* (What is Property?).

1867 Simon Schwendener (1829–1919) proposes in the Swiss Natural History Society annual meeting held in Rheinfelden (Switzerland) the dual hypothesis to explain the nature of lichens, indicating that they are an association of two organisms, a fungus and an alga, behaving as "master and slave."

1873 Johannes Reinke (1849–1931) refers to the relationship between the fungi and the algae in lichens as a *consortium.*

1875 Pierre-Joseph van Bénéden (1809–1894) introduces the *mutualism* concept for the animal kingdom in the work *Les Commensaux et les Parasites dans le Règne Animal* (*The Commensals and the Parasites in the Animal Kingdom*).

1877 Albert Bernhard Frank (1839–1900) introduces the term *symbiotismus* in a publication on the biology of lichens. This concept is similar to the symbiosis one introduced one year later by Anton De Bary.

1878 Heinrich Anton De Bary (1831–1888) introduces the concept of *symbiosis* (from Greek, meaning "living together") as "the living together of unlike named organisms" in a communication entitled "Ueber Symbiose" (On Symbiosis) during a meeting at Cassel (Germany) of the Congress of German Naturalists and Physicians. De Bary used this term when discussing the presence of the cyanobacteria in the leaf cavity of *Azolla* and also about the nature of lichens and the role of the alga and fungus in this association.

1883 Andreas Schimper (1856–1901) reports on the nature and growth of starch grains showing that they arise in specific organelles, which he named *chloroplasts*. He also noted the proliferation of these organelles through division, suggesting their symbiotic origin.

1885 Albert Bernhard Frank introduces the term "micorrhizen" *mycorrhiza* (fungus root) in a paper entitled "Ueber die auf Wurzelsymbiose beruhende Ernährung gewisser Bäume durch unterirdische Pilze" (On the Nourishment of Trees Through a Root Symbiosis with Underground Fungi) in the *Berichte der Deutschen Botanischen Gesellschaft*, to describe the mutualistic associations between soil fungi and plant roots.

1893 Roscoe Pound (1870–1964) publishes in the journal, *The American Naturalist.* "Symbiosis and Mutualism." based on the communication with the same title read at the Botanical Seminar of the University of Nebraska on December 17, 1892.

1893 Shosaburo Watasé (1862–1929) gives the lecture "On the Nature of Cell-Organization" before the Biological Club of the University of Chicago, on February 7 of this year, where he defends the idea of the eukaryotic cell as a symbiotic community, and published the following year in the *Biological Lectures of Marine Biological Laboratory of Woods Hall.*

1897 Albert Schneider publishes in the *Minnesota Botanical Studies,* "The Phenomena of Symbiosis," and redefines symbiosis as "a contiguous association of two or more morphologically distinct organisms, not of the same kind, resulting in a loss or acquisition of assimilated food-substances."

1899 Herbert Spencer introduces in his revised and enlarged second volume of *The Principles of Biology* the idea of symbiosis as a division of labor, a synthesis of a complementary physiological functions, resulting from early divergence in the history of life.

1902 Petr Kropotkin (1842–1921) publishes *Mutual Aid. A Factor of Evolution*. In this work, Kropotkin argues that despite the Darwinian concept of the survival of the fittest, co-operation rather than conflict is the main factor in the evolution of species. The book was written while he was in exile in England.

1904 Theodor Heinrich Boveri (1862–1915) suggests that the nucleated cells arose from a symbiosis of two kinds of single plasma-structures, Monera, in a fashion that a number of smaller forms, the chromosomes, established themselves within a larger one which is called the cytosome. In conclusion, the chromosomes would be independent elementary organisms that live symbiotically in the cytoplasm. This idea was further deeply developed by Constantin Merezhkowsky.

1905 Constantin Sergeevich Merezhkowsky (1855–1921) publishes the article "Uber Natur und Ursprung der Chromatophoren im Pflanzenreich" (On the Nature and Origin of Chromatophores in the Plant Kingdom) where, for the first time, coherent scientific arguments show that plastids arose from free-living cyanobacteria.

1907 Andrey Sergeevich Famintsyn (1835–1918), a Russian botanist contemporary of Merezhkowsky, publishes "On the Role of Symbiosis in the Evolution of Organisms," where the author developed the idea that symbiosis has an important evolutionary, or even adaptative, meaning.

1909 Publication of "The Theory of Two Plasms as Foundation of Symbiogenesis, New Doctrine on the Origin of Organisms" in Russian. The German version is published one year later. Constantin Merezhkowsky writes the work during his stay at Kazan University, introducing the concept of symbiogenesis as "The origin of organisms by the combination or by the association of two or several beings which enter into symbiosis." In this paper, he introduces not only the new concepts in the symbiogenesis field, but he also develops some important ideas about the origin of life, namely related to the role of extremophiles in that scenario. A new classification of the living world is proposed using symbiotic criteria.

1910 Frederick Keeble (1870–1952) publishes *Plant-Animals. A Study in Symbiosis*, a study of the biology of two marine worms, *Convoluta roscoffensis* and *Convoluta paradoxa*, and their algae symbionts.

1913 Hermann Reinheimer publishes *Evolution by Co-operation. A Study in Bio-economics.*

1915 Hermann Reinheimer publishes *Symbiogenesis: The Universal Law of Progressive Evolution*, reinforcing the idea that natural co-operation was as strong a force in evolution as Darwinian natural selection.

1918 Paul Portier (1866–1962) publishes *Les Symbiotes*. In this work, Portier develops the idea that all organisms are constituted of an association of different beings. In the case of mitochondria, he argues that those organelles are symbiotic bacteria that the author calls "symbiotes."

1920 Constantin Merezhkowsky publishes in the *Bulletin de la Société des Sciences Naturelles de l'Ouest de la France (Nantes)*, "La Plante Considerée comme un Complexe Symbiotique" (The Plant Considered as a Symbiotic Complex) where the author develops his previous ideas on the symbiotic origin of chloroplasts and nucleus. In opposition to all the current views at the time, Merezhkowsky defends that chloroplasts have not evolved from mitochondria or protoplasm, but from free-living cyanobacteria.

1920 *Symbiosis: A Socio-physiological Study of Evolution* is published by Hermann Reinheimer. In the book, the author points out the importance of the specific interrelations in the development of organisms as a whole, giving us a holistic perspective of organismal evolution.

1921 Constantin Merezhkowsky commits suicide in a room of the Hotel des Familles in Geneva, Switzerland, after several years of exile (January 9).

1921 Paul Buchner (1886–1978) publishes his first book entitled *Tier und Pflanze in Intracellular Symbiose (Animals and Plants in Intracellular Symbiosis)*.

1922 Maurice Caullery (1868–1958) publishes *Le Parasitisme et la Symbiose*, translated into English in 1952 with the title *Parasitism and Symbiosis.*

1923 George H.F. Nuttall (1862–1937) publishes in the journal, *The American Naturalist,* the article, "Symbiosis in Animals and Plants."

1923 Lemuel Roscoe Cleveland (1892–1969) publishes in the *Proceedings of the National Academy of Sciences* the article "Symbiosis between Termites and their Intestinal Protozoa" referring for the first time to the symbiotic nature of the intestinal flagellates of termites.

1923 Ivan Emmanuel Wallin (1883–1969) publishes in *The American Naturalist*, "The Mitochondria Problem," emphasizing the symbiotic origin of these organelles against the cytoplasmic point of view. He joined the University of Colorado in 1918 and the next year became professor of anatomy, a position he held for 32 years.

1924 Boris Kozo-Polyansky (1890–1957) publishes in Russian the monograph "A New Principle of Biology: An Essay on the Theory of Symbiogenesis." In this work, Kozo-Polyansky tries to integrate the symbiogenesis theory with the Darwinian one.

1927 Ivan Wallin publishes *Symbionticism and the Origin of Species*, where the author defends the importance of symbiotic mechanisms in evolution, with emphasis on the symbiotic origin of mitochondria. Wallin also emphasizes the importance of microsymbiosis in this process, pointing out the idea that "Bacteria, which are popularly associated with disease, may represent the fundamental causative factor in the origin of species."

1952 Joshua Lederberg (1925–2008) publishes an article in the journal *Physiological Reviews* entitled "Cell Genetics and Hereditary Symbiosis," where he introduces the term *plasmid* to describe extranuclear genetic structures that can reproduce independently. In the same article, he defends a symbiogenic approach to the origin of mitochondria and chloroplasts, pointing out the similarities between known bacterial symbionts and those organelles.

1962 The definitive proof of DNA in chloroplasts is made by Hans Ris (1914–2004) and Walter Plaut (1931–) suggesting that chloroplasts originate from endosymbiotic cyanobacteria as was postulated by Constantin Merezhkowsky. The work titled "Ultrastructrure of DNA-Containing Areas in the Chloroplast of *Chlamydomonas*" is published in *The Journal of Cell Biology.*

1963 The First International Conference on Symbiosis titled "Symbiotic Associations" takes place in London (April), held by the Society for General Microbiology in its Thirteenth Symposium.

1963 René Dubos (1901–1982) and Alex Kessler publish in the *Proceedings of the 1st International Conference on Symbiosis* the article "Integrative and Disintegrative Factors in Symbiotic Associations."

1963 Margit Nass and Sylvan Nass found DNA fibers in mitochondria, reinforcing the symbiotic origin of these organelles. These results are published in two papers of *The Journal of Cell Biology.*

1967 Lynn Margulis (1938–2011) publishes in the *Journal of Theoretical Biology* the article "On the Origin of Mitosing Cells." In this paper, a theory of the origin of the discontinuity between eukaryotic and prokaryotic cells is presented. Mitochondria, basal bodies of the flagella and chloroplasts, are considered to have derived from free-living cells, and the eukaryotic cell is seen as the result of the evolution of ancient symbioses.

1967 At the same time that Margulis' 1967 article was published, an oft-forgotten short paper by the Norwegian microbiologist Jostein Goksoyr (1922–2000) appeared in *Nature,* providing a similar endosymbiotic theory for the origin of eukaryotic cells.

1969 Ivan Wallin submitted a short paper titled "Symbioticism in the Light of Recent Cytological Investigations" to *Science* magazine. This paper was rejected without any comments.

1970 Lynn Margulis publishes the book, *Origin of Eukaryotic Cells: Evidence and Research Implications for a Theory of the Origin and Evolution of Microbial, Plant and Animal Cells on the Precambrian Earth,* in sequence with her previous article. Using information from cellular and molecular biology, she promotes the serial endosymbiotic theory for the origin of the eukaryotic cells.

1972 Kwang W. Jeon publishes in the journal, *Science,* a short article entitled "Development of Cellular Dependence on Infective Organisms: Micrurgical Studies in Amoebas" about the role of intracellular symbionts on cellular divergence and variation.

1975 James Lovelock (1919–) and Lynn Margulis propose the Gaia hypothesis, supporting the idea that Earth is a complex self-regulatory, flexible living system.

1976 Richard Dawkins (1941–) writes *The Selfish Gene*, redefining the concept of symbiosis to include relations between individuals of the same species. He also introduces says that there is no selection for "the good of species" and "we are gigantic colonies of symbiotic genes."

1979 Liya N. Khakhina publishes in Russian the book, *Problema Simbiogeneza: Istoriko-Kritichesky Ocherk Issledovany Otechestvennykh Botanikov*, translated into English in 1992 as *Concepts of Symbiogenesis. A Historical and Critical Study of the Research of Russian Botanists*, and edited by Lynn Margulis and Mark McMenamin, an important contribution to the knowledge of the history of symbiosis research in Russia.

1981 Lynn Margulis publishes *Symbiosis in Cell Evolution: Life and its Environment on the Early Earth*. In this book, the author presents a modern synthesis of the mechanisms and processes of cell evolution, offering a coherent explanation of how eukaryotic cells evolved from bacterial ancestors by a series of symbioses. In this sense, the origin of the eukaryotic cell is perceived as a special case of a general phenomenon, the evolution of microbial associations.

1982 Christian de Duve (1917–2013) suggests that peroxisomes arose from aerobic bacteria that were adopted as endosymbionts before mitochondria.

1985 Douglas H. Boucher (1950–) edits *The Biology of Mutualism. Ecology and Evolution*. This book develops the point of view that the mutually beneficial interactions between species are just as important as competition and predation, and how mutualisms affect population dynamics and community structure.

1987 David C. Smith and Angela E. Douglas publish *The Biology of Symbiosis*. This important textbook was primarily aimed at filling a gap in the symbiosis literature to base a course in the field for the biology curricula at the university level.

1988 The Microcosmos Project begins. This project co-ordinated by Douglas Zook and Lynn Margulis at the University of Boston aims at the use of the microorganism world for a more earth-conscious approach to education, with particular interest in co-operative biological systems and the maintenance of species diversity.

1991 The book *Symbiosis as a Source of Evolutionary Innovation: Speciation and Morphogenesis* is published. It is edited by Lynn Margulis and René Fester.

1991 Francisco Carrapiço (1951–) publishes in the journal *Plant and Soil* the article "Are Bacteria the Third Partner of *the Azolla-Anabaena* Symbiosis?" presenting data showing that bacteria existing in the *Azolla* leaf cavities and megasporocarps follow a developmental pattern identical to the cyanobacteria *Anabaena azollae* and can be considered the third partner of the symbiotic association.

1994 Jan Sapp (1954–) writes *Evolution by Association. A History of Symbiosis*, an important scientific landmark in the history of symbiosis theory.

1994 Angela Douglas publishes *Symbiotic Interactions*, considering that "The common denominator of symbiosis is not mutual benefit but a novel metabolic capability, acquired by one organism from its partners."

1996 Peter Corning (1935–) publishes in the *Journal of Evolutionary Theory* the article "The Co-operative Gene: On the Role of Synergy in Evolution," an important contribution to understanding evolution in a more synergistic and cooperative way.

1997 The International Symbiosis Society (ISS) is founded on April 15 at the Second International Symbiosis Congress in Woods Hole, United States.

1998 Lynn Margulis publishes *Symbiotic Planet. A New View of Evolution*, a personal and autobiographical journey to the science and symbiosis world.

1998 Douglas Zook in the article, "A New Symbiosis Language," published in the *ISS Symbiosis News*, proposes a new definition for symbiosis: "Symbiosis is the acquisition and maintenance of one or more organisms by another that results in novel structures and metabolism. Some symbiotic evolution may involve partner genetic exchanges."

1999 William Martin and Klaus V. Kowallik publish in the *European Journal of Phycology* the annotated English translation of Merezhkowsky's (1905) paper "Uber Natur und Ursprung der Chromatophen im Pflanzenreich" (On the Nature and Origin of Chromatophores in the Plant Kingdom).

2000 Surinder Paracer and Vernon Ahmadjian write *Symbiosis. An Introduction to Biological Associations.*

2000 Rosmarie Honegger publishes in the journal, *The Bryologist,* the article, "Simon Schwendener (1829–1919) and the Dual Hypothesis of Lichens."

2000 Marc-André Selosse writes *La Symbiose: Structures et Fonctions, Rôle Ecologique et Évolutif.*

2002 The book *Cyanobacteria in Symbiosis* is edited by Amar N. Ray, Birgitta Bergman, and Ulla Rasmussen. It is a reference work in the field of plant–cyanobacteria interactions and nitrogen biological fixation.

2002 Joseph Seckbach (1934–) edits *Symbiosis: Mechanisms and Model Systems*, providing in a clear and broad way the inter- and multidisciplinary dimension of the interspecific relationships, and their mechanisms of work and evolution.

2002 Lynn Margulis and Dorion Sagan (1959–) publish *Acquiring Genomes. A Theory of the Origins of Species*. In this work, the authors point out that the acquisition of new genomes involving symbiogenic processes is the main driving force in evolution, not random mutations, and include a solid criticism of neo-Darwinism.

2002 Jan Sapp, Francisco Carrapiço, and Mikhail Zolotonosov (1954–) publish in the journal of *History and Philosophy of the Life Sciences* the article, "Symbiogenesis: The Hidden Face of Constantin Merezhkowsky," revealing the controversial dimension of his life and work.

2003 Jan Sapp introduces the terms *symbiomics* and *symbiome* in his new book *Genesis. The Evolution of Biology*, revealing a new approach to the understanding of this science in an evolutive perspective, reinforcing its symbiogenic component. In this work, the author points out an important and innovative idea that

Every eukaryote is a superorganism, a symbiome composed of chromosomal genes, organellar genes, and often other bacterial symbionts as well as viruses. The symbiome, the limit of the multicellular organism, extends beyond the activities of its own cells. All plants and animals involve complex ecological communities of microbes, some of which function as commensals, some as mutualists, and others as parasites, depending on their nature and context.

Acknowledgments I am grateful to Nathalie Gontier and Maria Helena Costa for their encouragement in the development of this work and to Helena Carrapiço and Ben Farrand for the English revision of the text. Also, a word of gratitude to Jonathan Bujak for helping to obtain information about Hermann Reinheimer related to the 1911 England Census, and to Padmavathi Jayajeevan for the excellent work in publishing this chapter.

References

Archibald JM (2011) Origin of eukaryotic cells: 40 years on. Symbiosis 54:69–86

Archibald J (2014) One plus one equals one. Symbiosis and the evolution of complex life. Oxford University Press, New York

Boucher DH (1985) The idea of mutualism, past and future. In: Boucher DH (ed) The biology of mutualism: ecology and evolution. Croom Helm Ltd., Oxford University Press, New York, pp 1–28

Bowler PJ (1975) The changing meaning of "evolution". J Hist Ideas 36:95–114

Brucker RM, Bordenstein SR (2012) Speciation by symbiosis. Trends Ecol Evol 27(8):443–451

Carrapiço F (2006a) Is the Azolla-Anabaena symbiosis a co-evolution case? General botany: traditions and perspectives. Materials of the International Conference, dedicated to 200th anniversary of the Kazan Botanical School (23–27 Jan 2006). Part I. Kazan, 2006, pp 193–195. Edited by Andrew Sitnykov and published by the Kazan University, Russia

Carrapiço F (2006b) The origins of life and the mechanisms of biological evolution. Proceedings of SPIE 6309: 630900-1-630900-5

Carrapiço F (2010a) Azolla as a superorganism. Its implication in symbiotic studies. In: Seckbach J, Grube M (eds) Symbioses and stress: joints ventures in biology, cellular origin, life in extreme habitats and astrobiology, vol 17. Springer, New York, pp 225–241

Carrapiço F (2010b) How symbiogenic is evolution? Theory Biosci 129:135–139

Carrapiço F (2012a) The symbiotic phenomenon in the evolutive context. In: Pombo O, Rahman S, Torres JM, Symon J (eds) Special sciences and the unity of science, logic, epistomology and the unity of science, vol 24. Springer Science+Business Media B.V, New York, pp 113–119

Carrapiço F (2012b) Beyond neo-Darwinism. Building a symbiogenic theory of evolution. Abstracts of the International Colloquium on Emergence and non-Fundamentalist Metaphysics. Lisbon, 14–16 May 2012. (published in Kairos, Journal of Philosophy & Science 12:47–53, 2015)

Corning PA (1996) The co-operative gene: on the role of synergy in evolution. Evol Theor 11:183–207

Corning PA (2005) Holistic darwinism. Synergy, cybernetics, and the bioeconomics of evolution. The University Chicago Press, Chicago

Corning PA (2014) Systems theory and the role of synergy in the evolution of living systems. Syst Res 31:181–196

Corning PA, Szathmáry E (2015) "Synergistic selection": a Darwinian frame for the evolution of complexity. J Theor Biol 371:45–58

De Bary A (1878) Ueber symbiose—Tageblatt 51 Versamml. Deutscher Naturforscher u. Aerzte, Cassel, pp 121–126

De Bary A (1879) Die erscheinung der symbiose. Vortrag auf der Versammlung der Deutschen Naturforscher und Aertze zu Cassel, Strasburg: Verlag von Karl J. Trubner, pp 1–30
Darwin C (1859) On the origin of species by means of natural selection or the preservation of favored races in the struggle for life. In: Murray J (ed), London
Dawkins R (1976) The selfish gene. Oxford University Press, Oxford
Frank AB (1877) Ueber die biologischen Verhältnisse des Thallus einiger Krustenflechten. Beitr Biol Pfl 2(2):123–200
Goksoyr J (1967) Evolution of eucaryotic cells. Nature 214:1161
Gontier N (2007) Universal symbiogenesis: an alternative to universal selectionist accounts of evolution. Symbiosis 44:167–181
Guerrero R, Margulis L, Berlanga M (2013) Symbiogenesis: the holobiont as unit of evolution. Int Microbiol 16:133–143
Guilliermond A (1918) Sur l'origine mitochondriale des plastides. C. R. Hebd Séanc. Acad Sci, Paris 167:430–433
Holldobler B, Wilson EO (2009) The superorganism: the beauty, elegance and strangeness of insect societies. W.W. Norton and Company, New York
Honegger R (2000) Simon Schwendener (1829–1919) and the dual hypothesis of lichens. Bryologist 103:167–183
Huxley J (1942) Evolution. The modern synthesis. George Allen & Unwin Ltd (eds). London
Kozo-Polyansky B (2010) Symbiogenesis. A new principle of evolution. In: Fet V, Margulis L (trans) Harvard University Press, Cambridge (from the Russian edition, 1924)
Khakhina LN (1992) Concepts of symbiogenesls. In: Marguhs L, McMenamin M (eds) A historical and critical study of the research of russian botanists. Yale University Press, New Haven
Kutschera U (2011) From the scala naturae to the symbiogenic and dynamic tree of life. Biol Direct 6(33):1–25. doi:10.1186/1745-6150-6-33
Kropotkin P (1902) Mutual aid. A factor of evolution. McClure Phillips, New York
Lipnicki IL (2015) The role of symbiosis in the transition of some eukaryotes from aquatic to terrestrial environments. Symbiosis 65:39–53. doi:10.1007/s13199-015-0321-7
Lamarck J-BM (1809) Philosophie zoologique, ou exposition des considérations relatives à l'histoire naturelle des animaux. Paris
Lumière A (1919) Le mythe des symbiotes. Masson et Cie. Ed., Paris
Margulis L (1990) Words as battle cries—symbiogenesis and the new field of endocytobiology. Bioscience 40(9):673–677
Margulis L, Sagan D (2002) Acquiring genomes. A theory of the origin of species. Basic Books, New York
Martin W, Kowallik KV (1999) Annotated english translation of Mereschkowsky's 1905 paper "Uber Natur und Ursprung der Chromatophoren im Pflanzenreich". Eur J Phycol 34:287–295
Mayr E (2001) What evolution is. In: Weidenfeld, Nicolson (eds) Basic books, London
Mayr E (1942) Systematics and the origins of species, from the viewpoint of a zoologist. Harvard University Press, Cambridge
Merezhkowsky C (1905) Uber natur und ursprung der chromatophoren im pflanzenreiche. Biol Centralbl 25(593–604):689–691
Merezhkowsky C (1909) The theory of two plasms as foundation of symbiogenesis. A new doctrine on the origins of organisms. Proceedings of studies of the Imperial Kazan University, vol 12. pp 1–102
Merezhkowsky C (1920) La plante considerée comme un complexe symbiotique. Bulletin de la Societé des Sciences Naturelles de l'Ouest de la France. 6:17–98
Muller HJ (1927) Artificial transmutaion of the gene. Science 66(1699):84–87
Nylander W (1896) Les lichens des environs de Paris. Typographie Paul Schmidt, Paris
Pereira L, Rodrigues T, Carrapiço F (2012) A symbiogenic way in the origin of life. In: Seckbach J (ed) Genesis—in the beginning. Precursors of life, chemical models and early. Biological evolution, cellular origin, life in extreme habitats and astrobiology, vol 22. Springer Science+Business Media, Dordrecht, pp 723–742. doi:10.1007/978-94-007-2941-4_36
Portier P (1918) Les symbiotes. In: Masson et Cie. (ed), Paris

Reid RGB (2007) Biological emergences. Evolution by natural experiment. MIT Press, Cambridge
Reif WE, Junker T, Hossfeld U (2000) The synthetic theory of evolution: general problems and german contribution to the synthesis. Theor Biosci 119:41–91
Reinheimer H (1913) Evolution by co-operation. A study in bio-cconomics. In: Paul KT (ed). Trubner & Co. Ltd., London
Reinheimer H (1915) Symbiogenesis. The universal law of progressive evolution. Knapp, Drewett and Sons Ltd (eds), Westminster
Reinheimer H (1920) Symbiosis. A socio-physiological study of evolution. Headley Brothers (eds), London
Reinke J (1873) Zur kenntniss des rhizoms von Corallorhiza und Epipogon. Flora 31:145–209
Rieppel O (2011) Evolutionary theory and the creation controversy. Springer, Heidelberg
Sagan L (1967) On the origin of mitosing cells. J Theor Biol 14:225–274
Sapp J (1994) Evolution by association: a history of symbiosis. Oxford University Press, New York
Sapp J (2003) Genesis. The evolution of biology. Oxford University Press, New York
Sapp J, Carrapiço F, Zolotonosov M (2002) Symbiogenesis: the hidden face of Constantin Merezhkowsky. Hist Phil Life Sci 24:421–449
Sciama Y (2013) Penser coopération plutôt que compétition. Sci Vie 1147:139–148
Selosse MA (2000) La symbiose: structures et fonctions, rôle écologique et évolutif. Vuibert (ed), Paris
Spencer H (1876) The principles of sociology. D. Appleton and Company (eds), New York
Taylor FJR (1974) Implications and extensions of the serial endosymbiosis theory of the origin of eukaryotes. Taxon 23(2/3):229–258
Taylor FJR (1976) Autogenous theories for the origin of eukaryotes. Taxon 25:377–390
Taylor FJR (1979) Symbioticism revisited: a discussion of the evolutionary impact of intracellular symbioses. Proc R Soc London B 204:267–286
Taylor J, Whalley JI, Hobbs AS, Battrick EM (1995) Beatrix potter 1866–1943. The artist and her world. Imago Publishing Limited, Hong Kong
Todes DP (1989) Darwin without Malthus. The struggle for existence in Russian evolutionary thought. Oxford University Press, New York
Tower WL (1916) Review of the book symbiogenesis. The universal law of progressive evolution. Am J Sociol:695–696
Van Bénéden PJ (1875) Les comensaux et les parasites dans le règne animal. Bibl Sci Int, Paris
Villarreal L, Ryan F (2011) Viruses in host evolution: general principles and future extrapolations. Curr Top Virol 9:79–90
Wallin IE (1923) The mitochondria problem. Am Nat 57(650):255–261
Wallin IE (1927) Symbionticism and the origin of species. Williams and Wilkins (eds), Baltimore
Wheeler WM (1911) The ant-colony as an organism. J Morphol 22(2):307–325
Wheeler WM (1928) The social insects, their origin and evolution. Harcourt, Brace & Company, New York
Wilson EO (1975) Sociobiology. The new synthesis. Harvard University Press, Cambridge
Wilson DS, Sober E (1989) Reviving the superorganism. J Theor Biol 136:337–356
Zilber-Rosenberg I, Rosenberg E (2008) Role of microorganisms in the evolution of animals and plants: the hologenome theory of evolution. FEMS Microbiol Rev 32:723–735

Novel Endosymbioses as a Catalyst of Fast Speciation

Vitor G. Faria and Élio Sucena

Abstract Many symbiotic bacteria complete their life cycle inside eukaryotic cells. In arthropods, facultative endobacteria such as *Wolbachia* and *Spiroplasma* influence enormously the ecology and evolution of their hosts. In the last decades, the idea that endosymbiotic co-evolution can lead to host speciation has been proposed and, in some instances, verified. However, although usually transmitted vertically, these bacteria can also change host through horizontal transmission. After this transfer and in a virtually instantaneous fashion, endobacteria can alter the fitness of their new host by modifying its response to the environment and/or manipulating its reproduction. In this light, horizontally transmitted endosymbionts could strongly influence the evolutionary path taken by their new hosts. Here, we argue that from this evidence emerges a testable five-step scenario for the appearance of novel host lineages.

Keywords Endosymbiosis • Speciation • Arthropod evolution • *Wolbachia* • Vertical transmission • Horizontal transmission

1 Endosymbiosis in a Symbiotic World

Symbiosis is the generic terminology to classify close and in general long-term biological interactions between organisms of different species, conferring a benefit or disadvantage to at least one of them (de Bary 1879). Many symbiotic

V.G. Faria (✉) · É. Sucena (✉)
Gulbenkian Institute for Science (IGC), Apartado 14, 2780-901 Oeiras, Portugal
e-mail: vfaria@igc.gulbenkian.pt

É. Sucena
Department of Animal Biology, Faculty of Science, University of Lisbon,
Edifício C2, Campo Grande, 1749-016 Lisbon, Portugal
e-mail: jesucena@fc.ul.pt

N. Gontier (ed.), *Reticulate Evolution*, Interdisciplinary Evolution Research 3,
DOI 10.1007/978-3-319-16345-1_4

relationships have been reported, revealing dynamic interactions pivotal to the evolution of species and their ecology (Douglas 1994; Nieberding and Olivieri 2007). Both within and between kingdoms, permanent or sporadic associations can be found linking organisms of different species in a range of habitats and environments (Moran 2006; Tarkka et al. 2009). Unlike ectosymbionts, which establish themselves in the host's body surface, endosymbionts are lodged in the host's tissues or organs, intra- or extracellularly (Douglas 1994). These associations persist across generations by vertical transmission (maternal and/or paternal) which is directly inherited [like endobacteria in fungi (Bianciotto et al. 2004) and a number of endosymbionts in invertebrates (Moran et al. 2008)] or by horizontal indirect transmission, where associations are formed de novo [as mycorrhization and root nodulation (Lima et al. 2009)]. Yet, endosymbiosis can be obligate or facultative (for one or both partners), according to the necessity for the presence of the symbiotic partner to the completion of the host's life cycle (Moran 2006).

Here, we will focus on intracellular symbiotic relationships between different multicellular hosts and facultative endobacteria, which can be horizontally transmitted to individuals of other populations or species. We will look in detail, mostly in insects, at the potential of endosymbiosis of facultative intracellular bacteria to enable sudden phenotypic change in novel multicellular hosts (White 2011) as well as to potentially facilitate rapid speciation processes through reproductive manipulations of the host (Hurst and Schilthuizen 1998).

Below, we will systematize a large yet disperse body of evidence, which in principle supports a five-step scenario where the establishment of new endosymbioses can lead to the emergence of new host (incipient) species. We will review the literature and show that (i) endosymbiosis is a common phenomenon; (ii) the presence of the endosymbiont in the host frequently affects its fitness; (iii) horizontal transmission of endosymbionts is likely; (iv) vertical transmission may ensue and thus lead to a stable phenotype; and (v) the presence of endobacteria in the new host induces significant phenotypic change with fitness consequences that may promote directional or disruptive selection. We will organize and link evidence, which support this five-step scenario and argue that the establishment of endosymbioses may culminate in fast speciation events.

2 Widespread Co-evolution Between Endosymbionts and Multicellular Hosts

When looking at the three most well-defined kingdoms of the Eukarya domain, we can relate them directly to the establishment of bacterial endosymbioses (de Duve 2007; Sapp 1994). Organelles that have co-evolved intracellularly until becoming mutually obligate from free-living prokaryotes ensure the main oxidative metabolism of fungi, plants and animals (Dyall et al. 2004; Osteryoung and Nunnari 2003).

In addition to the bacterial endosymbionts which gave rise to organelles of eukaryotic cells, endosymbiotic relationships can be found in all kingdoms of the Eukarya domain (in higher or lower frequency) and it is likely that many are yet to be discovered, together with the full range of phenotypic changes they may cause (Moran and Wernegreen 2000; Moran 2006; Faria and Sucena 2013). In fungi, we can find many symbiotic relationships with intracellular bacteria (Tarkka et al. 2009). The majority of them include species of the Phylum Glomeromycota, which is composed of arbuscular mycorrhizal fungi (Parniske 2008) and Geosiphon (Gehrig et al. 1996). In other phyla, there are at least two other associations described: the basidiomycete *Laccaria bicolor* (Bertaux et al. 2003) and *Rhizopus microsporus* of the Phylum Zygomycota (Partida-Martinez and Hertweck 2005). In plants, most relationships established with bacteria occur via nodulation. The bacteria *Frankia* and *Rhizobia* are the most recurrent ones, establishing themselves in the root and forming nodules, mutualistically exchanging nutrients with plants (Oldroyd and Downie 2008). Moreover, several bacteria species are responsible for follicular nodulation in plants of the Myrsinaceae and Rubiaceae families, a type of association less frequent and less studied than those of the root (Lersten and Horner 1976). In animals, many of the endosymbiotic relationships involve bacteria as illustrated by the enormous abundance of reported associations between endobacteria and invertebrates (Ruby 2008). Within this group, the most studied are the endosymbionts of arthropods, mainly insects.

Some endobacteria are presented as obligate for the host, resulting from close co-evolution with the host species and, thereafter, a corresponding diversification (Baumann 2005). Usually, these endosymbionts, also called primary endosymbionts, lodge in a bacteriome and produce essential nutrients for the host (Werren and O'Neill 1997). On the other hand, facultative (or secondary) endobacteria are found in several cells of various host tissues, being able to infect organisms that already have obligate endobacteria (Moran et al. 2008) In addition to being transmitted vertically, and unlike obligate endosymbionts that are entirely dependent on the host for perpetuation, facultative bacteria are also occasionally transmitted horizontally within and between host species and typically show a short evolutionary history with the current host (for review see Moran et al. 2008). We argue that this horizontal transmission phenomenon can bring new hosts into a process of rapid speciation with high impact in the evolution of host species (Fig. 1).

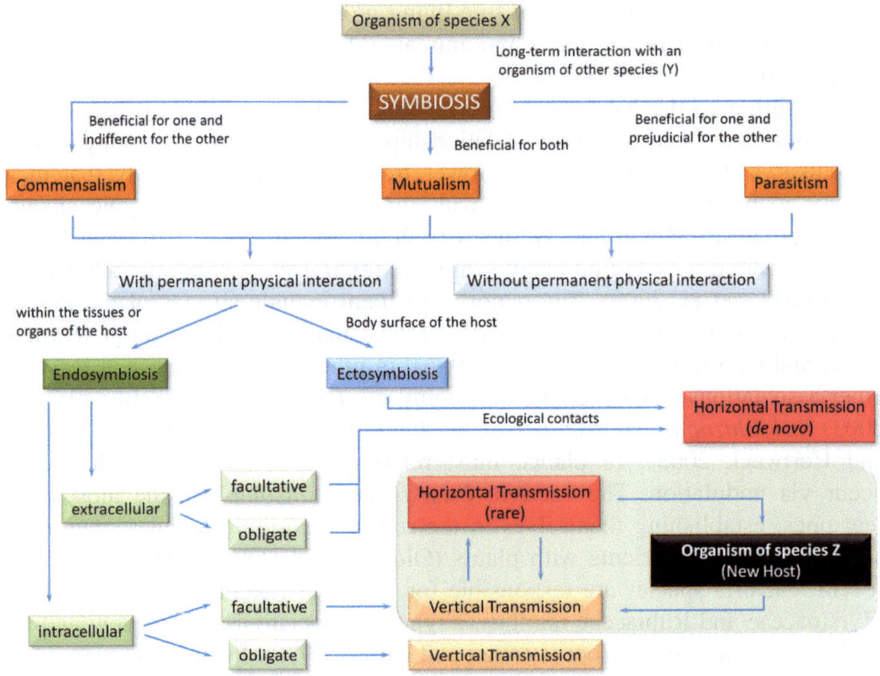

Fig. 1 Symbiotic relationships and the potential emergence of a novel host lineage. When two organisms of different species are stably related in nature, we are in the presence of a symbiotic relationship. These can exist in several combinatorial outputs between the interacting agents, with or without permanent physical interactions. Both ectosymbiosis and extracellular endosymbiosis are indirectly maintained by intraspecific horizontal transmission, where symbiosis occurs de novo in future generations through intimate environmental contacts. In intracellular endosymbiosis, we have maternal and/or paternal vertical transmission of obligate endosymbionts, which are necessary partners for the host's development and reproduction. Here, the partners have strong co-evolution and the endosymbionts' diversification is consistent with the diversification of host populations. Facultative endosymbionts, despite being transmitted by vertical transmission, present occasional horizontal transmission within and between host species. This horizontal transmission of intracellular endosymbionts to a new host may create an organism bearing an immediate novelty, which is now subjected to new environmental selective pressures and may strive in a new lineage

3 Fitness Consequences of the Presence of Facultative Endosymbionts

Many bacteria that complete their life cycle within eukaryotic cells constitute a fully polyphyletic group that exerts a wide range of effects on their hosts (Moran and Wernegreen 2000). One of the most extreme consequences of this symbiotic interaction is the manipulation of the host's reproduction, an important factor in several evolutionary processes, namely in ecologic specialization and speciation (Engelstadter and Hurst 2009; Tsuchida et al. 2004).

The intracellular bacterium *Wolbachia* is the most pandemic symbiont in arthropods and is predominantly transmitted through the female germ line. *Wolbachia* exhibits an extraordinary ability to alter the host's reproduction to selectively favour infected females, thus facilitating its maternal transmission. *Wolbachia* causes four distinct reproductive phenotypes in a range of arthropod orders: feminization, where genetic males develop as females through *Wolbachia*'s interference with thc sex-determination pathway; parthenogenesis, where males are no longer required for reproduction through disruption the host's cell cycle by the bacterium; male killing, where infected males are eliminated to the advantage of surviving *Wolbachia*-infected female siblings; and cytoplasmic incompatibility (CI) that reduces or prevents infected males from producing viable zygotes with females with the same infection *status* (for review see Werren et al. 2008). CI manipulation, the most frequently found *Wolbachia*-induced phenotype, creates an incompatibility between sperm and egg by the alteration of the pronuclear envelope breakdown speed, resulting in the loss of sperm chromosomes following fertilization (Tram and Sullivan 2002). In *Aedes albopictus* mosquitos, *Wolbachia*-infected females are at a reproductive advantage relative to uninfected females due to both CI and a fitness increase (longevity, fecundity and egg hatch) associated with *Wolbachia* infection (Dobson et al. 2004). In *D. mauritiana,* infection with *Wolbachia* increases fecundity substantially through a boost of cell division and decrease of apoptosis of germ line stem cells (Fast et al. 2011). In other bacterial groups, the helical gram-positive bacterium *Spiroplasma* or the bacterium *Cardinium* can also confer a variety of fitness effects and induce host phenotypic alterations by reproductive manipulation (Engelstadter and Hurst 2009; Zchori-Fein et al. 2001).

Facultative endosymbionts can also influence their hosts' defences against natural enemies (Gil-Turnes et al. 1989; Hurst and Hutchence 2010) and specialization to different plant species (for review see Oliver et al. 2010). In the pea aphid, *Acyrthosiphon pisum*, the endosymbiotic association with facultative bacteria confers resistance to attack by the parasitoid wasp, *Aphidius ervi*, causing high mortality of developing parasitoid larvae (Oliver et al. 2003). Subsequently, it was shown that one of the common facultative symbionts of *A. pisum*, the bacterium *Regiella insecticola*, has a major effect on the resistance of the host to a fungal pathogen and lowers its rate of transmission (Scarborough et al. 2005). Recently, some studies have demonstrated that the presence of *Wolbachia* can also increase the fitness of the host. In *Drosophila melanogaster*, infection with *Wolbachia* increases resistance to RNA viruses such as *Drosophila* C virus, a natural pathogen of *Drosophila* (Teixeira et al. 2008; Hedges et al. 2008). Furthermore, it was shown that *Spiroplasma* protects *Drosophila neotestacea* against the sterilizing effects of a parasitic nematode, underscoring the potential impact of facultative endosymbioses in the ecological distribution and population dynamics of the host species (Jaenike et al. 2010; Jaenike and Brekke 2011). These data support the notion that the response of a host to environmental conditions also depends on its resident endobacteria.

4 Secondary Endobacteria are Horizontally Transmitted Between Hosts

As stated above, facultative endobacteria are mostly vertically transmitted to the progeny. However, since in many cases, there is no concordance between the phylogeny of bacteria and their hosts, and there is indication of horizontal transmission (Vavre et al. 1999; Thao et al. 2000; Russell et al. 2003; Ahmed et al. 2013). It is conceivable that in an environment inhabited by organisms infected and non-infected with bacteria, given enough time, high densities and reiterated contacts, the probability of horizontal transmission of symbionts is not negligible (Gehrer and Vorburger 2012; Le Clec'h et al. 2013). Moreover, several studies have demonstrated that some microbial symbionts retain a generalized ability to infect multiple hosts (Schilthuizen and Stouthamer 1997; Heath et al. 1999; Huigens et al. 2004; Duron et al. 2010).

In *Drosophila*, the only heritable endosymbionts described thus far are *Wolbachia* and *Spiroplasma* (Mateos et al. 2006). Recently, a phylogenetic analysis of *Spiroplasma* from several *Drosophila* species revealed at least five independent introductions of four phylogenetically distinct *Spiroplasma* haplotypes, indicating imperfect vertical transmission in host populations and likely horizontal transmission (Haselkorn et al. 2009). Likewise, *Wolbachia* molecular phylogenies are not concordant with those of their hosts, supporting occasional events of horizontal transmission (Werren and O'Neill 1997; Jiggins et al. 2002; Baldo et al. 2008). Additionally, it has been demonstrated that *Wolbachia* is able to establish itself as a stable and vertically transmitted infection upon transfer into the hemolymph of uninfected *D. melanogaster* females (Frydman et al. 2006).

Parasitoid insects constitute a prime candidate for acting as vectors of *Wolbachia* horizontal transmission. Some studies revealed extensive similarities between the *Wolbachia* strains found in parasitoids and their hosts (Vavre et al. 1999), strongly supporting the hypothesis of natural *Wolbachia* transfer into other species. Another putative vector for horizontal transmission of endosymbionts is parasitic mites. Indeed, ectoparasitic mites have been shown to transfer *Spiroplasma poulsonni* from infected *D. nebulosa* to *D. willistoni* whose females will, subsequently, transmit the infection to their offspring (Jaenike et al. 2007). Thus, endosymbiotic facultative bacteria show a clear propensity to establish promiscuous relationships with various intra- and interspecific hosts.

5 Endosymbiont-Associated Traits are Transferred to the New Host and Maintained by Bacterial Vertical Transmission

As we have seen, even though we currently do not fully understand the ecological mechanisms for horizontal transmission of facultative endosymbionts, there is ample evidence that it occurs. In this section, we will provide evidence that these endosymbionts may bring instant metabolic or internal morphological novelty to their novel host, usually the same phenotypic alteration that was induced in the previous host (Huigens et al. 2004; Veneti et al. 2004; Braig et al. 1994). Additionally, the endobacteria which change host species, undergoing strong selection for their permanence in the new host (Vallet-Gely et al. 2008), can ensure the evolutionary sustainability by maintenance or acquisition of stable vertical transmission (Mira and Moran 2002; McGraw et al. 2002). Multiple independent lines of evidence support this scenario.

Three species of vertically transmitted Gammaproteobacteria from different aphid host species can infect, spread and induce variation in fitness of the host, when microinjected into a new aphid host (the pea aphid *Acyrthosiphon pisum*), as well as sustain stable vertical transmission to its offspring (Russell and Moran 2005). Recent data reinforce the potential of facultative endosymbioses in modifying the aphid phenotype. Leonardo and Mondor have shown that the endosymbiont *Regiella insecticola* can manipulate polyphenic development by changing the number of winged versus non-winged individuals under crowding, as well as the time of sexual maturation (Leonardo and Mondor 2006). More recently, an interspecific transfection of this endosymbiont from the pea aphid to the vetch aphid *Megoura crassicauda* has proven sufficient to confer the ability to utilize clover as a host plant (Tsuchida et al. 2011). In yet another example with the pea aphid, it has been shown that the presence of an endosymbiont of the genus *Rickettsiella* is sufficient to change body colour and may affect host fitness by influencing interactions with both predators and other endosymbionts (Tsuchida et al. 2010).

Further, when male-killing *Spiroplasma* from coccinellid beetles was artificially injected into a series of naive arthropod species, this bacterium colonized host tissues and was vertically transmitted in all cases tested. Moreover, both the bacteria's efficiency of transmission and its ability to distort offspring sex ratios in novel hosts were unaffected in the case of transfers to the native genus and reduced or incomplete in more distantly related species (Tinsley and Majerus 2007).

In *Wolbachia*, in yet another case of putative reiterated horizontal transmission, it was shown that a male-killing *Wolbachia* strain has consistent phenotypic effects in *Drosophila borealis* and its closely related species (Sheeley and McAllister 2009). In other examples, as in the butterfly *Hypolimnas bolina*, the male-killing effect of *Wolbachia* presence is suppressed without significant reduction in bacterial load. In this case, *Wolbachia* induces CI in the surviving males (Hornett et al. 2008). Similarly, the wCauA strain of *Wolbachia*, which induces CI in the lepidopteran *Cadra cautella*, causes male killing upon transfer to *Ephestia kuehniella* (Sasaki et al. 2002, 2005). Another example comes from the interaction between *Wolbachia* and *Trichogramma* where uninfected immature wasps that acquired *Wolbachia* while inside the host egg displayed a parthenogenetic phenotype (Huigens et al. 2004). These examples suggest plasticity in the deployment of *Wolbachia*'s large arsenal of host reproduction manipulation strategies upon horizontal transfer and consequently in its adaptation to novel hosts.

In summary, evidence is abundant for the stabilization of de novo endosymbionts through vertical transmission upon seemingly rare episodes of horizontal transfer. Furthermore, in many of these instances, such newly established relationships will have instantaneous phenotypic effects with impact in the fitness of the host, thus having the potential to drive evolutionary change.

6 Host Speciation and Endosymbiont-Induced Novelties

As seen above, when the genetic system of a bacterial species is combined with that of another (arthropod) species through horizontal transmission, the new symbiotic partnership may create novel forms of coping with selective pressures in the environment. In particular, when a mechanism of reproductive manipulation is brought by endobacteria from a former host, the endosymbiont may trigger a rapid speciation in the new host (Bordenstein 2003; Coyne and Orr 2004).

The presence of *Wolbachia* in two closely related species of parasitic wasps severely reduces the frequency of hybrid offspring through bidirectional CI in interspecific crosses and precedes the evolution of other postmating reproductive barriers (Bordenstein et al. 2001). In addition, bidirectional CI in host populations may: (i) substantially reduce gene flow (Telschow et al. 2002); (ii) reinforce genetic divergence by association between nuclear alleles and respective microbe infection state (Telschow et al. 2005); (iii) increase behavioural isolation from the *Wolbachia*-infected species; and/or (iv) lead to behavioural isolation between populations of the uninfected species (Jaenike et al. 2006). Yet, other mechanisms may contribute to gene flow reduction between infected and uninfected individuals such as assortative mating and oviposition site preference (Vala et al. 2004). Also, incipient isolation is observed between the sister species *Drosophila recens* and *Drosophila subquinaria*, via the combined action of CI, prezygotic isolation and hybrid sterility (Shoemaker et al. 1999; Jaenike 2007). Taken together, these results support the view that facultative endosymbionts may, directly and indirectly, contribute to reproductive isolation and promote speciation of their hosts (for review, see Thompson 1987; Coyne and Orr 2004; Engelstadter and Hurst 2009; Brucker and Bordenstein 2012).

7 Closing the Circle: From a Different Organism to a New Lineage

In intracellular bacteria, the mechanism of vertical transmission is essential to the unity of the symbiotic complex and for the co-evolution of increased benefits for both species (which ultimately may transform the bacterium into an organelle). Prior to this, ecological interactions may foster the transfer of these bacteria between hosts, within or across species. This horizontal transmission creates

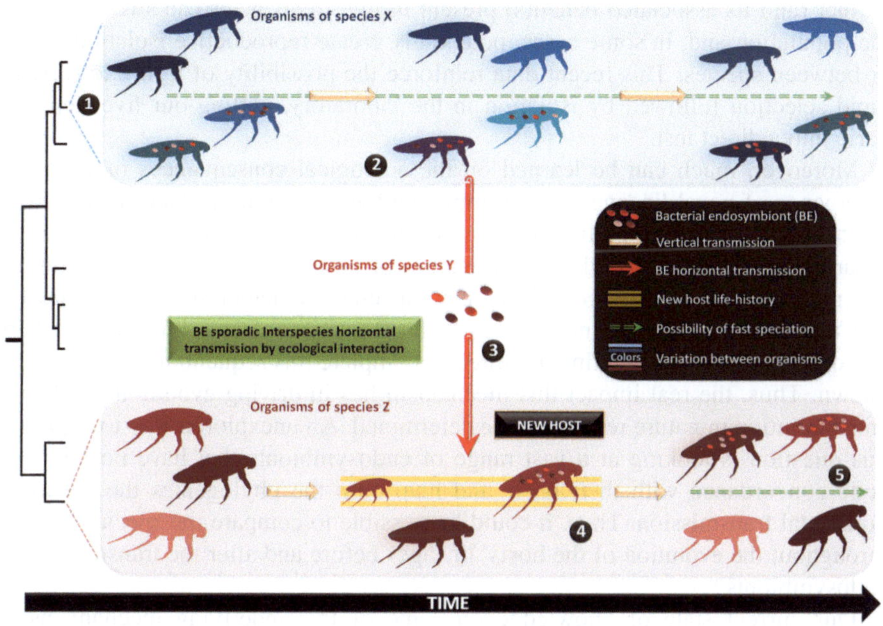

Fig. 2 Five-step scenario for the fast emergence of new lineages. (*1*) A population of species X contains infected and uninfected individuals that inherit bacterial endosymbionts (BE) of the Y species by vertical transmission (sexual—maternal and/or paternal—or asexual). (*2*) The presence of BE can bring reproductive modifications and/or metabolic advantages to the host. (*3*) Ecological interactions (e.g. predation, cannibalism or parasitic vectors, such as wasps and mites) may facilitate the horizontal transmission of the BE to a new species. (*4*) The endosymbiont will impact the phenotype and fitness of its new host, and if this transmission takes place during the reproductive age of the host, there may be stable vertical transmission to the next generation. (*5*) Through a mechanism of sexual manipulation (or others), which may have co-evolved with the previous host, BE may induce reproductive modifications on its new host, leading to rapid speciation

points of contact between evolutionary paths and produces new synergistic combinations of phenotypic variation between organisms (morphological and/or metabolic) with direct fitness impacts and adaptive potential. Indeed, it is reasonable to assume that, as in many of the examples presented above, in some cases, the mere presence of the endosymbiont will contribute to reproductive isolation and promote speciation of its host. In these circumstances, this cyclic chain of bacterial transmission would contribute to catalyse evolution by creating organisms with new phenotypes, which would be founders of new lineages (Fig. 2).

A recent report has confirmed that endosymbionts can combine developmental modifications and reproductive manipulations, which translate into high fitness gains. In a six-year period in nature and in few generations in the laboratory, *Rickettsia* bacteria were able to increase their prevalence from a small percentage of individuals to a rampant infection in *B. tabaci* populations (sweet potato whiteflies) (Himler et al. 2011). Although the selective pressure that causes this difference in fitness is unknown, this report illustrates that an endosymbiotic partner (and its associated benefits) present in low frequencies can sweep through the population and, in some cases, potentially create reproductive isolation within or between species. This recent data reinforce the possibility of trying to recreate rapid selection followed by isolation in the laboratory, putting our five-step scenario into a direct test.

Moreover, much can be learned on the ecological consequences of the rapid emergence of novel lineages: for example, on the evolution of plant lineages with the appearance of new pollinator species; on rapid changes in the food chain in a particular habitat; or on the dissemination of new strategies such as evolutionary induction of parthenogenesis in new species. Nonetheless, and despite the fact that each of the steps necessary for the formation of new lineages can happen quickly, the minimum time required to complete this sequential scenario is not known. Thus, the real impact this mechanism has in driving evolutionary change and speciation in nature remains to be determined. An unexplored way to approach this question is looking at a vast range of endosymbionts that have no phylogenetic concordance with their hosts and map onto the phylogenies the events of horizontal transmission. Thus, it could be possible to compare the speciation rates throughout the evolution of the hosts' lineages before and after the transmission of endosymbionts.

Our current state of knowledge on some of the underlying mechanisms of reproductive manipulation, developmental change and behavioural modulation by facultative endosymbionts is paving the way to approach putative processes of rapid speciation in the laboratory upon endosymbiont horizontal transmission. We argue that the time is right to test experimentally the real potential of the role of facultative endosymbionts in speciation through the controlled manipulation of partners and their relationships in customized novel endosymbioses.

Acknowledgments The authors wish to thank Filipa Vala, Margarida Matos, Sara Magalhães, Thiago Carvalho, Patrícia Beldade, Pedro Lima, Inês Trancoso and Alexandre Leitão for discussions and for the critical reading of the manuscript. This work was supported by Fundação para a Ciência e a Tecnologia (FCT), Portugal (PPCDT/BIA-BDE/60950/2004 and POCTI/0664/2004).

Glossary

Arthropods Arthropods belong to the phylum of Arthropoda and include insects and other animal that are characterized by an exoskeleton, a segmented body part and jointed appendages.

Aphids Aphids are small sap-sucking insects belonging to the Aphidoidea, and include plant lice as well as green-, black- and whiteflies.

Endosymbionts versus ectosymbionts Endosymbionts are all organisms that live on the surface of their host, while ectosymbionts are all organisms that live inside their host (in the gastrointestinal tract, airways, lymphatic systems).

Obligate symbionts versus facultative symbionts Obligate symbionts entertain a symbiotic association with their host that is either necessary for the symbiont or the host or both, while facultative symbionts are not necessary for either the symbiont or the host's survival.

References

Ahmed MZ, De Barro PJ, Ren SX, Greeff JM, Qiu BL (2013) Evidence for horizontal transmission of secondary endosymbionts in the *Bemisia tabaci* cryptic species complex. PLoS ONE 8:e53084

Baldo L, Ayoub NA, Hayashi CY, Russell JA, Stahlhut JK, Werren JH (2008) Insight into the routes of *Wolbachia* invasion: high levels of horizontal transfer in the spider genus *Agelenopsis* revealed by *Wolbachia* strain and mitochondrial DNA diversity. Mol Ecol 17:557–569

Baumann P (2005) Biology bacteriocyte-associated endosymbionts of plant sap-sucking insects. Annu Rev Microbiol 59:155–189

Bertaux J, Schmid M, Prevost-Boure NC, Churin JL, Hartmann A, Garbaye J, Frey-Klett P (2003) In situ identification of intracellular bacteria related to *Paenibacillus* spp. in the mycelium of the ectomycorrhizal fungus *Laccaria bicolor* S238N. Appl Environ Microbiol 69:4243–4248

Bianciotto V, Genre A, Jargeat P, Lumini E, Becard G, Bonfante P (2004) Vertical transmission of endobacteria in the arbuscular mycorrhizal fungus *Gigaspora margarita* through generation of vegetative spores. Appl Environ Microbiol 70:3600–3608

Bordenstein SR (2003) Symbiosis and the origin of species. Insect Symbiosis

Bordenstein SR, O'Hara FP, Werren JH (2001) *Wolbachia*-induced incompatibility precedes other hybrid incompatibilities in *Nasonia*. Nature 409:707–710

Braig HR, Guzman H, Tesh RB, O'Neill SL (1994) Replacement of the natural *Wolbachia* symbiont of *Drosophila simulans* with a mosquito counterpart. Nature 367:453–455

Brucker RM, Bordenstein SR (2012) Speciation by symbiosis. Trends Ecol Evol 27:443–451

Coyne JA, Orr HA (2004) Speciation. Sinauer Associates, Sunderland

de Bary A (1879) The phenomenon of symbiosis. Strasbourg

de Duve C (2007) The origin of eukaryotes: a reappraisal. Nat Rev Genet 8:395–403

Dobson SL, Rattanadechakul W, Marsland EJ (2004) Fitness advantage and cytoplasmic incompatibility in *Wolbachia* single-and superinfected *Aedes albopictus*. Heredity 93:135–142

Douglas AE (1994) Symbiotic interactions. Oxford University Press, New York

Duron O, Wilkes TE, Hurst GDD (2010) Interspecific transmission of a male-killing bacterium on an ecological timescale. Ecol Lett 13:1139–1148

Dyall SD, Brown MT, Johnson PJ (2004) Ancient invasions: from endosymbionts to organelles. Science 304:253–257

Engelstadter J, Hurst GDD (2009) The ecology and evolution of microbes that manipulate host reproduction. Annu Rev Ecol Evol Syst 40:127–149

Faria VG, Sucena E (2013) *Wolbachia* in the malpighian tubules: evolutionary dead-end or adaptation? J Exp Zool B Mol Dev Evol 320:195–199

Fast EM, Toomey ME, Panaram K, Desjardins D, Kolaczyk ED, Frydman HM (2011) *Wolbachia* enhance *Drosophila* stem cell proliferation and target the germline stem cell niche. Science 334:990–992

Frydman HM, Li JM, Robson DN, Wieschaus E (2006) Somatic stem cell niche tropism in *Wolbachia*. Nature 441:509–512

Gehrer L, Vorburger C (2012) Parasitoids as vectors of facultative bacterial endosymbionts in aphids. Biol Lett 8:613–615

Gehrig H, Schussler A, Kluge M (1996) Geosiphon pyriforme, a fungus forming endocytobiosis with *Nostoc* (cyanobacteria), is an ancestral member of the Glomales: evidence by SSU rRNA analysis. J Mol Evol 43:71–81

Gil-Turnes MS, Hay ME, Fenical W (1989) Symbiotic marine bacteria chemically defend crustacean embryos from a pathogenic fungus. Science 246:116–118

Haselkorn TS, Markow TA, Moran NA (2009) Multiple introductions of the *Spiroplasma* bacterial endosymbiont into *Drosophila*. Mol Ecol 18:1294–1305

Heath BD, Butcher RD, Whitfield WG, Hubbard SF (1999) Horizontal transfer of *Wolbachia* between phylogenetically distant insect species by a naturally occurring mechanism. Curr Biol 9:313–316

Hedges LM, Brownlie JC, O'Neill SL, Johnson KN (2008) *Wolbachia* and virus protection in insects. Science 322:702

Himler AG, Adachi-Hagimori T, Bergen JE, Kozuch A, Kelly SE, Tabashnik BE, Chiel E, Duckworth VE, Dennehy TJ, Zchori-Fein E, Hunter MS (2011) Rapid spread of a bacterial symbiont in an invasive whitefly is driven by fitness benefits and female bias. Science 332:254–256

Hornett EA, Duplouy AM, Davies N, Roderick GK, Wedell N, Hurst GD, Charlat S (2008) You can't keep a good parasite down: evolution of a male-killer suppressor uncovers cytoplasmic incompatibility. Evolution 62:1258–1263

Huigens ME, de Almeida RP, Boons PA, Luck RF, Stouthamer R (2004) Natural interspecific and intraspecific horizontal transfer of parthenogenesis-inducing *Wolbachia* in *Trichogramma wasps*. Proc Biol Sci 271:509–515

Hurst GDD, Hutchence KJ (2010) Host defence: getting by with a little help from our friends. Curr Biol 20:R806–R808

Hurst GDD, Schilthuizen M (1998) Selfish genetic elements and speciation. Heredity 80:2–8

Jaenike J (2007) Spontaneous emergence of a new *Wolbachia* phenotype. Evolution 61:2244–2252

Jaenike J, Brekke TD (2011) Defensive endosymbionts: a cryptic trophic level in community ecology. Ecol Lett 14:150–155

Jaenike J, Dyer KA, Cornish C, Minhas MS (2006) Asymmetrical reinforcement and *Wolbachia* infection in *Drosophila*. PLoS Biol 4:e325

Jaenike J, Polak M, Fiskin A, Helou M, Minhas M (2007) Interspecific transmission of endosymbiotic *Spiroplasma* by mites. Biol Lett 3:23–25

Jaenike J, Unckless R, Cockburn SN, Boelio LM, Perlman SJ (2010) Adaptation via symbiosis: recent spread of a *Drosophila* defensive symbiont. Science 329:212–215

Jiggins FM, Bentley JK, Majerus ME, Hurst GD (2002) Recent changes in phenotype and patterns of host specialization in *Wolbachia* bacteria. Mol Ecol 11:1275–1283

Le Clec'h W, Chevalier FD, Genty L, Bertaux J, Bouchon D, Sicard M (2013) Cannibalism and predation as paths for horizontal passage of *Wolbachia* between terrestrial isopods. PLoS ONE 8:e60232

Leonardo TE, Mondor EB (2006) Symbiont modifies host life-history traits that affect gene flow. Proc Biol Sci 273:1079–1084

Lersten NR, Horner HT (1976) Bacterial leaf nodule symbiosis in angiosperms with emphasis on Rubiaceae and Myrsinaceae. Bot Rev 42:145–214

Lima PT, Faria VG, Patraquim P, Ramos AC, Feijo JA, Sucena E (2009) Plant-microbe symbioses: new insights into common roots. BioEssays 31:1233–1244

Mateos M, Castrezana SJ, Nankivell BJ, Estes AM, Markow TA, Moran NA (2006) Heritable endosymbionts of *Drosophila*. Genetics 174:363–376

McGraw EA, Merritt DJ, Droller JN, O'Neill SL (2002) *Wolbachia* density and virulence attenuation after transfer into a novel host. Proc Natl Acad Sci USA 99:2918–2923

Mira A, Moran NA (2002) Estimating population size and transmission bottlenecks in maternally transmitted endosymbiotic bacteria. Microb Ecol 44:137–143

Moran NA (2006) Symbiosis. Curr Biol 16:R866–R871

Moran NA, Wernegreen JJ (2000) Lifestyle evolution in symbiotic bacteria: insights from genomics. Trends Ecol Evol 15:321–326

Moran NA, McCutcheon JP, Nakabachi A (2008) Genomics and evolution of heritable bacterial symbionts. Annu Rev Genet 42:165–190

Nieberding CM, Olivieri I (2007) Parasites: proxies for host genealogy and ecology? Trends Ecol Evol 22:156–165

Oldroyd GE, Downie JA (2008) Coordinating nodule morphogenesis with rhizobial infection in legumes. Annu Rev Plant Biol 59:519–546

Oliver KM, Russell JA, Moran NA, Hunter MS (2003) Facultative bacterial symbionts in aphids confer resistance to parasitic wasps. Proc Natl Acad Sci USA 100:1803–1807
Oliver KM, Degnan PH, Burke GR, Moran NA (2010) Facultative symbionts in aphids and the horizontal transfer of ecologically important traits. Annu Rev Entomol 55:247–266
Osteryoung KW, Nunnari J (2003) The division of endosymbiotic organelles. Science 302:1698–1704
Parniske M (2008) Arbuscular mycorrhiza: the mother of plant root endosymbioses. Nat Rev Microbiol 6:763–775
Partida-Martinez LP, Hertweck C (2005) Pathogenic fungus harbours endosymbiotic bacteria for toxin production. Nature 437:884–888
Ruby EG (2008) Symbiotic conversations are revealed under genetic interrogation. Nat Rev Microbiol 6:752–762
Russell JA, Moran NA (2005) Horizontal transfer of bacterial symbionts: heritability and fitness effects in a novel aphid host. Appl Environ Microbiol 71:7987–7994
Russell JA, Latorre A, Sabater-Munoz B, Moya A, Moran NA (2003) Side-stepping secondary symbionts: widespread horizontal transfer across and beyond the Aphidoidea. Mol Ecol 12:1061–1075
Sapp J (1994) Evolution by association: a history of symbiosis. Oxford University Press, New York
Sasaki T, Kubo T, Ishikawa H (2002) Interspecific transfer of *Wolbachia* between two lepidopteran insects expressing cytoplasmic incompatibility: a *Wolbachia* variant naturally infecting Cadra cautella causes male killing in Ephestia kuehniella. Genetics 162:1313–1319
Sasaki T, Massaki N, Kubo T (2005) *Wolbachia* variant that induces two distinct reproductive phenotypes in different hosts. Heredity 95:389–393
Scarborough CL, Ferrari J, Godfray HC (2005) Aphid protected from pathogen by endosymbiont. Science 310:1781
Schilthuizen M, Stouthamer R (1997) Horizontal transmission of parthenogenesis-inducing microbes in Trichogramma wasps. Proc Biol Sci 264:361–366
Sheeley SL, McAllister BF (2009) Mobile male-killer: similar *Wolbachia* strains kill males of divergent *Drosophila* hosts. Heredity 102:286–292
Shoemaker DD, Katju V, Jaenike J (1999) *Wolbachia* and the evolution of reproductive isolation between *Drosophilla recens* and *Drosophila subquinaria*. Evolution 53:1157–1164
Tarkka MT, Sarniguet A, Frey-Klett P (2009) Inter-kingdom encounters: recent advances in molecular bacterium-fungus interactions. Curr Genet 55:233–243
Teixeira L, Ferreira A, Ashburner M (2008) The bacterial symbiont *Wolbachia* induces resistance to RNA viral infections in *Drosophila melanogaster*. PLoS Biol 6:e2
Telschow A, Hammerstein P, Werren JH (2002) The effect of *Wolbachia* on genetic divergence between populations: models with two-way migration. Am Nat 160(Suppl 4):S54–S66
Telschow A, Hammerstein P, Werren JH (2005) The effect of *Wolbachia* versus genetic incompatibilities on reinforcement and speciation. Evolution 59:1607–1619
Thao ML, Clark MA, Baumann L, Brennan EB, Moran NA, Baumann P (2000) Secondary endosymbionts of psyllids have been acquired multiple times. Curr Microbiol 41:300–304
Thompson JN (1987) Symbiont-induced speciation. Biol J Linn Soc 32:385–393
Tinsley MC, Majerus ME (2007) Small steps or giant leaps for male-killers? Phylogenetic constraints to male-killer host shifts. BMC Evol Biol 7:238
Tram U, Sullivan W (2002) Role of delayed nuclear envelope breakdown and mitosis in *Wolbachia*-induced cytoplasmic incompatibility. Science 296:1124–1126
Tsuchida T, Koga R, Fukatsu T (2004) Host plant specialization governed by facultative symbiont. Science 303:1989
Tsuchida T, Koga R, Horikawa M, Tsunoda T, Maoka T, Matsumoto S, Simon JC, Fukatsu T (2010) Symbiotic bacterium modifies aphid body color. Science 330:1102–1104
Tsuchida T, Koga R, Matsumoto S, Fukatsu T (2011) Interspecific symbiont transfection confers a novel ecological trait to the recipient insect. Biol Lett 7:245–248

Vala F, Egas M, Breeuwer JA, Sabelis MW (2004) *Wolbachia* affects oviposition and mating behaviour of its spider mite host. J Evol Biol 17:692–700
Vallet-Gely I, Lemaitre B, Boccard F (2008) Bacterial strategies to overcome insect defences. Nat Rev Microbiol 6:302–313
Vavre F, Fleury F, Lepetit D, Fouillet P, Bouletreau M (1999) Phylogenetic evidence for horizontal transmission of *Wolbachia* in host-parasitoid associations. Mol Biol Evol 16:1711–1723
Veneti Z, Toda MJ, Hurst GD (2004) Host resistance does not explain variation in incidence of male-killing bacteria in *Drosophila bifasciata*. BMC Evol Biol 4:52
Werren JH, O'Neill SL (1997) The evolution of heritable symbionts. Influential passengers. Oxford University Press, Oxford
Werren JH, Baldo L, Clark ME (2008) *Wolbachia*: master manipulators of invertebrate biology. Nat Rev Microbiol 6:741–751
White JA (2011) Caught in the act: rapid, symbiont-driven evolution: endosymbiont infection is a mechanism generating rapid evolution in some arthropods-but how widespread is the phenomenon? BioEssays 33:823–829
Zchori-Fein E, Gottlieb Y, Kelly SE, Brown JK, Wilson JM, Karr TL, Hunter MS (2001) A newly discovered bacterium associated with parthenogenesis and a change in host selection behaviour in parasitoid wasps. Proc Natl Acad Sci USA 98:12555–12560

Historical and Epistemological Perspectives on What Horizontal Gene Transfer Mechanisms Contribute to Our Understanding of Evolution

Nathalie Gontier

Abstract Since the 1990s, results coming in from molecular phylogenetics necessitate us to recognize that Horizontal Gene Transfer (HGT) occurs massively across all three domains of life. Nonetheless, many of the mechanisms whereby genes can become transferred laterally have been known from the early twentieth century onward. The temporal discrepancy between the first historical observations of the processes, and the rather recent general acceptance of the documented data, poses an interesting epistemological conundrum: Why have incoming results on HGT been widely neglected by the general evolutionary community and what causes for a more favorable reception today? Five reasons are given: (1) HGT was first observed in the biomedical sciences and these sciences did not endorse an evolutionary epistemic stance because of the ontogeny/phylogeny divide adhered to by the founders of the Modern Synthesis. (2) Those who did entertain an evolutionary outlook associated research on HGT with a symbiotic epistemic framework. (3) That HGT occurs across all three domains of life was demonstrated by modern techniques developed in molecular biology, a field that itself awaits full integration into the general evolutionary synthesis. (4) Molecular phylogenetic studies of prokaryote evolution were originally associated with exobiology and abiogenesis, and both fields developed outside the framework provided by the Modern Synthesis. (5) Because HGT brings forth a pattern of reticulation, it contrasts the standard idea that evolution occurs solely by natural selection that brings forth a vertical, bifurcating pattern in the "tree" of life. Divided into two parts, this chapter first reviews current neo-Darwinian "tree of life" versus reticulate "web of life" polemics as they have been debated in high-profile academic journals, and secondly, the historical context of discovery of the various means whereby genes are transferred laterally is sketched. Along the way, the reader is introduced to how HGT contradicts some of the basic tenets of the neo-Darwinian paradigm.

N. Gontier (✉)
AppEEL—Applied Evolutionary Epistemology Lab, University of Lisbon,
Lisbon, Portugal
e-mail: nlgontier@fc.ul.pt

N. Gontier (ed.), *Reticulate Evolution*, Interdisciplinary Evolution Research 3,
DOI 10.1007/978-3-319-16345-1_5

Keywords Tree of life · Web of life · Horizontal Gene Transfer · Transformation · Transduction · Conjugation · Gene transfer agents · Modern Synthesis · Extended Synthesis · Biomedical sciences

> Imagine that in a coffee house you brush up against a guy with green hair. In so doing, you acquire that part of his genetic endowment, along with perhaps a few more novel items. Not only can you now transmit the gene for green hair to your children, but you yourself leave the coffee shop with green hair. Bacteria indulge in this sort of casual, quick gene acquisition all the time. (Margulis and Sagan 2000: 93)

1 Introduction

The concept of "horizontal gene transfer" (HGT) and also the recognition that HGT occurs abundantly across all three domains of life have only been brought to the attention of the wider evolutionary community from the early 1990s onward (Gogarten et al. 1989; Hilario and Gogarten 1993; Doolittle 1999; Rivera and Lake 2004; Syvanen and Kado 1998). This wider recognition correlates with advances made in molecular phylogenetics (Woese et al. 1990), in particular with prokaryotic systematics, but the results also extend to eukaryotic life forms. Molecular phylogenetic reconstructions have brought forth numerous inconsistencies, incongruences, and anomalies in the traditional "tree of life." Whole-genome sequencing techniques of various species belonging to all three domains of life evidence that species genomes contain significant amounts of "foreign DNA," i.e. genes that are neither shared with their ancestral lineages, nor the result of random mutations of existing genes. In short, they are not acquired from parental species and thus not the result of genealogical or reproductive descent with modification. Rather, these foreign genes are acquired outside the genealogical descent line, by means of HGT (Goldenfeld and Woese 2007; Zhaxybayeva and Doolittle 2011: R243–4).

HGT refers to various processes by which biological individuals can acquire genes coming from outside the germ line and the means by which genes are exchanged either between distinct organisms with different genealogical histories, or between distinct genomes present in the same organism. Genes can be transferred between prokaryotes, between prokaryotes and eukaryotes (in both directions), between eukaryotes, and between viruses and pro- and eukaryotes.

In prokaryotes, HGT occurs via bacterial *transformation*, phage-mediated *transduction*, plasmid transfer via *bacterial conjugation*, via *Gene Transfer Agents* (GTAs), or via the movement of *transposable elements* such as insertion sequences. The genes that underlie significant prokaryotic metabolic pathways such as energy metabolism, cofactor/vitamin metabolism, and antibiotic resistance were mostly acquired by HGT (Iwasaki and Takagi 2009). Because these genes play such a crucial role in prokaryotic evolution, several authors claim that it is the main way in which evolutionary novelty arises in these microorganisms (Lopez and Bapteste 2009; Doolittle 2005; Fournier et al. 2011; Ragan et al. 2009; Zhaxybayeva and Doolittle 2011).

In eukaryotes, HGT is mediated by processes such as endosymbiosis, phagocytosis and eating, infectious disease, and hybridization or divergence with gene flow, which facilitates the movement of *mobile genetic elements* such as transposons and retrotransposons between different organisms (Keeling and Palmer 2008; Arnold 2008; Ryan 2004, 2009).

Although HGT has been dubbed "biology's next revolution" (Goldenfield and Woese 1997), most of the processes by which HGT takes place were already discovered in the early twentieth century. There are merely fine lines to be drawn between HGT and endosymbiosis or processes of infectious heredity, and data on HGT were originally interpreted from within a general symbiosis theory. Bacterial conjugation and phage-induced transduction of bacterial DNA were dubbed instances of "hereditary symbiosis" and "infective heredity" by their discoverer Lederberg (1952). Also today, a significant amount of scholars continue to consider HGT, endosymbiosis, and infectious heredity as aspects of a larger, symbiogenetic (Gontier 2006, 2007; Margulis 1970; Moran and Jarvik 2010; Ryan 2006, 2009; Sapp 1994, 2003, 2004), or a more general reticulate evolutionary theory (Andam et al. 2010; Doolittle and Bapteste 2007; Zhaxybayeva and Doolittle 2011; Keeling and Palmer 2008).

Many of the observations made on HGT furthermore track back to, and associate with major milestones and advances made within standard evolutionary theory. Research on bacterial transformation tracks back to the identification of genes as the bearers of genetic material. The study of bacterial conjugation and transduction is associated with increasing insight into cell cytology, cytoplasmic inheritance, and evo-devo which are milestones in evolutionary thinking that first developed during the "eclipse of Darwin." Research on mobile genetic elements associates with epigenetics.

The discrepancy in time between the first observations of HGT in the early twentieth century and the wider acceptance and recognition of its abundant occurrence in the 1990s, as well as the remarkable associations of advances in knowledge on reticulate evolution with the standard milestones of evolutionary thought, raise a series of interesting anthropological and epistemic problems. Why have these data been ignored for so long by mainstream evolutionary scholars? Why have they not been incorporated into the Modern Synthesis? And why are these data argued to contradict the Modern Synthesis?

Divided into two parts, this chapter first investigates the historical factors that contributed to the wider recognition of HGT, and secondly, the various mechanisms by which HGT occurs are sketched against their general context of discovery.

2 Reticulate Evolution and Webs of Life

Tree diagrams are nowadays the most common means by which the evolutionary descent of species is illustrated. The concept of evolution is currently so intertwined with these tree diagrams that both laypeople and scientists alike often find it difficult to think about evolution without envisioning phylogenetic tree images. Tree of life imagery serves as an educational aid where, by analogy with natural trees, evolutionary tree diagrams depict life as having originated from one single trunk that bifurcates into branches that in turn split into twigs whereupon the leaves grow. The leaves represent the species, the twigs are analogous to the genera, the branches to the phyla, and the trunk symbolizes the last and single universal common ancestor of life.

Tree metaphors and treelike structures were first drawn in pre-evolutionary times to depict the genealogical descent of natural phenomena. These non-evolutionary genealogical tree diagrams were later adopted by Charles Darwin and Ernst Haeckel to understand and depict evolutionary descent relationships of biological species. If evolution is a fact of life, then how does one depict the "evolutionary descent with modification" of the various species that ever existed? How do we illustrate life's early origins and extinctions? Do speciations from unicellular organisms to multicellular life forms entail some kind of "progress," and "linear arrangement," or do "the bottom of branches … appear like circles" (Darwin 1837/1838: 1–27)? Darwin pondered about these questions in his Notebook B that he filled in 1837–38. Inspired by familial pedigree thinking and genealogical tree models that illustrate the natural history of languages (Gontier 2011), on the one hand, Darwin favored a "tree of life metaphor" because species do not "really pass into each other." On the other hand, he wondered whether "The tree of life should perhaps be called the coral of life, base of branches dead; so that passages cannot be seen," but he added that such an imagery "offers contradiction to constant succession of germs in progress" and "makes it excessively complicated" (Darwin 1837/1838: 25–6).

Eventually, both in his notebook and also in the *Origin*, the tree diagram made it. The famous "I think" diagram depicts a first hypothetical branching diagram that Darwin used to hypothesize about species relatedness and speciation, and in his *Origin of Species*, Darwin (1859) made ample use of the "tree of life" metaphor in his Chap. 4, the chapter wherein he advanced his views on natural selection. The *Origin* also contains a famous diagram on how species hypothetically speciate over time by means of natural selection. Nonetheless, the first evolutionary "tree of life" that depicts *actual*, chronological evolutionary-descent relationships between species was first drawn by Haeckel (1866).

Darwin thought of other ways to depict evolutionary descent relations, and it is no coincidence that the tree metaphor and branching diagrams were favored. Evolution by means of natural selection is conjectured to occur either when diverging species gradually split off from existing branches (cladogenesis), or when existing species gradually and linearly evolve into new species (anagenesis). Cladogenesis naturally brings forth a bifurcating and ramificating pattern (Doolitte and Bapteste 2007), while anagenesis, though conceived to be gradual, nonetheless entails a break between the parental and the newly evolved species.

Scholars coming from different evolutionary research fields have increasingly come to question the utility and accuracy of tree diagrams in depicting the evolution of various strands of life. Advocates of punctuated equilibria theory (Eldredge and Gould 1972) have long reported that tree diagrams do not adequately portray the often rapid speciation events of eukaryotes. Because the tree of life tends to have a maximum species diversity at the end of the tree (Gould 1986), it does not adequately depict the many (mass-)extinction events that have occurred throughout life's evolution (for discussions, see the fourth issue of the 2010 edition of *Evolution, Education and Outreach*, and Serrelli and Gontier 2015).

Critique has also come from symbiologists and microbiologists. As early as 1905, Constantin Mehrezkowsky provided a prolegomena for a symbiogenetic double-origin theory of life, a view he fully developed and illustrated with a phylogenetic reconstruction in the 1910 translation of his 1909 work (depicted in Carrapiço's chapter, this volume). He assumed that life evolved from two separate "*Plasmaarten*" (life forms), *Mycoides Plasma* (Mykoplasma) and *Amöboides Plasma* (Amoeboplasma). Both life forms were conjectured to have evolved separately, and afterwards, they engaged in a "primary symbiosis." Mereschkowsky (1910: 280, my translation):

> Until now, there was the general conviction, that the tree of life was a single one. The task set forth in this work, is to demonstrate that there are two trees of life, and that each tree originated on its own and independently from the other one, and this probably happened in different periods of earth's history. These trees partly developed on their own and independently from one another and partly stringed together and closely grew and developed together. Both trees are responsible for the diversity of the organic beings. The idea of a unity of organic nature has to be abandoned in favor of the idea of nature's duality.

A couple of years later, in 1915, Hermann Reinheimer was the first to introduce the concept of a "web of life" (Carrapiço, this volume), to describe the multiple cases of symbiosis and symbiogenesis that occur in the animal and plant kingdoms. And also Margulis has pioneered in developing "tree of life" and 5-kingdom iconographies that include the symbiogenetic mergings that underlie the evolution of the eukaryotic kingdoms (Whittaker and Margulis 1978; Margulis 1998, 1991; Margulis and Schwart 1997).

Both Lederberg (1952: 425) and Sapp (1994, 2009) have argued that most tree diagrams and evolutionary theories in general present a "sterile" view of evolution. And today, critique on the tree of life is based on incoming data on massive HGT in prokaryotes as evidenced by molecular phylogenetic reconstruction techniques. Current phylogenetic depictions of the evolution of life increasingly attempt to include the many reticulate means by which the micro-organismal world evolves. Such phylogenetic reconstructions look more like a "web" (Doolittle 1999), "network" (Gogarten 2000; Kunin et al. 2005), "net" (Williams et al. 2011), "ring" (Rivera and Lake 2004), or "cobweb" of life (Ge et al. 2005), which connects the splitting branches of the tree at the level of the roots, the trunk, the numerous branches, twigs, and nods. The emerging network-like diagrams draw interconnecting evolutionary lines within and across life's *three domains*, i.e., the Archaea, Bacteria, and Eukarya (Woese et al. 1990), and the connecting lines also crisscross with *viruses*, i.e., genetic agents traditionally conceived as non-living structures.

In this part, we first sketch the historical context wherein biochemical, molecular elements became used as markers to infer genealogical descent. Secondly, we investigate when the concepts of "reticulate evolution" in general and "HGT" in particular became associated with molecular phylogenetic reconstructions of the "tree" and "web" of life, and we end with briefly sketching the polemics that underlie tree versus web of life iconographies.

2.1 Molecular Phylogenetics and the Origin of Life

In the early 1960s, Pauling and Zuckerkandl proposed that "chemical paleogenetics" could aid paleontology and systematics in reconstructing the natural genealogies of species by comparing the uniform, constant rate of "semantide changes" ("DNA, RNA, and polypeptides") in related species (Zuckerkandl and Pauling 1962). In 1965, they wrote their seminal papers "Evolutionary divergence and convergence in proteins" and "Molecules as Documents of Evolutionary History" (Zuckerkandl and Pauling 1965a, b), wherein they called this uniform, constant rate of semantide change "the molecular evolutionary clock." Their work was foundational for neutral evolution theory (genetic drift), as it was developed by Motoo Kimura, and it launched the field of molecular systematics where the rate of molecular change in protein and gene sequences is used to deduce genealogical relationships and speciation events, an approach they characterize as "…the most rational, universal, and informative molecular phylogeny" because "… in macromolecules of these types there is more history in the making and more history preserved than at any other single level of biological organization" (Zuckerkandl and Pauling 1965b: 360).

Molecular phylogenies lend insight into evolutionary history in the following three ways, they give: "(1) the approximate time of existence of a molecular ancestor common to the chains that are being compared; (2) the probable amino-acid sequence of this ancestral chain; and (3) the lines of descent along which given changes in amino-acid sequence occurred" (Zuckerkandl and Pauling 1965b: 360). The method itself, however, cannot provide exact dates in time. To give an exact estimate of divergence in geological time, scholars need to calibrate and compare their results with the fossil record (for a history and discussion, see Morgan 1998; Morange 2000; San Mauro and Agorreta 2010).

Ever since, scholars have reconstructed ancestral-descent relationships by comparing the sequences of proteins (Fitch and Margoliash 1967) and DNA and RNA sequences (Sanger and Coulson 1975). Technological advances in molecular genetics, such as the polymerase chain reaction or PCR technique (Mullis 1983), and more recent shotgun sequencing (Messing et al. 1981; Staden 1979), high-throughput sequencing, and barcoding (Hebert et al. 2003), enable comparisons ranging from single-nucleotide sequences to whole genomes. Comparative studies allow the various species to become "rooted" into common ancestors that trace back to the very origins of life on earth (de Magalhães et al. 2010; Pettersson et al. 2009; Schuster 2008).

Molecular phylogenies are currently providing the primary tools to classify prokaryotic life. Prokaryotic organisms rarely fossilize, and before the advent of molecular genetics, scholars were limited to reconstructing morphological phylogenies. Due to increasing possibilities to sequence large data sets, including whole genomes and "metagenomes," molecular phylogenetics has from the 1970s onward provided bacteriologists with a means to identify the genetic diversity and relatedness of living prokaryotic beings, as well as to infer, from these comparisons, their evolutionary emergence in time. At present, these techniques even enable microbiologists to identify and distinguish bacterial species currently known only by their genetic sequence because scholars have so far been unable to isolate them for morphological, electron microscopic study (Eisen 2007; Chen and Pachter 2005; Handelsman et al. 1998; Hugenholz et al. 1998).

In recent years, worldwide, large-scale projects have been set up such as ToLweb—the Tree of Life Web Project (http://tolweb.org); the NSF-funded AToL—Assembling the Tree of Life (http://www.phylo.org/atol) and GoLife—Genealogy of Life (http://www.nsf.gov/funding/pgm_summ.jsp?pims_id=5129&org=DEB) projects; iTOL—the Interactive Tree of Life Project (http://itol.embl.de); and the Tree Thinking Group (http://www.tree-thinking.org). These all share the ambitious goal to once and for all determine every species' evolutionary ancestry and place on the tree of life.

One of these projects already originated in the late 1970s. With the goal to build the universal tree of life, Woese and Fox (1977; Fox et al. 1980) began comparing specific sections of genetic material that is present in both unicellular and multicellular life. They selected subunits of ribosomal RNA (rRNA) as biochemical markers and focused on 16S rRNA for prokaryotes and 18S rRNA for eukaryotes. Ribosomal RNA is a type of RNA that enables protein synthesis. Incoming results led Fox and Woese to undo the classic distinction of life into pro- and eukaryotes, and instead, they proposed that prokaryotes (or Monera) should be divided into two separate kingdoms: Archaebacteria and Eubacteria. Eukaryotes represented a third "urkingdom" that has a "chimeric nature," because it contains genes coming from both lines. The consequence was that there exist three "urkingdoms" or monophyletic lines of descent instead of two: the Archaebacteria, Eubacteria, and Eukaryota.

> This tripartite division of extant life is incompatible with the conventionally accepted view in which living systems are divided into two basic phylogenetic categories, prokaryotes and eukaryotes. However, the eukaryotic cell is now recognized to be a genetic chimera, whose evolutionary origins we do not yet understand. (Fox et al. 1980, 458)

In 1990, Woese and colleagues would take things one step further and demonstrate that there is sufficient reason to completely separate "Eubacteria" from "Archaebacteria" and to classify them as distinct domains of life (Bacteria and Archaea) that, together with the eukaryotes (Eukaryota), delineate the "three-domain hypothesis" or three-domain classification of life (Woese et al. 1990). The new comparisons also evidenced that Bacteria stand quite on their own and that Archaea are more closely related to eukaryotes, the third domain. The three-domain classification also came with an rRNA-based tree of life (Woese et al. 1990: 4578) where LUCA, the last universal common ancestor wherefrom these domains evolved, i.e., the "universal" and single root of the tree, still remained to be identified (Lawton 2009).

In later publications, Woese (1998) and other scholars (Doolittle 1999, 2005; Martin 1999; Villarreal 2006) increasingly came to question whether one such universal common ancestor will ever be found, and instead, Woese raised the possibility that the roots of the tree are multiple. Also, Margulis (1991) has conjectured that although all eukaryotes probably share a last eukaryotic common ancestor (LECA), a single origin for prokaryotic life forms is less likely.

The original rRNA trees did not take HGT into account. In fact, to enable Woese to draw his early trees, he had to ignore any such events. However, later work (Woese 1998) made it clear that HGT is quite common in prokaryotes which made him join the previous claims made by scholars such as Gogarten et al. (1989, Hilario and Gogarten 1993) that HGT occurs abundantly (Goldenfeld and Woese 2007). Ever since, scholars have been developing means to identify the numerous instances of HGT by applying molecular phylogenetic reconstruction techniques.

Nonetheless, mainstream phylogenetic reconstructions continue to ignore HGT events. In 2006, in association with the iTOL project, a team lead by Bork (Ciccarelli et al. 2006) introduced a new "tree." To fit all life forms, they had to turn the tree into a circle, which was quite innovative, but the scholars also consciously ignored HGT data. HGT has long been considered a seldom event in evolutionary history. In this view, it does not threaten the common tree iconography of evolution because one can look for "core genes" shared by all major taxa. Bork's iconography, for example, was based upon 31 orthologous "core genes" present in all 191 examined lineages coming from both pro- and eukaryotic species.

HGT scholars criticized Bork's tree for representing a "tree of one percent" (Dagan and Martin 2006). Doolittle (2009: 221) contended that

> Enthusiastic TOLers see this TOC, no matter how little of the actual phenotype-determining information or history the organisms they wish to classify it encompasses, or how extreme the algorithm used to derive it, as a triumph of the Darwinian method and a vindication of their belief in the TOL. It is, in their view, the genealogy upon which Darwin thought classification could safely, and ultimately must, rest. This is I think a misreading of history and a non-trivial re-formulation of the goals of phylogenetic practice.

Because these phylogenetic reconstructions attempt to find the universal "roots" of the tree of life, this research also extends toward fields such as exobiology and astrobiology, i.e., fields that study the origin of life on this and possible other planets. Researching the origin of life does not form a basic tenet of the Modern Synthesis. The standard neo-Darwinian paradigm provides a theory of biogenesis: It explains how existing life brings forth new life forms by means of natural selection. Thomas Henry Huxley, for example, called research on abiogenesis, which in his epoch associated with theories on spontaneous generation and epigenetics, to a halt. Research on abiogenesis, or how life evolves out of inorganic physical, and biochemical particles, developed mostly after the foundation of the Modern Synthesis. Theories on the RNA world, "spontaneously generated" or self-organizing autocatalytic biomolecular networks, proteinoid microspheres, etc., evolved in association with increasing knowledge of the biochemical elements that build the living cell, and in association with molecular phylogenetic reconstructions. An RNA world, for example, was already envisioned by Woese (1967); the article by Fox et al. (1980) demonstrated that the first life forms were chemoautotrophs instead of heterotrophs, and it were molecular sequences of RNA and DNA viruses that demonstrated that these genetic agents most likely evolved before life (Villareal and Defilipps 2000; Villarreal and Witzany 2010).

2.2 First Usage of the Terms HGT and Reticulate Evolution

Evidence for HGT dates back to discoveries of bacterial transformation, conjugation, and transduction, but concepts such as "HGT" and "reticulate evolution" date to later periods in time. HGT between the eukaryotic nuclear genome and cellular organelles such as mitochondria and chloroplasts was first hypothesized to occur by early symbiologists including Margulis (1970), and genetic exchange was later confirmed by Wolf and Delguidice (1987) and Gray et al. (1989). More evidence for naturally occurring gene transfer between organisms was reported by Trevors et al. (1987), Coughter and Stewart (1989), Daniels et al. (1990), Doolittle et al. (1990), and Gupta and colleagues (Gupta and Signh 1994; Golding and Guptha 1995).

One of the first usages of the concepts "HGT" and "reticulate evolution" comes from Ambler et al. (1979), Hartley (1980), Busslinger et al. (1982), and Champion et al. (1980). Champion, in a review paper on the evolution of gram-negative *Pseudomonas fluorescens* bacteria, wrote that:

> Bacteria can acquire new phenotypic characters either from other bacteria by *Horizontal gene transfer* or through manipulation of their own genetic material (vertical evolution). … In addition, it has long been recognized that a group of bacterial strains are usually delineated from other groups, not by the exclusive possession of a single or several traits, but by possessing a particular set of traits. These facts have been taken to indicate a *reticulate mode of evolution* in which the potential uniqueness of any group of strains has been undermined from extensive *horizontal exchange of genetic material* from closely to distantly related groups. The pseudomonads may be a good example of this phenomenon. (Champion et al. 1980: 506, my italics)

In 1984, Syvänen (1984a and also see Syvänen 1984b, 1986 and 1987) speculated that conserved regions in mammalian beta-globin possibly resulted from "cross-species gene exchange," and a year later, he wrote a seminal and very interesting article on "Cross-species gene transfer; implications for a new theory of evolution," wherein he hypothesized that:

> … genes are transferred and expressed among all species, and that such exchange is facilitated by, and can help account for, the existence of the biological unities, from the uniform genetic code to the cross-species similarity of the stages of embryological development. If this idea is correct, the uniformity of the genetic code would allow organisms to decipher and use genes transposed from chromosomes of foreign species, and the shared sequence of embryological development within each phylum would allow the organism to integrate these genes, particularly when the genes affect complex morphological traits. The cross-species gene transfer model could help explain many observations which have puzzled evolutionists, such as rapid bursts in evolution and the widespread occurrence of parallelism in the fossil record. (Syvänen 1985: 333)

Peter Gogarten and colleagues (Gogarten et al. 1989; Hilario and Gogarten 1993) focused on ATPase genes as genetic markers for molecular phylogenetic reconstructions and suggested that these genes were acquired by "Horizontal Gene Transfer," which made the authors start to debunk the single tree of life before Woese did. ATPase genes encode for proteins and enzymes that play a crucial role in cell membranes by enabling the uptake of foreign material (Gogarten et al. 1989; Hilario and Gogarten 1993). The ATPase genes differ between all three "urkingdoms of life" (Fox et al. 1980), and are therefore a good genetic marker to root the tree. The comparison of ATPase genes also groups Archaea closer to Eukaryota and both are more distant from Bacteria, which converges with, and also confirms, Woese's et al. (1990) classification of life into three separate domains based upon comparisons of rRNA subunits. ATPase genes also give discrepancies in what regards gene versus species trees and Hilario and Gogarten (1993: 118) therefore concluded that:

> The finding that genes were exchanged between distantly related species implies that a single gene phylogeny can no longer be readily interpreted as a species tree. To determine the evolution of species more than one gene tree should be considered. (Hilario and Gogarten 1993: 118)

Ever since, scholars who study HGT have been introducing new metaphors and visualizations that capture the reticulate evolutionary pattern brought forth by HGT. Already in 1999, Ford W. Doolittle provided a now classic reticulate image of the "web" that sought ways to visualize the massive HGT, and later, in 2005, he expanded his image in order to include the symbiogenetic acquisition of chloroplasts and mitochondria in eukaryotic life forms (Doolittle 1999, 2005; Doolittle and Bapteste 2007; Bapteste et al. 2005, 2009). Rivera and Lake (2004) have introduced a "ring of life" that depicts the chimeric origin of the eukaryotic genome; Dagan and Martin (2009) have provided networks that depict both the horizontal and vertical exchanges between distinct microbial lineages; in 2010, Luis Villarreal (Villarreal and Witzany 2010) provided a first attempt to root the tree of life with viruses, and he tried to illustrate the susceptibility of all three domains of life to "viral colonization" (and also see Mindell and Villarreal 2003).

2.3 *Was Darwin Wrong?*

The new reticulate icons of evolution are often treated with gigantic suspicion. In 2009, the January 21st issue of the *New Scientist* magazine featured a cover titled "Darwin was wrong: Cutting down the tree of life." The front page of the magazine featured a tree of life drawn by the Russian artist Yulia Brodskaya. In that tree, some branches crossed and the trunk was divided into several different lineages. In the journal, Lawton (2009) wrote an article on HGT titled "Axing Darwin's tree; The tree of life is an iconic image, but it could be time to fell it" and the editorial was titled "Uprooting Darwin's tree." The tree, the editorial, and the article received enormous media and scholarly attention, and most of the reactions were negative. Dennett et al. (2009: 25) wrote a very angry letter which opened as follows:

> What on earth were you thinking when you produced a garish cover proclaiming that 'Darwin was wrong' …? First, it's false, and second, it's inflammatory. And, as you surely know, many readers will interpret the cover not as being about Darwin, the historical figure, but about evolution. … You have made a lot of extra, unpleasant work for the scientists whose work you should be explaining to the general public. We all now have to try to correct all the misapprehensions your cover has engendered.

Indeed, emotions ran high. Dennett and his cowriters pointed the finger in dispraise, arguing that the cover, the editorial, and the article gave green light to creationists and that it undermined scientific evolutionary thought.

These criticisms were out of proportion. For one, trees are the number one icon for many world religions, and they were also used in pre-evolutionary societies to depict non-evolutionary, abstract and logical, or genealogical descent relations of divine and earthly phenomena. Debunking tree images therefore hardly feeds into creationist thought (Gontier 2011). Secondly, reticulate evolution can be proven by an enormous amount of data, and these theories in no way lend credibility to the ideas of creationism. Not trying to incorporate these findings into educational imagery, now that would be against science. Moreover, the tree that featured on the cover of *New Scientist* is still quite conventionally looking. There are many more extravagant "tree of life" images circulating around in science these days that do not even slightly resemble an actual, natural tree. It is therefore highly interesting to see such emotional responses made "ex auctoritate" when a public image such as the tree of life is being criticized. Nonetheless, evolution is no longer synonymous with natural selection, and the reticulate evolutionary mechanisms deserve their educational tools.

As early as 2004, the *American Journal of Botany* dedicated a special issue to the tree of life of plants (see especially the paper by Palmer et al. 2004). The overall message conveyed by the issue was that the early symbiogenetic origin of plants, as well as their numerous hybridization events, disable one to straightforwardly draw the tree of plants as a branching pattern wherein lineages solely split into new ones.

In response to the outbursts of some of the "hardcore" neo-Darwinians, scholars working on HGT and symbiogenesis have been stirring up debate in the August 2009 issue of the *Philosophical Transactions of the Royal Society, B: Biological Sciences*, which featured a theme issue titled "The network of life: genome beginnings and evolution" (Ragan et al. 2009). Epistemological aspects of the tree of life debate were discussed in the September 2010 issue of *Biology and Philosophy* (O'Malley et al. 2010). In 2011, Gribaldo et al. edited a special issue for the journal *Research in Microbiology* on "Archaea and the tree of life," and O'Malley and Koonin (2011) edited an issue titled "Beyond the Tree of Life" for *Biology Direct*.

As can be deduced from the issue's titles, the authors pled for a replacement of the tree of life image by a "network" or "web" of life. During horizontal evolution, evolutionary lineages can cross, melt, and dissolve into one another. By analogy, the roots of the tree, and even its distinct branches, can cross or melt together, and an increasing amount of scholars acknowledge that life probably evolved from multiple roots that evolved into various trunks.

Critique also came from virology. Trees of life include extant and extinct biological species, but should it end there? Virologists are increasingly suggesting that viruses should be included in phylogenetic reconstructions as circling around the tree of life, where the existing roots, trunks, branches, and twigs are constantly "colonized" by viral agents. The code words by which viruses can enter the tree of life iconography are again lateral gene transfer, symbiogenesis, and infective heredity.

> … The 'Tree of Life' concept has been severely undermined and cannot apply to … large scale HGT processes … or explain the role of viruses … . Yet a tree-like structure of genetic evolution is observed in all domains of life, including most viruses. Thus HGT is colonizing an existing tree from non-ancestral (viral) sources. However, 'Tree-thinking' which explains tree growth by ancestral variation and natural selection continues to be vigorously defended leading many to dismiss the prokaryotes as 'odd-balls' that evolve differently from other life. Evidence now compels us to revise our definition and vision of the Tree of Life to include viruses. … Reticulate evolution and symbiosis apply to all life and must now be incorporated into our conceptual framework … . (Villarreal and Witzany 2010: 699)

If we include viruses, should we also include the overall abiotic environment? Darwin (1837/38: 23–24), for example, already wondered how the environment, divided into air, land, and water, could be brought into tree of life imagery, and how a similar environment would cause for affinity in the major branches of life's tree.

> Would there not be a triple branching in the tree of life owing to three elements air, land and water, and the endeavour of each ~~one~~ typical class to extend his domain into the other domains, and subdivision three more, double arrangement. — if each main stem of the tree is adapted for these three elements, there will be certainly points of affinity in each branch.

In 2014, in a special issue on "The tree of life in ecosystems: evolution of plant effects on carbon and nutrient cycling" published in the *Journal of Ecology*, an additional requirement was added to the tree of life; namely, such an iconography should be able to feature the various biochemical cycles as well as the hierarchical relations life endorses with the biotic and abiotic environment (Cornelissen and Cornwell 2014). Also in 2014, Kathleen Scott called out for contributions to a special issue for the journal *Life*, titled "Modern Phylogeny: The Three Domains of Life" wherein she was aiming for chapters that include metabolic cycles and physiological capabilities such as photosynthesis and mutagenesis, which also need phylogenetic reconstructions and overall integration into our evolutionary descent imagery (the call can be read at http://www.mdpi.com/journal/life/special_issues/phylogeny, and the issue is currently forthcoming).

The incoming data can no longer be banned from evolutionary iconographies. Nonetheless, one can wonder whether patterns of evolution can or cannot be inferred from the tree of life imagery (for a discussion, see the 2008 special issue for the *Journal of Systematics and Evolution edited by* Hong et al. 2008). It seems a logical and scientific necessity to demand that a universal "tree of life," or more general educational aids that visualize the evolution of life, should be able to at minimum fit in all forms of life, all time periods, and it should adequately depict all types or modes of descent. This in turn raises interesting questions on whether evolutionary descent iconographies should be able to provide insight into the major mechanisms by which life evolves, the hierarchies of life and its major transitions, as well as what shapes such depictions should take on.

Debating such questions merges fluently with the ongoing debates on the adequacy of the Modern Synthesis and the necessity to extend its scope. Including data acquired from molecular phylogenies, exobiology, virology, and ecology implies an inclusion of fields that were marginalized during the formation of the Synthesis.

2.4 Conclusion to Part 1

Because there are so many special issues and numerous papers that debate these epistemic questions, I have chosen to guide the reader to the literature rather than to give an in-depth analysis myself. Over the past years, scholars have been trying to put more and more ranks on the tree of life, and they have been trying to visualize and incorporate the numerous evolutionary mechanisms whereby life evolves in order to make evolutionary iconographies truly universal. On the other hand, scholars have questioned the possible to draw one universal tree of life that is able to illustrate all of life's complexity, thereby arguing that each rank should have its own tree or network of life.

At present, it is unclear whether all of life's complexity can indeed be depicted into one tree, web, or one iconic image. Illustrating the common, and not so common, descent of life can only be done right if we take into account the different evolutionary theories and mechanisms that explain life's descent. Moreover, the drawing of the tree highly depends on the conceptual classification framework one uses: Different species concepts, different definitions of life, different evolutionary theories, and different evolutionary mechanisms provide different evolutionary diagrams.

Throughout history, the tree has strongly contributed to our understanding of the evolutionary process. Science is associated with images, and these images therefore help in the dissemination and acceptance of ideas. It therefore becomes all the more important that our scientific illustrations convey the right messages, and the origin of life out of non-life, dissipative structures, autocatalytic adaptive systems, RNA worlds, symbiogenesis, HGT, viral colonization, and hybridization are most certainly among those messages.

3 Mechanisms of HGT

> Medical microbiology had a life of its own, but it was almost totally divorced from general biological studies. Pasteur and Koch were scarcely mentioned by the founders of cell biology and genetics. Instead, bacteriology was taught as a specialty in medicine, outside the schools of basic zoology and botany. Conversely, bacteriologists scarcely heard of the conceptual revolutions in genetic and evolutionary theory. (Lederberg 2003: 287)

While concepts such as "Horizontal Gene Transfer" and "reticulate evolution" date back to the 1980s and 1990s, bacterial transformation, conjugation, and transduction were already observed in the early decades of the twentieth century. Many of these observations track back to, and associate with major milestones and advances made within standard neo-Darwinian evolutionary theory. However, these discoveries are rarely featured in historical reviews on the onset of evolutionary thought. Instead, such reviews will guide their reader through a set of historical milestones that include the following: the introduction of cell theory by Mathias Schleiden and Theodor Schwann in the late 1830s; the introduction of natural selection theory by Charles Darwin in 1859; the temporary "eclipse of Darwinism" in the late nineteenth century due to advances made in ecology, symbiology, and epigenetics (that includes what we today call "evolutionary developmental biology"); the rediscovery and synthesis of Mendelian hereditary laws with Theodor Boveri and Walter Sutton's chromosome theory of inheritance, Darwin's natural selection theory, and various aspects of mutation theory; the rise of theoretical population genetics; the foundation of the Modern Synthesis in the 1940s; the discovery of the structure of hereditary material in the early 1950s; the subsequent development of molecular genetics; and the current plea to extend the Modern Synthesis in order to integrate both presynthetically (but marginalized) and postsynthetically evolved theories.

In this part, we detail how insights into the mechanisms that underlie HGT correlate with these major advances in evolutionary thought. The discovery of bacterial transformation correlates with the discovery of DNA as the bearer of hereditary material. Insights into bacterial conjugation and phage-induced transduction of bacterial DNA associate with increasing knowledge into cell cytology, cytoplasmic inheritance, and evo-devo. And transduction and knowledge on mobile genetic elements bring us to the epigenetic era. Though ignored by the standard evolutionary framework, even more puzzling is that discoveries that currently enable us to understand HGT have from the very onset been recognized as major breakthroughs in the biomedical sciences. In fact, one way by which one can detail the history of HGT is by guiding the reader through the various Nobel Prizes that have been awarded in the category of Physiology or Medicine from the early twentieth century onward. Why was there this discrepancy between the biomedical and evolutionary sciences?

For one, evolutionary biology is a diachronically oriented research field: It studies the natural history of species, and therefore, it is directed toward the past. In contrast, the biomedical sciences' epistemic stance is futuristic: By trying to understand the current causes of disease, they try to find cures that will remedy disease in the future. That is why they do not form part of the evolutionary sciences, neither academically speaking in what regards the division of the sciences, nor epistemologically speaking in what regards their theoretical outlook. It is only recently, partly due to the wider recognition of HGT, that the biomedical and evolutionary sciences are becoming synthesized.

Secondly, the Weismann (1885) barrier put an end to neo-Lamarckian evolutionary theories that developed during the "eclipse of Darwin." Ontogenetically acquired traits were no longer contended to feed back into the germ line. The founders of the Modern Synthesis therefore drew clear barriers between ontogeny and phylogeny. HGT, on the contrary, associates with "infective heredity," i.e., the study of diseases and foreign DNA that are acquired during the individual's life span.

Thirdly, because the phylogenetically oriented neo-Darwinian field was out of epistemic reach for biomedical scholars, instead, they associated their discoveries with more ontogenetically oriented disciplines, namely symbiology, ecology, developmental biology, and epigenetics (what we today designate as "evolutionary developmental biology"). These research schools all developed in the late nineteenth century, and they did so in close association with one another, because these fields study the various lifetime-interactions species engage in, either with one another or with the abiotic environment. For that very same reason, these latter disciplines have evolved outside or in the margins of the standard evolutionary framework.

As a philosopher of science, my aim here is not to review insights on the biochemical, molecular structures that underlie HGT processes, but to explain the basics in simple terms, as well as to briefly situate the discoveries in time, thereby highlighting the major implications they have for general evolutionary theory.

3.1 History of Infectious Disease: The Origin of HGT Research in Symbiology, Ecology, Developmental Biology, and the Biomedical Sciences

> Genetics, symbiotology, and virology have a common meeting place within the cell. There is much to be gained by any communication between them which leads to the diffusion of their methodologies and the obliteration of semantic barriers. (Lederberg 1952: 32)

As already mentioned in the introduction to this volume, there is only a fine line to be drawn between studies on symbiosis, infectious heredity, and HGT. This becomes especially striking when one reconstructs the historical origins of lateral transfer studies.

Symbiology, medicine, bacteriology, virology, and overall microbiology are fields that developed their frameworks in the late nineteenth-century period that has been designated by Huxley (1942) as the "eclipse of Darwinism." The eclipse of Darwinism is a period in history that demarcates a demise in the adherence to natural selection theory in favor of more ecological, symbiotic, and ontogenetic (including cell cytological) frameworks. Symbiotic research paralleled ecological research and both developed outside Darwinian and neo-Darwinian theory (Sapp 1994). Ecology, for example, only became integrated in the 1960s, and even today, symbiology remains unintegrated.

From the mid-nineteenth century onward, symbiologists such as de Bary (1861) and Van Beneden (1873, 1875) had pointed toward parasitic symbiotic microorganisms as the cause of plant and animal diseases. The "golden age" of bacteriological and microbiological fields as well as the advent of the biomedical sciences with the "germ theory of disease" are thus ultimately driven by insights into symbiosis.

The biomedical sciences are an outgrowth of (1) increased knowledge of bacteria, protozoan microorganisms, and eventually also viruses as causal agents of disease (Beijerinck 1898; Cohn 1875; d'Herelle 1917; Iwanowski 1892; Koch 1876, 1882; Pasteur 1880; Laveran 1880; Lewis 1879; Mayer 1886; Twort 1915) and (2) knowledge on immunology and its associated researches on vaccination therapies, serology, and chemotherapies (Ehrlich 1877, 1879a, b; Jenner 1798). From its onset, the biomedical sciences have emphasized the possibility of horizontal exchange of biochemical substances between various organisms and horizontal transmission of these substances outside of the germ line, during various stages of organismal development.

As early as 1717, Mary Wortley Montagu introduced an immunization technique against smallpox (Variola) in England. During extensive stays in Turkey, she learned that rubbing the scabs from individuals infected with smallpox against carved skin of healthy individuals leads to the development of only a mild form of smallpox. Somehow, there must therefore have been a form of horizontal transmission from the infected pustules to the healthy individuals, which rendered the latter less vulnerable to the disease. Her work led to the introduction of inoculation and injection techniques to develop immunity against various diseases and these techniques eventually found all research on vaccinations. Soon after, scholars would conduct similar experiments with less virulent cowpox which proved to also cause immunity against Variola (Case and Chung 1997), and eventually, Jenner (1798) would introduce a first cowpox-based "vaccine" against smallpox.

Such inoculation and vaccination experiments, which, in hindsight, are per definition based upon the artificially induced horizontal transfer of biochemical substances, or the artificially induced endosymbiosis of foreign cells into a host, became the primary means by which scholars identify and study disease.

In 1884, Robert Koch, one of the founders of the germ theory of disease, published his etiology of tuberculosis that proves that the *Tubercle bacillus* is the disease-causing agent of tuberculosis. Koch (1884) was able to cultivate pure strains on blood serum and he proved that inoculation in guinea pigs caused disease. From this work, he derived 4 postulates that serve as testing devices to identify microorganisms as disease-causing agents: (1) The microorganism must be present in sick organisms and absent in healthy ones; (2) doctors have to be able to isolate the disease-causing organisms from infected organisms and grow pure strains of them; (3) inoculation of isolated cultures must cause disease in healthy organisms; and (4) after inoculation, the microorganisms must be found in the infected host, and they must be identical to the originally identified pathogens.

In 1886, D.E. Salmon and Theobald Smith, the first Americans to study bacteria as disease-causing agents, developed a new means to induce immunity against contagious diseases by injecting whole, heat-killed cells of virulent strains in healthy individuals. And four years later, two students of Koch, Behring and Kitasato (1890), developed immunizing techniques against Diphtheria and Tetanus.

Immunology became a subfield of the medical sciences due to advanced knowledge on human blood and the plasma it contains. Immunology encompasses the field that studies the reactions of the body against infectious agents and unwanted substances. The injection of foreign agents such as bacteria or viruses into the skin or the veins of organisms causes the body to react with an immune response. This immune response involves the production of antibodies (mostly consisting of white blood cells) that attack the antigens of the pathogen. Antigens are chemical substances found on the surface of infectious agents. When the pathogens are rendered harmless, the body has created a "memory" of the infectious agent: It maintains the specifically generated antibodies which ensure protection against future encounters with the pathogen.

Scholars first learned about such complex interactions between antibodies and antigens through the works of Paul Ehrlich and Ilya Ilyich Mechnikov

(Élie Metchnikoff). Ehrlich (1877, 1879a, b) first identified the various types of human blood cells that exist, thereby founding the field of serology. In 1891, he discovered antibodies that cause immunity against plant toxins, work that led him to develop one of the first theoretical frameworks to understand immunology (Ehrlich 1898, 1900; Ehrlich and Morgenroth 1902). His "side-chain theory" understands the workings of serum on toxins (antibodies on antigens) as a chemical reaction whereby the molecules of the serum and the toxin bind to one another. Most importantly, he thought that immunology was inheritable and horizontally transmittable, in mammals especially via lactation from mother to offspring (Ehrlich 1892a, b).

In 1882, Mechnikov experimentally introduced thorns from a tangerine tree into the larvae of starfish and observed that specific cell types surrounded the thorns, which made him assume they protect the larvae by eating the foreign material. In 1884, he was able to confirm his idea, by observing that "eating cells" destroyed fungal spores that he introduced in *Daphnia* (a freshwater flea). Later in his career, he discovered that mammalian white blood cells engulf and kill the Anthrax bacterium. Carl Friedrich Claus proposed Mechnikoff to call the cells phagocytes (eating cells), and Mechnikov introduced the term "phagocytosis" to describe the process of elimination by eating and conjectured that it lies at the basis of cellular immunity (see Karnovsky 1981; Tauber 2003; Tan and Dee 2009 for a discussion). Today, we know that phagocytes are a group of motile white blood cells, and mast and dendritic cells that play crucial roles as primary defenders of immunity. They eat harmful pathogens and eliminate debris. Phagocytosis is furthermore the mechanism suggested today to underlie primary, secondary, and tertiary endosymbiosis, whereby the acquisition of the symbionts is understood as a failed digestion of the independently evolved cells (for a discussion, see Zook, this volume).

New microscopes enabled a better visualization of the microorganisms, and symbiology theory (especially parasitism), more so than natural selection theory, facilitated a better theoretical conceptualization of the acquisition and development of infectious diseases, while immunizing techniques such as inoculation experiments and vaccination therapies enabled protection against disease. Similar inoculation experiments underlie the discovery of bacterial transformation by Frederick Griffith in 1928.

Investigating immunological processes entails a recognition that symbiotic and coevolutionary relations have evolved between hosts and pathogens, their antibodies, and antigens. And this recognition necessitates research in ontogeny because such interactions take place during the life span of the individual.

At the turn of the twentieth century, von Faber (1912) took the issue one step further by theorizing that ontogenetically acquired, parasitic, and beneficial symbiotic relationships can become hereditary. Von Faber's notion of "erbliche Zusammen leben" became translated by Cowles (1915) as "hereditary symbiosis" which became understood as a major driving force of symbiogenesis by scholars such as Buchner (1921), Wallin (1927), and Lederberg (1952). In fact, Lederberg, who discovered bacterial conjugation and phage-mediated transduction, understood both as instances of "hereditary symbiosis" and "infective heredity."

3.2 Bacterial Transformation

Transformation was first observed by Griffith (1928) in the context of medical research on the nature and cause of pneumonia. Griffith conducted various experiments whereby he inoculated healthy organisms with virulent strains of the bacteria that cause pneumonia and in the process, he discovered that some bacteria transformed. The study of the underlying patterns and mechanisms whereby bacteria transform also played a crucial role in the discovery of genes as carriers of hereditary information (Avery et al. 1944). Before we turn to history, we first outline the basics of bacterial transformation.

Bacterial transformation is a type of HGT whereby living bacteria take up naked genes from their surroundings, including genes coming from decomposing bacteria (Chen and Dubnau 2004; Dubnau 1999; Downie 1972; Redfield et al. 1997; Sisco and Smith 1979). The acquired genes can range from small DNA fragments such as transposons to plasmids, and even the donor's entire bacterial chromosome can become absorbed by the recipient (Akamatsu and Taguchi 2001). Uptake of plasmids or an entire chromosome occurs mostly under artificial laboratory conditions, during genetic engineering experiments, while the lateral acquisition of small DNA fragments occurs abundantly in natural settings (Mandel and Higa 1970; Johnsborg et al. 2007).

The acquired DNA fragments can be used for DNA repair (Hoelzer and Michod 1991); it can function as a nutritional source (Finkel and Kolter 2001); the acquired genes can become part of the bacterial genome, or it can integrate into possible plasmids already residing inside the bacterium (Fig. 1). The acquisition and insertion of DNA fragments into the bacterial genome enable genome growth, and the integrated DNA often not merely changes the bacterium's genetic makeup, it also changes functional metabolism.

How and why bacteria take up foreign DNA particles is still not completely understood (Chen and Dubnau 2004; Dubnau 1999). Bacterial types such as *Streptococcus pneumonia* appear to have a natural competence to take up foreign DNA, an ability that is biochemically "programmed" in their genes. This competence relates to the morphology of their bacterial envelope and membrane, as well as their pili (filaments attached to the cell's surface) (Sisco and Smith 1979; Redfield et al. 1997). Pili are also involved in bacterial conjugation, discussed in the next part.

Transformation is a costly biochemical process, and bacteria mostly engage in DNA uptake when they find themselves in a state of starvation, under harsh environmental conditions, or when they contain damaged DNA (Engelmoer and Rozen 2011). In addition, transformation is one of the means by which bacteria such as *E. coli* naturally acquire resistance genes to antibiotics (Anderson 1968; Cohen and Miller 1969, 1970; Cohen et al 1972). Stanley Cohen, Annie Chang, and Leslie Hsu demonstrated that "the introduced R-factor DNA can persist in such cells as an independently replicating plasmid, and can express both the fertility and antibiotic resistance functions of the parent R factor" (Cohen and Miller 1970: 2110). "R factors" stand for "resistance transfer factor" or "antibiotic resistance factors."

Bacteria can also integrate DNA from bacterial viruses. Mandel and Higa (1970), for example, demonstrated that in the laboratory, *E. coli* bacteria can take

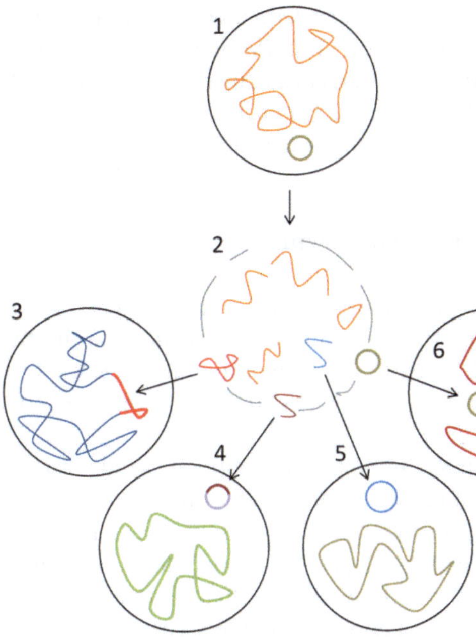

Fig. 1 Examples of bacterial transformation. *1* A bacterium with its bacterial genome and a plasmid. *2* The bacterium dies, the cell membrane and the bacterial chromosome disintegrate, and some fragments and the plasmid are released from the dead bacterium. *3* A DNA fragment is absorbed by a recipient cell and becomes integrated into the bacterial chromosome. *4* A transposon carrying antibiotic resistance gene(s) is absorbed by a recipient cell, and the transposon becomes integrated into the bacterial plasmid. *5* A DNA fragment is incorporated into a recipient bacterial cell, but the DNA is not integrated into the bacterial genome; instead, either the DNA fragment is broken down and used as a nutritional source or the DNA fragment remains in the cell as extrachromosomal DNA. *6* The plasmid from the donor cell becomes integrated into the recipient bacterium

up genes coming from the *lambda* bacteriophage. And already in 1951, Victor Freeman reported on HGT from a bacteriophage to an avirulent *Corynebacterium diphtheria* and indicated that such transfer renders the bacteria virulent:

> Regardless of the fact that the underlying mechanism is not understood, the knowledge that avirulent cultures of *C. diphtheriae* can become virulent in the presence of specific bacteriophage is of importance to any consideration of the many perplexing problems that have confronted bacteriologists and epidemiologists interested in the study of diphtheria. If the virulence of the *diphtheria bacillus* should prove dependent not only on its toxigenic ability and its invasive power but also on the degree of its association with a specific bacteriophage, then some of the difficulties involved in understanding the complex problems of bacterial metabolism and immunity as they occur in the diphtheria case or carrier might be partially solved. (Freeman 1951: 686)

Bacterial transformation raises interesting questions on adaptive environmental responses as well as biochemical "communication." Experiments demonstrate an increase in competence to transform when the bacteria are somehow threatened, and the mere possibility to take up foreign genes depends upon the biochemical recognition of exogenous DNA in its surroundings and biochemical capacities to transport and insert the foreign DNA particles. Insofar as bacteria seem to prefer to take up naked DNA similar to their own genetic makeup, it must involve some kind of biochemical recognition of this similarity. Furthermore, upon death, many bacteria release their genes to the surroundings, and one can wonder whether such an act requires a higher-order, group explanation: Does such release resemble some kind of "altruistic group behavior"?

3.2.1 Griffith's Inoculation Experiments

Around the turn of the twentieth century, Pneumococci (*Streptococcus pneumoniae*) became indicated in causing lobar pneumonia in humans. Neufeld (1902) and Neufeld and Händel (1910), working at the Koch Institute, had classified various strains into 3 different types. In 1917, Avery and colleagues added a fourth type (Downie 1972: 2).

With the goal to abstract immune sera (antibodies) against pneumococci, the British microbiologist and physician Frederick Griffith (1877–1941) abstracted the various bacterial types from the mucus coming from the lower airways (sputum) of infected humans that had developed lobar pneumonia. He subsequently "grew" these bacterial strains in the belly of mice to then abstract them again after which he tested whether these bacteria correspond to the types he found in the human sputum. Afterward, he combined the sputum with the type serum and injected them into healthy mice to see how they react to various strains of the bacterial types.

What was striking was that the mice that developed inflammation would often die from a bacterial strain different from the one it was injected with. Contrary to his contemporaries, who assumed the fixity of bacterial types, Griffith's research led him to conclude that the bacterial types underwent modification: They were able to "transform" and acquire new virulent functions. Such transformation, Griffith (1928: 139) furthermore noticed, was "the property of the whole strain in each case."

But what exactly happened? Several explanations were possible: Either the various bacterial types and strains merely represent several stages of the same individual, much like a caterpillar is a stage in the life cycle of a butterfly; or the cultures were not pure; or infection with one type might make infection with another type more likely; or perhaps during reproduction one type would randomly mutate to another type; or perhaps environmental conditions played a role because transformation of type was more likely to occur under the skin than in the bloodstream.

To find answers, Griffith (1928: 133–6) conducted further experiments and observed that when avirulent pneumococci were grown on a blood agar plate for 24 hours, they would develop colonies that appear morphologically "Rough" to the observer. Instead, virulent colonies would have a "Smooth" glistering appearance. Nonetheless, there were circumstances in which he inoculated mice with an attenuated Rough strain (R strain), and yet, they would die from sepsis. When these individual's blood was inspected, they would contain Smooth strains (S strain) with "well-marked capsules" (a substance that surrounds the bacterial envelope). The well-marked capsules, we now know, are coats made up of sugar and proteins that protect the virulent strains against recognition by the host's antibodies. Rough strains are rough because they do not have this protective shield, and the infected body therefore recognizes the intruder and renders it harmless.

So at some point, a conversion occurred and Griffith (1928: 141–158) subsequently investigated "whether an avirulent R pneumococcus can be transformed into the virulent S form by growth in the body of mice." Several of these experiments went as follows (Fig. 2). First, he cultured pure R and pure S strains. When he inoculated healthy mice with living attenuated non-encapsulated R strains, the mouse stayed healthy, thereby confirming that the R strain is avirulent. When he inoculated mice with the living encapsulated and virulent S strains, the mice would die, thereby confirming that the S strain is virulent. When he inoculated mice with dead S strains that he killed through heating, the mice would also remain healthy. But, when he combined living R strains with the heat-killed S strain and injected them into mice, the mice would die. Inspection of the blood of these dead mice would evidence the presence of living S strains.

He therefore concluded that transfer of substances from the dead S strain to the living R strain happened, enabling the R strain to transform into the virulent S strain. Griffith (1928: 166): "… the attenuated organisms actually make use of

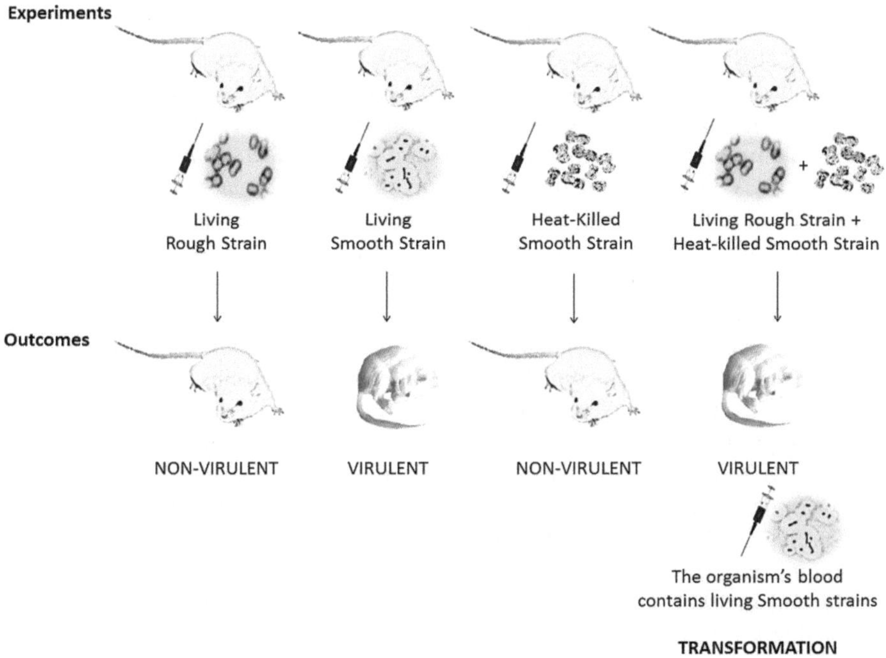

Fig. 2 Schematic of Griffith's transformation experiments. Living Rough pneumococcal strains and heat-killed Smooth strains are harmless for the mice, which proves that strains with this morphology are non-virulent. Living Smooth pneumococcal strains are virulent and cause death. A combination of living Rough avirulent strains with heat-killed Smooth strains made the mice die. When he examined the blood of the dead mice, he found living Smooth and thus virulent strains. The Rough strains had transformed from being avirulent to being virulent, and Griffith speculated that the Rough strains had acquired a "transforming factor" from the dead virulent strain

the products of the dead culture for the synthesis of their S antigen. An R strain is most readily transformed into the S variety when the killed culture used is of the same serological type as that from which the R strain was derived."

He also noted that such transformation occurred when inoculation happened in the skin or the belly. In the bloodstream, the R strains are immediately rendered harmless by immune responses of the body, thereby disabling any transformation to a virulent S strain. But, Griffith noted that such rough morphology is adaptive for the bacteria, because they can lie latent in the body and switch to a virulent stage when the body is infected with new, virulent strains (Griffith 1928: 172).

This is the principle of lateral transfer by transformation. Living bacteria can snatch compounds of dead bacteria. Griffith, however, did not know what the nature was of the transforming factor and thought that the capsule, which is made up of proteins, had something to do with the transformation from R to S strains.

3.2.2 The Avery–MacLeod–McCarty Experiments

Griffith's experiments were confirmed by several scholars including Neufeld, Levinthal, and Bauerhenn, as well as by scholars from the Rockefeller Institute for Medical Research (Avery et al. 1944: 137). Oswald Avery was a medical doctor who had been studying the capsules that surround the virulent strains, because he thought that these capsules were responsible for disease. From 1928 to 1931, work done by Martin H. Dawson and Richard Sia, members of Avery's laboratory, enabled the "conversion" of R into S forms in a test tube, in vitro. More specifically, they accomplished the growth of "R cells in a fluid medium containing anti-R serum and heat-killed encapsulated S cells" (Avery et al. 1944: 137).

While Griffith had induced transformation in various types and strains of pneumococci, in all of Avery, MacLeod, and McCarthy experiments, they focused on the transformation of an attenuated R type II strain (itself derived from a virulent S culture of pneumococcus type II) to virulent type III S strains (Avery et al. 1944: 139). Their experiments were successful.

The reproduction of the experiments subsequently enabled the Rockefeller scholars to begin isolation studies on the "transforming factor" in 1935, work that was first performed by James Lionel Alloway and later by M. MacLeod, McCarthy, and Avery himself. In a series of experiments, the scholars used various techniques that specifically and selectively break down DNA, RNA, proteins, and lipids, and in 1944, in a now famous article, they revealed that the transforming factor that enables non-capsular, avirulent R strains to transform into capsular S forms, must be made up of DNA. They concluded as follows:

> Equally striking is the fact that the substance evoking the reaction and the capsular substance produced in response to it are chemically distinct, each belonging to a wholly different class of chemical compounds. The inducing substance … appears to be a highly polymerized and vicious form of sodium desoxyribonucleate [deoxyribonucleate]. On the other hand, the Type II capsular substance, the synthesis of which is evoked by this transforming agent, consists chiefly of a non-nitrogenous polysaccharide constituted of glucose-glucuronic acid units

> linked in glycosidic union. The presence of the newly formed capsule … confers in the transformed cells all the distinguishing characteristics of Pneumococcus Type III. Thus, it is evident that the inducing substance and the substance produced in turn are chemically distinct and biologically specific in their action and that both are requisite in determining the type specificity of the cell of which they form part. (Avery et al. 1944: 152)

Particularly, the last point is important. The scholars underlined that DNA is the enabler of capsule formation (madc up of proteins), but DNA itself is made up of a substance different from the substance that makes up the capsule. For the authors, this suggested that DNA carries hereditary information, information that is specific and differential. It does not provide the material to make the capsule; rather, it carries the information on how to make it. Their work countered the then prevailing notion, brought forth by scholars such as Phoebus Levene, that DNA was a simple and repetitive structure made up of the same elements. They further noted how odd they found it to reach these conclusions based upon the study of "immunological techniques."

Their work, and also Griffith's, presents a prototypical example of how empirical evidence often precedes theory. As Downie (1972: 2) notes, one of the reasons Griffith conducted so many experiments to confirm his results was probably because he was "conditioned to believe that bacteria existed in immutable types." And also Avery was at first reluctant to accept the incoming results from his collaborators, who continued their research during a leave of absence due to thyroid intoxication by their senior.

Avery, MacLeod, and McCarthy's work was nonetheless well received and even popularized by Scientific American (Morange 2000: 33). But although the work and the results were recognized, it remained difficult for scholars to think through the consequences. Because both Griffith and Avery's team artificially induced transformation under laboratory settings, it was assumed that such transformation did not occur in natural settings. The authors themselves noticed that: "Transformation of types has never been observed to occur spontaneously and has been induced experimentally only by the special techniques outlined earlier in this paper" (Avery et al. 1944: 140). The idea that hereditary information can be transmitted horizontally under natural conditions during ontogeny (terms they never used to describe these phenomena) was too far away from the idea that hereditary information is transmitted vertically during sexual reproduction.

That it is indeed DNA that carries hereditary information was later confirmed by Hershey and Chase (1952) at the Cold Spring Harbor Laboratory, in their famous experiments with T2 bacteriophages and *E. coli* bacteria. They used different radioactive isotopes to color the proteins that make up the bacteriophage's capsid and the DNA that makes up the core of the phage's head, in order to track which parts enter the bacterial cell upon infection and which parts enable the production of new phages. They concluded that it is the phage's DNA that becomes injected upon infection, while the protein coat remains attached to the surface of the bacterium. And the phage's progeny mainly inherits genetic material rather than proteins.

> We have shown that when a particle of bacteriophage T2 attaches to a bacterial cell, most of the phage DNA enters the cell, and a residue containing at least 80 per cent of the

> sulfur-containing protein of the phage remains at the cell surface. This residue consists of the material forming the protective membrane of the resting phage particle, and it plays no further role in infection after the attachment of phage to bacterium. These facts leave in question the possible function of the 20 per cent of sulfur-containing protein that may or may not enter the cell. We find that little or none of it is incorporated into the progeny of the infecting particle, and that at least part of it consists of additional material resembling the residue that can be shown to remain extracellular. Phosphorus and adenine … derived from the DNA of the infecting particle, on the other hand, are transferred to the phage progeny to a considerable and equal extent. We infer that sulfur-containing protein has no function in phage multiplication, and that DNA has some function. (Hershey and Chase 1952: 54)

3.3 Bacterial Conjugation

A *plasmid* is an extrachromosomal, circular-shaped, double-stranded DNA molecule. In other words, plasmids do not form part of the bacterial chromosome, but reside in the cytoplasm of the bacterial cell. Here, they can replicate autonomously from the latter. Plasmids are central agents for HGT by means of bacterial conjugation (Thomas 2000; Griffith et al. 2000), whereby one copy of the double-stranded DNA molecule that makes up the plasmid is laterally transferred from a donor bacterium to a recipient. Bacterial conjugation requires cell-to-cell contact between two bacterial organisms, and these organisms have a different morphology: One has a plasmid and a sex pilus (a sexual appendage), and the other has not. Pili are hairlike filaments made up of pilin proteins that extend from the bacterial cell surface (Proft and Baker 2009). There exist different types of pili: Some enable motility in which case they are called common pili or fimbriae (Mattick 2002), while others enable conjugation. When the pili enable bacterial conjugation, they are called conjugative pili or sex pili. The genes required to produce pili are encoded in the conjugative plasmid.

During bacterial conjugation in *E. coli*, one such sex pilus of the donor bacterium attaches itself to the recipient bacterium and pulls the latter cell closer (Fig. 3). The membranes join, and the lateral exchange occurs. During the exchange, one strand of the double-stranded DNA molecule that forms the plasmid of the donor cell is passed on to the recipient. Thus, the plasmid itself does not, in its entirety, move from the donor to the recipient cell. What happens is that the double-stranded DNA of the plasmid is cleaved, and a single-stranded plasmid is transferred to the recipient. After the transfer, both the donor and the recipient synthesize the complementary strand of the plasmid DNA, thereby rebuilding the plasmid into a double-stranded DNA molecule. After conjugation, and thus after acquisition of the plasmid, the recipient cell is also able to engage in bacterial conjugation.

Bacterial conjugation was first discovered in the *E. coli* K-12 strain by Lederberg and Tatum (1946). The *E. coli* K-12 strain is a laboratory strain that was abstracted from a human patient's stool (Bachmann 1972), and ever since, it has been cultivated in various laboratories around the world. When Lederberg and Tatum first reported on bacterial conjugation, they did not know how the mechanism of transfer worked. Their experiments on "direct hereditary interaction of one

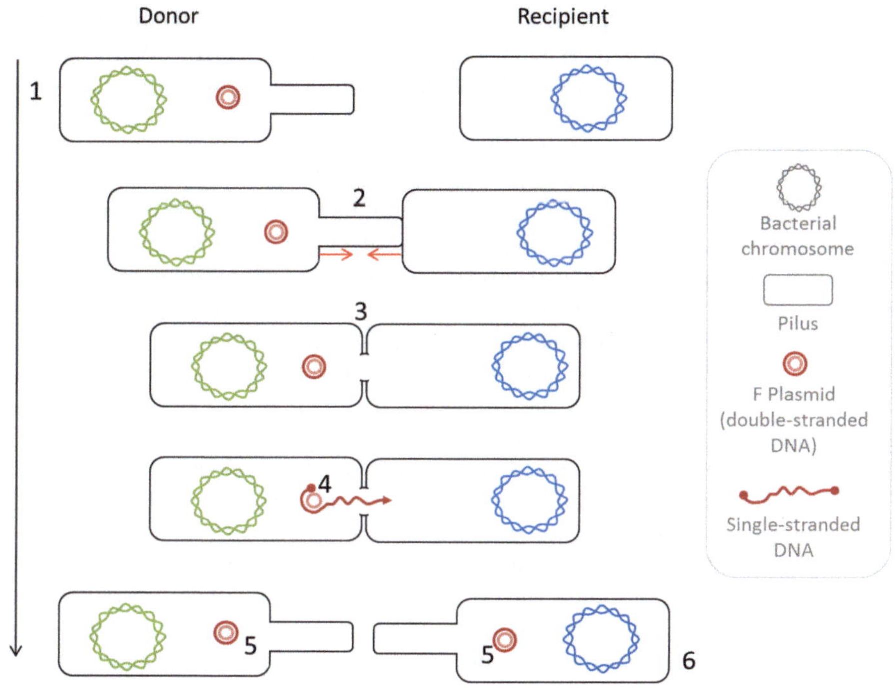

Fig. 3 Steps in bacterial conjugation of *E. coli*. *1* On the *left*, a donor bacterium (F + cell) carrying an F plasmid. *Right*, a recipient cell (F−) that does not have an F plasmid. *2* The conjugative pilus of the donor bacterium attaches to the recipient bacterium and pulls the latter closer. *3* The cells become connected. *4* A single DNA strand of the plasmid is transferred. *5* Both the donor and the recipient synthesize the complementary DNA strand which enables the restoration of the double-stranded plasmid molecule. *6* Because of the transfer, the recipient cell now also becomes an F + cell that can donate plasmids to other F− cells

bacterial type with another" involved mixing two mutated strains of *E. coli* K-12, each deficient in the synthesis of certain proteins and enzymes disabling them to grow. One strain could produce what the other lacked, so the strains were complementarily insufficient for their lack of synthesis of these elements. However, when the two strains were mixed together, new colonies emerged. So the bacteria appeared to have exchanged what the other lacks to reestablish growth, and the authors presumed that this occurred via bacterial "recombination" (bacterial mating) or "hybridization," though a "complete analogy cannot be drawn at present between the inheritance of bacterial characters and the Mendelian processes of higher forms" (Lederberg and Tatum 1946: 681).

But what had been exchanged and how? Did one or both strains engage in a form of "nutritional symbiosis" (Lederberg and Tatum 1946: 766), or did they exchange genes in order to repair their DNA (Griffiths 2000)?

Major insights into the mechanisms were brought forth by the Lederberg couple as well as by Luigi-Luca Cavalli Sforza, William Hayes, François Jacob, and

Ellie Wollman, first independently and later in close collaboration (Holloway and Broda 1996). Seven years later, in a letter to Nature, Hayes (1952) reported that bacterial conjugation was unidirectional, going from a donor to a recipient. Lederberg et al. (1952) and, a year later, Hayes (1953) discovered the "F Fertility factor" that enables the initiation of bacterial conjugation in *E. coli*. It was Esther Lederberg who first coined the term "Fertility Factor."

In *E. coli*, the F factor can also integrate into the bacterial chromosome through homologous recombination, a process first described by Cavalli Sforza (1950). As a factor that can both exist independently in the cytoplasm and integrate into the bacterial chromosome, the F plasmid was considered an example of an "episome," a term first coined by Jacob and Wollman (1958; Wollman and Jacob 1955). Integration of the F factor into the bacterial chromosome results in a "High-frequency recombination cell" or Hfr cell (Williams, this volume).

Plasmids were considered by Lederberg (1952: 403) to "comprise part of the genetic determination of the organic whole," and he understood such elements to evolve in a non-Mendelian fashion, by "infective heredity" (Lederberg 1952: 413) which made them examples of "hereditary symbiosis" (Lederberg 1952: 415 1955).

Finally, bacterial conjugation has mostly been studied in gram-negative bacteria such as the *E. coli*, but it also occurs in gram-positive bacteria (Grohmann et al. 2003). Gram-positive bacteria are furthermore able to transfer the entire double-stranded plasmid DNA (Grohmann et al. 2003; Scott and Zähner 2006). Heinemann and Sprague (1989) have additionally demonstrated that genetic transfer from conjugative *E. coli* to eukaryotic yeast organisms (*Saccharomyces cerevisiae*) can happen which proves that transfer can occur from pro- to eukaryotes.

3.3.1 Bacterial Conjugation and the Acquisition of Antibiotic Resistance

Bacterial conjugation was the first suggested mechanism by which bacteria acquire resistance to antibiotics. Antibiotics are substances produced by certain bacteria that have detrimental effects on other bacteria. When antibiotic substances penetrate the bacterial cell, it can target cellular components such as ribosomes, proteins, or the bacterial cell wall and induce decomposition leading to bacterial cell death. Antibiotic resistance occurs when bacteria evolve genes that disrupt the function of antibiotic substances, by disabling penetration of the bacterial cell wall, or by altering its target. Bacteria have also evolved means to simply push the antibiotic substances out of the cell, and they have evolved enzymes that decompose the antibiotic substances (Andersson and Hughes 2010; Allen et al. 2010).

By the 1950s, several antibiotics, including Chloramphenicol, Tetracycline, Streptomycin, and Sulfonamide, had become widely fabricated and administered to sick patients (Lederberg 2003: 288). And in a little less than 10 years, the first reports came in that bacteria had acquired resistance to these antibiotics.

These reports came from Japan where Ochiai et al. (1959), Akiba et al. (1960), and Mitsuhashi et al. (1961) had discovered naturally occurring antibiotic resistance in *Shigella* bacteria. These are gram-negative bacteria that include *S. dysenteriae* bacteria responsible for dysentery and they are also closely related to *Salmonella* and *E. coli*. The scholars observed that several dysentery patients became non-responsive to the four different types of antibiotics.

In the feces of the dysentery patients, they also found *E. coli* bacteria that were equally resistant to these antibiotics. The case studies brought to light that the acquired drug resistance of *Shigella* often occurred against all four types of antibiotics, even when patients were only administered one type of antibiotics. Neither the antibiotics themselves nor the interaction between the antibiotics and the *Shigella* bacteria could therefore underlie the acquired resistance. Instead, it was suggested that *Shigella* bacteria acquired antibiotic resistance from resistant *E. coli* strains that exist together in the intestinal canal of dysentery patients. The scholars tested the hypothesis and were able to artificially induce (in vitro): "(1) multiple-resistant clones of *Shigella* by mixing the cultures of the drug-sensitive *Shigella* and the multiple-resistant *E. coli*, and (2) mutual transfer of resistance between *Shigella* and *Escherichia*" (Akiba et al. 1960: 225).

At the time, all three canonical forms of HGT among bacteria had been reported, and based upon the principle of exclusion, the Japanese scholars theorized that the means by which antibiotic resistance was acquired was by bacterial conjugation.

> From the results obtained [...] transformation is not believed to be responsible for this phenomenon. The authors have not succeeded in the isolation of phage from resistant strains which is infective to give resistance to sensitive recipients. From this result, transduction is hardly considered as an essential mechanism. Therefore, recombination or conjugation may be the most probable mechanism involved. (Akiba et al. 1960: 226)

The Japanese scholars' knowledge on antibiotic resistance was transferred to a general Western audience by Watanabe (1971), who furthermore associated the acquisition of antibiotic resistance genes with Lederberg's notion of "infectious heredity" which in turn was inspired by symbiology jargon. The drug resistance genes, dubbed "Resistance factors" or "R factors," were thought of as "episomes": elements that can either replicate autonomously or integrate into the bacterial chromosome:

> We have found ... that the multiple drug resistance factors are carried and transferred by an episome ... . Multiple drug resistance is, therefore, an example of "infective heredity ..." (Watanabe 1971: 87)

As such, they provided another example for the very existence of episomes. Other early reports on transferrable antibiotic drug resistance include the works by Andersön (1968) and Jones and Sneath (1970).

The transmission of antibiotic resistance genes via bacterial conjugation demonstrates how rapid evolution by means of HGT can be. In less than 10 years after the first worldwide administrations of antibiotics, *Shigella* bacteria were able to acquire and spread this resistance. Some of these resistance genes already existed within the bacteria involved, and other resistance genes have evolved since the massive introduction of antibiotics. The current standard paradigm assumes that genetic mutations are "random," and it considers such random mutations to result from "copying errors." Whether or not the evolution of resistance genes is "random" or "directed" remains a topic of considerable research, but the rapidness by which transfer occurs across species outnumbers any suspicion of randomness.

Today, all canonical forms of HGT have been indicated in the acquisition of bacterial resistance. In the case of transduction, this also implies that resistance can be passed on from bacteriophages to bacteria or vice versa, and besides viruses, also fungi and other microorganismal pathogens have acquired resistance for the antibiotics administrated against them.

This again demonstrates intra- and interspecific exchange which in turn indicates an intricate coevolutionary and symbiotic way of living.

3.4 Phage-Mediated Transduction

> When *Salmonella typhimurium* is grown in the presence of a variety of mildly deleterious agents, especially weakly lytic phages, it produces a filterable agent (FA) capable of transferring hereditary traits from one strain to another. Individual filtrates may transduce many different traits, but no more than one in a single bacterium. The activities of a filtrate parallel the characteristics of the donor cells. Nutritional, fermentative, drug resistance, and antigenic characters have been transduced. The new characters are stable after many generations of subcultures. … (Zinder and Lederberg 1952: 697)

Bacteria can become infected by specific viruses called bacteriophages, a word that is often abbreviated as phages. When a bacteriophage attaches to the surface of bacteria, it is able to penetrate the membrane and inject its phage DNA (Fig. 4).

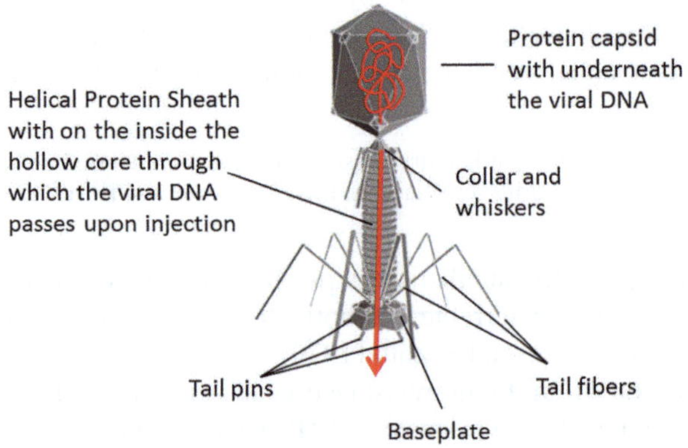

Fig. 4 Morphological structure of a bacteriophage. The phage's DNA is located inside the head of the virus. When the phage infects a bacterium, only the DNA becomes injected into the bacterium. The DNA travels through the helical protein sheath and passes through the bacterial membrane, while the remaining protein-based structure of the phage remains outside the bacterial cell

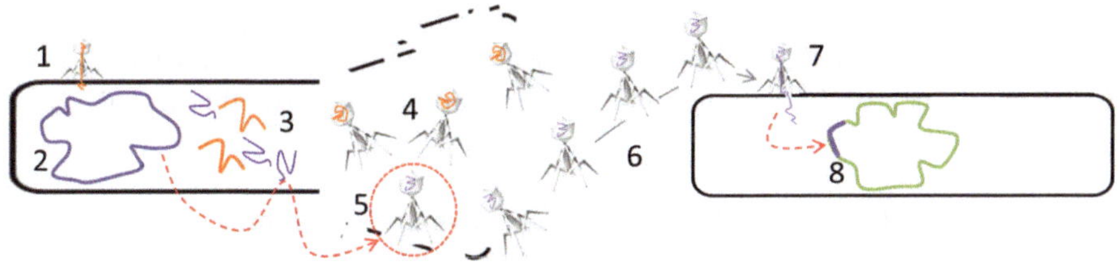

Fig. 5 General transduction or lateral transfer of bacterial genes via bacteriophages. *1* A bacteriophage injects its genes into a bacterium; *2* the bacterial chromosome; *3* phage reproduction causes the bacterial chromosome to disintegrate (lytic phase); *4* the bacterial cell dies, and new virulent phages are released; *5* one phage accidentally packaged bacterial genes instead of phage genes; *6–7* this defective phage (the transducing phage) travels to another bacterium and injects the acquired bacterial genes; *8* the bacterial genes become integrated into the genome of the recipient bacterium

When the bacteriophage's genes enter a bacterial cell, either they can take on a dormant, non-virulent "prophage" phase (a phase called the lysogenic cycle), or they can make use of the bacterial metabolism to synthesize new phage particles (a phase called the lytic cycle) (Lwoff 1953).

As avirulent, latent prophages, either they integrate into the bacterial genome, or they can reside inside the cell's cytoplasm as integrons or as circular DNA that resembles plasmids. When bacteria carry prophages, they are called lysogenic bacteria. When the phage remains virulent, or during the transition from a latent and lysogenic to a virulent and lytic phase, new virus particles are produced and the bacterial cell undergoes lysis: The cell bursts, and the new viral particles are released into the environment where they can infect surrounding bacterial cells.

Bacterial transduction (Fig. 5) is a form of HGT of bacterial genes that is mediated by bacteriophages (Zinder and Lederberg 1952; Zinder 1992). During the production of new bacteriophages, a bacteriophage accidentally packages bacterial genes instead of bacteriophage genes. When these transducing phages (the with bacterial genes infected phages) are released, they can infect other bacteria, and transfer the aquired bacterial genes. Upon infection with a transducing phage, the bacterial DNA from the donor is incorporated into the bacterial DNA of the recipient via homologous recombination.

In short, during general transduction, bacteriophages accidentally serve as vectors for the transportation and transmission of bacterial DNA. In genetic engineering, phages are one of the preferred means to introduce foreign DNA elements into cells.

Scholars distinguish "generalized transduction," where any gene can become transferred between bacteria via phages, from "specialized transduction" (Griffiths et al. 2000). During specialized transduction, the bacterial genes that are packaged and transferred are specific, and also the location where the transducing genes insert inside the recipient bacterial genome is highly site-specific. An example of specialized transduction involves the transduction of galactose-fermenting genes by the lambda phage in *E. coli* bacteria (Morse et al. 1956). Lambda phages always insert themselves on a specific site of the bacterial chromosome of *E. coli*, a site that lies right next to the bacterium's galactose-fermenting genes. When the phage transitions to the lytic phase, it packages these specific bacterial genes, which subsequently become the subject of transduction toward other *E. coli* genes.

3.4.1 Bacteriophages and Prophages: Parasites or Symbionts?

Understanding the mechanism of phage-mediated transfer of bacterial genes was greatly facilitated by increased knowledge on bacteriophages (Twort 1915; d'Herelle 1917–1922) as well as the lysogenic and lytic cycles of bacteriophages which first became described by Lwoff and Gutmann in 1950. In this section, we briefly review these advances.

By the end of the nineteenth century, microscopic studies enabled the visualization of bacteria. Bacteria became recognized as agents of disease, and experimental studies enabled their isolation which in turn facilitated their examination. But scholars such as Iwanowski (1891) and Beijerinck (1898) became convinced that besides bacteria, there must exist other pathogenic agents, at the time invisible to the observers. Both scholars came to this knowledge by studying the tobacco mosaic disease, a disease in tobacco plants that, we now know, is caused by the tobacco mosaic virus. Techniques that usually enabled the isolation of bacteria were ineffective in isolating the pathogen that causes the tobacco mosaic disease, and both Iwanowski and Beijerinck therefore speculated about the existence of another contagious substance. Beijerinck called this "contagium vivum fluid" or "living contagious liquid" a virus.

By 1915, Twort discovered that bacteria can be killed or "lysed," and he was able to filter the agents responsible and to experiment with them. In 1917, Felix d'Herelle called the "bacterial parasites" bacteriophages and held them responsible for bacterial cell death.

The existence of bacteriophages in turn induced polemic debates between various scholars about where bacteriophages stem from: Are they naturally part of bacteria? That is, are they part of the genetic endowment of bacteria and are they passed on to future generations? Do bacteria spontaneously "grow" bacteriophages? Or are bacteriophages exogenous structures that infect their bacterial hosts? Another intensely debated subject was whether or not bacteriophages are always "parasitic" and "pathogenic" (d'Herelle 1917), leading to bacterial cell death, or whether bacteriophages could entertain a more mutualistic and commensal "symbiotic" relationship with their host (Burnet 1934), where they switch to a non-virulent form that enables them to entertain a more harmonious relationship.

The debates were settled by Lwoff and Gutmann in 1950, who showed that bacteriophages are the result of an initial infection, but the infectious agents can entertain both virulent and thus parasitic and non-virulent, symbiotic relations with their host. Lwoff and Gutmann (1950) cultured 19 generations of with phage-infected *Bacillus megaterium* bacteria, and not once were they able to isolate free bacteriophages from their cultures which proved the existence of "vegetative" phages.

These non-active and non-virulent phages were called "prophages," and the scholars also distinguished between the lysogenic and lytic, virulent phase. The lytic phase always leads to the death of the bacterial host, while lysogenic bacteria (bacteria that harbor a prophage) transmit the prophage during division. In other words, "Lysogeny is the *hereditary power* to produce bacteriophage. A lysogenic bacterium is a bacterium possessing and transmitting the power to produce bacteriophage" (Lwoff 1953: 271, my italics, original italic deleted) and a "Prophage is the form in which lysogenic bacteria perpetuate the power to produce phage. Its multiplication is correlated with bacterial reproduction. It seems to be located at a specific site of a bacterial chromosome and to behave in crosses as a bacterial gene" (Lwoff 1953: 272, original italic deleted).

Lwoff and Gutmann came to their discoveries at the Pasteur Institute in France, and at this institute, two of their colleagues, Jacques Monod and François Jacob, were discovering the gene regulatory network that underlies galactose fermentation in *E. coli*, the first ever to become described.

Phage-mediated transduction was first observed in *Salmonella* bacteria by Zinder and Lederberg (1952). This is two years after Lwoff and Gutmann (1950) described the lysogenic and lytic phases of bacteriophages, and the same year wherein Hershey and Chase (1952) conducted their experiments. They demonstrated that bacteriophages carry their hereditary material, DNA, inside their head, and that this substance is injected upon infection, while the protein capsid is left behind at the surface. The DNA helix would be described a year later, and genetic material was still assumed to reside both inside and outside the chromosomes of organisms.

Zinder and Lederberg had turned to *Salmonella typhimurium* bacteria that are closely related to *E. coli*, with the purpose to investigate whether also *Salmonella* bacteria engage in bacterial conjugation. They did not find evidence for such conjugation, where multiple traits are transferred at once, but found that single traits can become transferred via a bacteriophage known by the name P22 (Lederberg 1956: 271). Around 30 different traits could be transferred via phages between *Salmonella* bacteria. These traits included resistance to antibiotics (Zinder 1955), flagellar antigens (Sakai and Iseki 1954), and genes involved in fermentation and sugar metabolism (Jacob 1955).

Lederberg and coworkers knew that bacteriophages were involved in the transmission of bacterial genes, as "passive" "vehicles" (Lederberg 1952: 37), but the exact chemical nature of bacteriophages remained obscure. Filtration and sedimentation experiments demonstrated that the substance was below 0.1 micron and it was therefore still beneath microscopic visibility of their time.

Lwoff's distinction between the lysogenic and lytic phases made Lederberg understand the division as being one between "lysogenic symbiosis" and "lethal parasites" (Lederberg 1956: 272). Lysogenic bacteria entertain "stable symbiotic associations" with their prophages (Lederberg 1952: 419), and because prophages can become transmitted to progeny during cell division, he understood it as an example of "hereditary symbiosis" or "infectious heredity."

Lwoff, on the other hand, preferred a more parasitic jargon. In 1965, he shared the Nobel Prize in Physiology or Medicine with François Jacob and Jacques Monod, and in his Nobel lecture, Lwoff used "master/slave" conceptualizations to characterize viruses as "intracellular parasites" that favor "war" over "peaceful coexistence":

> The virus is necessarily an *intracellular parasite*. The genetic material of a virus has thus entered the cell. The cellular and viral molecules will confront each other, and the fate of the two partners will be decided. Two extreme cases may present themselves. Either the virus will multiply in the cell or else the cell will *enslave* the virus. Quite naturally, investigation was first directed toward the *total war*, which offers greater attraction for the combative intellect than *peaceful coexistence*. (Lwoff 1965, my italics)

Lwoff furthermore noted that "the development of the prophage into bacteriophage is a "mortal disease" because the "prophage is a potentially lethal factor." As already noted in the introduction to this volume, parasitic jargon is nonetheless symbiotic jargon.

It should also be noted that not all bacteriophages can take on a dormant form (Guttman et al. 2002, Chap. 10). Phages such as T2 and T4 are always virulent, and infection always leads to the death of the infected bacteria. Not all phages integrate into the bacterial genome either. Some, such as the P1 phage (Lennox 1955), remain part of the cell cytoplasm, where they take on a circular morphological form that resembles a plasmid or an integron. Alternatively, they integrate into the bacterial chromosome. In both cases (as plasmid or integron, or as an integrated part in the bacterial genome), the prophage can become transmitted vertically over future generations.

Transitions from lysogenic to lytic phases are a relatively rare process: Release of bacteriophages happens in round and about one cell per million in a normal culture (Lederberg 1956: 274) and appears to increase under stress conditions, such as the exposure to ultraviolet light (Jacob 1955). The integration of one prophage also protects the bacterium from becoming infected with other bacteriophages.

3.4.2 Transduction and Evo-Devo

Studies on transduction also contributed to the operon concept and the recognition of the existence of gene regulatory networks. The first gene regulatory network was described by François Jacob and Jacques Monod in what regards the genotype-to-phenotype mappings of lactose breakdown or "galactose fermentation" in *E. coli* (Summers 2006). Today, we know that a distinction can be made between "structural" and "regulatory" genes (Jacob et al. 1960: 30, 31). Structural genes are genes that encode for specific traits, while regulatory genes are genes that can orchestrate the activation and deactivation of structural genes. Regulatory genes encode for proteins that return to the helix where they turn other genes on or off, thereby underlying overall ontogenetic development of bodily form and function.

That a difference exists between regulatory and structural genes was first suggested by Barbara McClintock, in the context of discoveries of transposable or "jumping" genes. And these assumptions were confirmed by Jacques Monod and François Jacob in the 1950s–1960s whom were the first to track how genes encode for metabolism. This first "genotype-to-phenotype" map explains how *E. coli* metabolizes lactose (milk sugar).

E. coli is a bacterium that naturally occupies the gut of mammals, and all newborn mammals in turn depend upon milk for successful survival. *E. coli* have evolved intricate coevolutionary relations with the mammals they inhabit. *E. coli* possesses enzymes that enable them to digest and metabolize lactose (the beta-galactosidase enzyme), and use of their genetically programmed lactose-metabolizing apparatus, i.e., galactose fermentation, is regulated. Research has shown that in an artificially induced, lactose-low environment, *E. coli* do not produce the beta-galactosidase enzyme, while exposure to lactose induces a rapid production of this enzyme. Within 3–5 min, the enzyme is produced about a thousand times faster. When the lactose is removed, the production comes to a halt (Guttman et al. 2002; Lodish et al.2000; Weickert and Adhya 1993).

In 1951, Esther Lederberg discovered that *E. coli* can become infected with a bacteriophage that she called *lambda*. Research on *lambda* provided the definite proof that viral DNA can become integrated into the bacterial genome as a prophage and the first proof that the prophage is transmitted together with the bacterial genome to offspring (Lederberg et al. 1951; Lederberg and Lederberg 1953). A couple of years later, Morse et al. (1956) demonstrated that transduction in *E. coli* occurs via the lambda phage. This type of transduction is called specialized transduction because the lambda phage transduces the specific cluster of bacterial genes involved in galactose fermentation, and it does so only among specific members of the *E coli K*-12 strain.

When F+ *E. coli* (*E. coli* bacteria that carry the F factor that enables bacterial conjugation) are crossed with F− *E. coli* that carry the lambda prophage, the cross always leads to lysogenic recipients, but transduction and recombination of the lambda phage are hardly ever induced when F+ (lambda) are crossed with F− *E. coli* strains (Griffiths et al. 2000). Furthermore, the integration of lambda as a prophage into the bacterial chromosome of *E. coli* always occurs at a specific region, in between *E. coli's gal* genes involved in the galactose fermentation process and the *bio* genes responsible for the synthesis of biotin vitamin. Integration of lambda into *E. coli* therefore became a medium by which Jacob and Monod could study the expression and regulation of *gal* genes.

3.5 Mobile Genetic Elements: Gene Transfer Agents, Transposons, Retrotransposons, and (Endogenous) Retroviruses

Eukaryotic genomes contain noncoding DNA sequences (genes that do not encode for functional proteins) that include repetitive elements such as terminal repeats (that include satellite DNA), tandem repeats, and interspersed repeats. Around 98 % of the human genome, for example, exists out of noncoding DNA, and about 2/3 of our genome consists of repetitive elements (de Koning et al. 2011). Especially the interspersed repeats contain (remnants of) mobile genetic elements, such as transposons and retrotransposons. Lateral exchange of mobile DNA might be the reason why there is so much repetition in organismal genomes. Bacteria also contain repetitive DNA (in the form of Insertion Sequences, Gene Transfer Agents and prophages) but relatively little non-coding DNA in comparison to the amounts found in eukaryotic genomes (Gil and Amparo 2012). Prophages, bacterial Insertion Sequences, and Gene Transfer Agents are also classified as mobile genetic elements.

Transposons were originally named "jumping genes" by McClintock (1950, 1953), and these "transposable elements" were first classified by Campbell et al. (1977). The concept of "mobile genetic elements" was first introduced in a book edited by Shapiro (1983) that reviewed the various means whereby genetic material can become "transposed" or relocated within and between genomes. Two other seminal review works were edited by Berg and Howe (1989) and Craig et al. (2002), and today, academic journals exist that exclusively dedicate themselves to the study of mobile genetic elements such as *Mobile DNA* and the Journal of *Mobile Genetic Elements*.

Plasmids, Gene Transfer Agents (GTAs), ribozymes (group I and II introns), and (bacterial) viruses are all examples of mobile genetic elements, and besides the latter, scholars distinguish between transposons, retrotransposons, and (endogenous) retroviruses. These are distinctive classes of genetic elements that can become the subject of "transposition" or relocation and lateral exchange within and between genomes. Scientists are currently mapping these various mobile genetic elements in order to find recurring structures, elements, patterns, and mechanisms whereby these elements are transmitted. These efforts are designated as the *mobilome* projects (Frost et al. 2005; Siefert 2009).

3.5.1 Gene Transfer Agents

Several bacteria contain Gene Transfer Agents (GTAs) in their genome. GTAs are bacteriophage-like elements that can become horizontally exchanged (Lang and Beatty 2000; Lang et al. 2012; Stanto 2007; Yen et al. 1979). GTAs present a fourth form of HGT among bacteria that works similar to transduction. The main difference with bacteriophage-induced transduction is that the GTAs appear to be defective prophages that are part of the bacterial genome. Nonetheless, upon lysis they can move location and become inserted into the genomes of recipient bacteria.

3.5.2 Transposable Elements

Transposons can move to another location within that same genome, or in prokaryotes, they can travel horizontally from the bacterial genome to a bacterial plasmid or vice versa, and in eukaryotes, they can travel from organellar DNA to nuclear DNA and vice versa. When transposable elements move location, it implies a form of HGT because transposition does not require division of the cell or replication of the entire genome. Jumping genes were first identified by Barbara McClintock (1942, 1950: 344–5) in chromosome 9 of maize. She called them "mutable" and "instable" genes that underlie "variegation" and "mosaicism." The position switches of jumping genes on the chromosome alter functionality of the genome

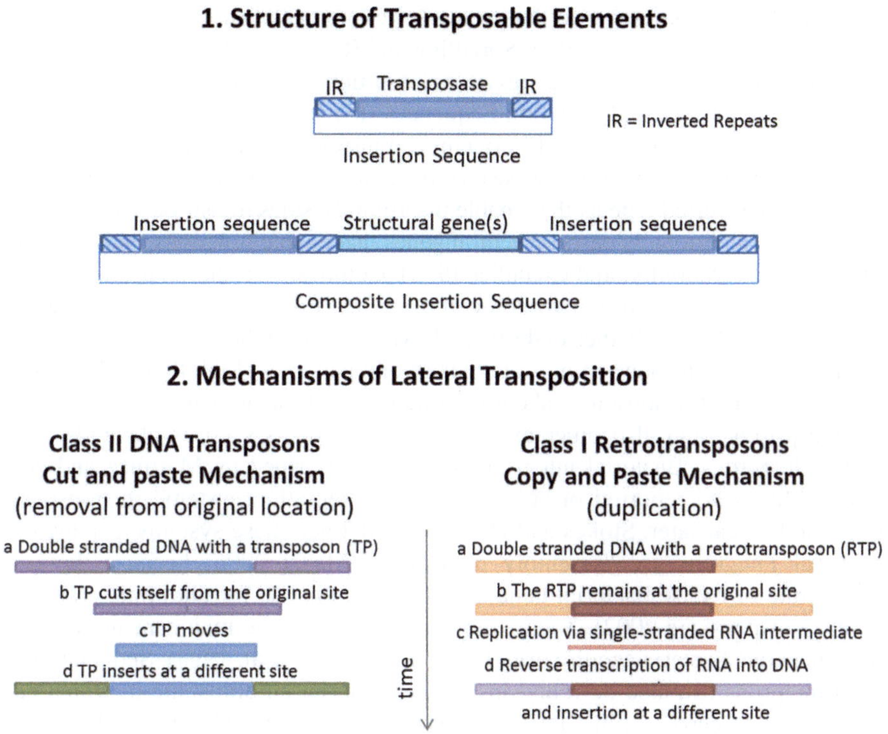

Fig. 6 *1* Schematic of simple and composite insertion sequences. *2* Schematic of the two main modes of transposition by which DNA transposons and retrotransposons can switch position inside or between genomes

by providing a loss or introduction of new traits. More specifically, "spontaneous translocations" can lead to deletion, duplication, loss, or introduction of functions.

There exist two main classes of transposable elements: *DNA transposons* and *retrotransposons or retroelements*. DNA transposons are found in both pro- and eukaryotes, while retrotransposons are mostly found in eukaryotes (but for some exceptions see Boeke 2003). As the name implies, DNA transposons (class II) are always made up of DNA and they *cut and paste* themselves into genetic sequences (Finnegan 1989). Transposition thus involves a complete removal of the transposon at the original site, followed by insertion at a new site (Fig. 6(2)). When the transposition occurs within the same genome, it can lead to a loss of function at the original site as well as a loss of function at the new site, especially when insertion interrupts an existing gene sequence. When transposition occurs from the genome to a bacterial plasmid or an organellar genome, the original genome is reduced in size. Transposable elements therefore have the potential to affect functional metabolism, and they can also interfere with successful survival, reproduction, and evolution. When P elements, for example, which are transposons found in *Drosophila*, are transmitted from male fruit flies to females that lack them, their offspring becomes sterile due to the large number of mutations such a cross induces (Kidwell and Novy 1979; Spradling and Rubin 1982).

Structurally, transposons can exist as insertion elements, first identified by Bukhari et al. (1977), and Insertion Sequences (IS) can be non-composite or composite (Berg and Howe 1989). IS contain a transposase-encoding region (i.e., the genes that encode for the transposase enzyme that enables the cutting), a region that is flanked by inverted repeats that enable insertion. Besides transposase genes, compositional IS additionally carry structural genes that enable functions such as antibiotic resistance (Mahillon and Chandler 1998). Composite IS that transfer structural genes are always flanked by two non-composite IS (Fig. 6(1)). Insertion sequences were first classified by Esther Lederberg (1981), and since the 1980s, numerous different IS have been found in pro- and eukaryotic genomes (Mahillon and Chandler 1998). However, transposons make up less than 2 % of the human genome.

Transposons, and also other mobile genetic elements such as plasmids, often contain integrons. Bacterial integrons were first described by Martinez and de la Cruz (1988) who reported on a "site-specific integration system" in transposon Tn21, and a year later, Stokes and Hall (1989) defined these systems as integrons. Bacterial integrons are "assembly platforms" (Mazel 2006) because they can incorporate or excise bacterial gene cassettes that encode for antibiotic resistance genes (Kovalevskaia 2002). Today, they are considered the major means whereby gram-negative bacteria acquire drug resistance (Barker et al. 1994). Sedentary bacterial integrons can also be found inside the bacterial chromosome in which case they are called chromosomal integrons (Mazel et al. 1998). These were first found in the gram-negative *Vibrio cholerae,* the bacterium that causes cholera, and their integron was called a "superintegron" because of its large size (Hall and Stokes 2004). There are over 50 known gene cassettes in gram-negative bacteria and five distinct classes of integrons (Barker et al. 1994; Kovalevskaia 2002; Hall 1997).

Retrotransposons (class I) are alternatively known as *transposons* via *RNA intermediates*, because they move about by *copying and inserting* themselves via RNA intermediates (Fig. 6(2)). The retrotransposon is not removed from the original site. Instead, it is transcribed into an RNA intermediate that subsequently moves to the new location where the RNA strand is reverse-transcribed into complementary DNA and inserted into the new site. Thus, after retrotransposition, the retrotransposable element is present in both the original site and the new site.

While DNA transposons leave gaps at the places where they cut themselves, leading to reduction in genome size and interruption of gene sequences at the insertion sites, retrotransposons enable genome growth by duplication of gene sequences.

Eukaryotic organisms contain many such gene duplications, and especially plants have huge genomes because of gene duplications. Maize genomes, for example, contain 50–70 % of retrotransposons, and the human genome is made up of 42 % of retrotransposons.

Both DNA transposons and transposons via RNA intermediates enable "genetic transformation" (Spradling and Rubin 1982) of the organismal genome they belong to. In other words, they change the genetic makeup of organisms and are therefore key players in evolution.

Retrotransposons can be further divided into LTRs, LINEs, and SINEs, a division that has to do with the length and position of the repetitive sequence. LTRs are longer than SINEs and LINEs. LTRs are Retrotransposons with Long Terminal Repeats, and they were first defined by Shine et al. (1977) in the avian sarcoma virus. LINEs are Long Interspersed Nuclear Elements first identified by Adams et al. (1980) in humans, and SINEs are Short Interspersed Nuclear Elements. The distinction between LINEs and SINEs was first made by Singer (1982). Both LTRs and LINEs use reverse transcriptase, while SINEs use RNA polymerase III for transcription and the latter make use of the more autonomous LINEs (Weiner 2002). Especially mammalian genomes, including human genomes, contain many retrotransposable LINE and SINE elements (Deininger and Batzer 2002). Some SINEs today are also classified as miniature inverted-repeat transposable elements (MITEs) (Bardaji et al. 2011; Zhang et al. 2000), which are especially numerous in flowering plants, but they are also found in insects and animals.

Especially retrotransposons violate the standard idea that information flows from DNA to RNA to proteins but not the other way round. This is because they are able to use RNA to assemble DNA. "Replicative transposition" was first described by Shapiro (1979), and the intermediate RNA is therefore sometimes called the "Shapiro intermediate."

3.5.3 Retroviruses

Several scholars (Wicker et al. 2007; Flavell 1981; Nelson and Hooley 2004; Ryan 2009; Temin 1980) classify retroviruses as a type of retrotransposons, namely LTRs, because retroviruses can be transformed into an LTR retrotransposon.

Retroviruses can integrate into the host's genome and become permanently transmitted to offspring. Human endogenous retroviruses are thought to make up 8 % of the human genome (Cotton 2001; Nelson and Hooley 2004; Khodosevich et al. 2002; Belshaw et al. 2004).

3.5.4 Mobile Genetic Elements and the Extended Synthesis

It is important to note that GTAs, DNA transposons, integrons, retrotransposons, and retroviruses are container terms for numerous families. At present, scholars are in the process of identifying (the members of) these families, and they are studying the biomolecular similarities, the various taxa they belong to, and the various taxa they can transfer to in order to gain insight into the underlying mechanisms and patterns. Besides these large classes of mobile genetic elements, there are still other kinds of mobile DNA and means by which genes can become exchanged between biological individuals, including exchange via intracellular nanotubes (Dubey and Ben-Yehuda 2011) and via release of DNA-containing membrane vesicles (Mashburn-Warren and Whiteley 2006; Chiura et al. 2011). At the same time, the scholars are involved in finding similarities and differences between the various identified transposable elements, in order to develop new classification systems (Wicker et al. 2007). These endeavors form part of the "mobilome projects" (Frost et al. 2005). Similarities in gene frequencies and genetic organization, enzymes that mediate transposition, similar terminal repeats, functional differences of the inserted transposable elements (whether they result in genome growth, mutation, or new features), and the particular taxa wherein they are found can all function to group the various mobile genetic elements into several classes (Mahillon and Chandler 1998).

These various mobile DNA families and their members are currently listed in DNA databases and DNA banks such as ENCODE (the ENCyclopedia Of DNA Elements, The ENCODE Project Consortium 2012, and for the official Web site, see http://www.genome.gov/encode/), but to my knowledge, no scholar has so far tried to visualize these elements into taxonomies or all-encompassing mobile DNA networks or trees that map onto the current networks and trees of life.

Transposable and retrotransposable elements contradict the standard neo-Darwinian view in many ways. For one, as McClintock (1941, 1950) already noted, during mitotic cell regeneration (development), genes can move position, leading to chromosomal breakage or deviant but stable chromosomal rearrangements in future somatic cells. She realized that the process is not unique to maize, rather:

> The same types of genetic instability appearing in the maize cultures have been described in many other organisms. The behavior of these new mutable loci in maize cannot be considered peculiar to this organism. The author believes that the mechanism underlying the phenomenon of variegation is basically the same in all organisms. (McClintock 1950: 345)

Chromosomal rearrangements can also affect the gametic cells, and when this happens, the chromosomal rearrangements can become permanently part of the progeny's genetic makeup and affect their successful survival, reproduction, and evolution. Chromosomal rearrangements were recognized by the founders of the Modern Synthesis to be one way by which genetic variation could be brought about, but for McClintock, it also implied that the location of genes on chromosomes is not fixed. Rather, genes can translocate at any moment in time, thereby significantly and rapidly impacting both ontogeny and phylogeny. McClintock also associated her research with macroevolutionary biology. In the early 1930s, she visited Goldschmidt (1940) in Germany who introduced the concepts of "macromutations" and "hopeful monsters," and jumping genes can be understood as responsible for both.

Secondly, for McClintock, translocation required an "activation" of the genes involved. That genes need to become activated implies they can be silent, and both imply differential gene expression and functionality in different periods in time. As such, she envisioned the existence of gene regulatory networks, where transposable elements function as "novel biological switches" (Cohen 1976), processes that today are studied from within evo-devo schools. McClintock (1950: 354):

> changes in quantity, quality or structural organization of heterochromatic elements may well alter the kind and/or degree of particular exchanges that occur, and in this way control the chromosome organization and the kind and the relative effectiveness of genic action. There can be little question that transpositions … occur and that the time of their occurrence in the development of a tissue is under precise control. This control is determined by the number of AC [activator] loci present and their organization and possibly their position in the chromosome complement.

Thirdly, besides to evo-devo, transposition links to epigenetics. Mobile genetic elements form part of "non-encoding" gene regions. In other words, they do not encode or translate into proteins that underlie functional anatomy and metabolism of the organism. Nonetheless, mobile DNA elements impact overall genome size and genome organization, and they are known to transcribe into RNA. As such, mobile DNA can rearrange and reshape existing genomes by enhancing, silencing, or promoting gene regulatory networks, and they have been implicated in various diseases. Reverse transcription, typical of retrotransposons, is a prototypical example of "epigenetic change" defined as post-translational or non-translational changes induced in the genetic code. Of course, many (retro-) transposable elements carry the genes required for their own copy/cut and pasting, but such genetic information is autonomous of the functional parts of the genome that underlies bodily form and functional metabolism. More specifically, mobile genetic elements underlie post-transcriptional RNA silencing/RNA interference/ quelling (Fire et al. 1998), as well as chromatin remodeling and DNA methylation. These are ways by which the genetic code can become permanently modified during ontogeny. Such ontogenetic modifications can become stable, and as such, they can become transmitted to future generations. The fact that many mobile genetic elements induce such changes implies that these epigenetic traits themselves are interchangeable and thus mobile between organisms (for a discussion, see Galun 2003). Transposons can be linked to pathogenicity and genome reduction, while retrotransposons potentially introduce genetic variation and innovation. Because transposons play an important part in genome size, they can causally influence evolution, by interferring with chromosomal compatibility.

Fourthly, because (retro-) transposable elements can have developmental and evolutionary functionality, it makes scholars question the existence of "Junk DNA" (Ohno 1972). Mobile genetic elements form part of the noncoding regions of DNA, and these noncoding regions were first designated as "Junk DNA" by Ohno (1972). Ohno is one of the molecular geneticists that theorized that gene duplications can be understood as "mutations" that contribute creatively to the evolutionary process (Ohno 1970). The idea that there exists "Junk DNA" in turn made scholars like Dawkins (1976) introduce the notion of "selfish replicators," themes that became repeated in exo- and astrobiological research schools as well as in research on transposable elements. In an April 1980 issue of the journal *Nature*, for example, Doolittle and Sapienza (1980) characterized transposable elements in particular and Junk DNA in general as "selfish genes," and in that same issue, Orgel and Crick characterized selfish DNA as the "ultimate parasite." Since then, many potential functions have been attributed to these "Junk" regions. According to Gregory (2007):

> Those who complain about a supposed unilateral neglect of potential functions for non-coding DNA simply have been reading the wrong literature. In fact, quite a lengthy list of proposed functions for non-coding DNA could be compiled …. Examples include buffering against mutations … or retroviruses … or fluctuations in intracellular solute concentrations …, serving as binding sites for regulatory molecules …, facilitating recombination …, inhibiting recombination …, influencing gene expression …, increasing evolutionary flexibility …, maintaining chromosome structure and behavior …, coordinating genome function …, and providing multiple copies of genes to be recruited when needed… .

According to the scholars involved in the ENCODE project, over 70 % of the noncoding DNA regions found in the human genome are transcribed and targeted by transcription factors of regulatory genes. In other words, according to these scholars, though they are not involved in the formation of anatomical form, there exists biochemical functionality (for a discussion, see Kellis et al. 2014).

Fifthly, because organisms can do without many of the inserted elements, because they are exchanged between various organisms, and because the acquisition of transposable elements has the potential to cause evolutionary innovation, several scholars understand especially LINEs and SINEs (Oshima and Okada (2005), as well as retroviruses and endogenous retroviruses (or LTRs) as symbionts (Ryan 2009; Villarreal 2008).

4 Conclusion

While HGT has only been recognized from the 1990s onward, the historical origin of lateral gene transfer mechanisms tracks back to the early twentieth century and converges with ongoing research on intracellular and hereditary symbiosis as well as the advent of the biomedical sciences and research on abiogenesis. For years now, vaccination therapies have relied on artificially induced symbiosis whereby foreign cells are brought together to create immunity. Any and all therapeutic cures proposed by medicinal doctors to treat disease, be they the abstraction and administration of antibiotics, or the introduction of chemotherapeutic agents, is always based upon horizontal exchange of substances during ontogeny. These therapies provide hope to eradicate disease, at minimum during the individual's life time, and preferably during all future generations to come (which, for example, was successfully induced with the worldwide eradication of the smallpox virus).

The discoveries of bacterial transformation, phage-mediated transduction, bacterial conjugation, and mobile genetic elements furthermore converge with the major milestones of neo-Darwinian theory. Artificially induced HGT of phages into bacterial cells, for example, has been the major tool whereby scholars have performed genetic linkage studies whereby they have come to learn how genes underlie metabolism and how metabolism underlies anatomical form. Yet, symbiogenesis and HGT, though known for so long and foundational for some of the major neo-Darwinian and biomedical claims, have hardly made it into the standard textbooks, which brings to light an almost schizophrenic ambivalence.

Why have these data served to proof neo-Darwinian claims, but why have the incoming data and applications of transformation, transduction, and conjugation not themselves been considered as instances of evolution? One of the major stumbling blocks has been the neo-Darwinian divide between ontogeny and phylogeny. Another is the neo-Darwinian demand to understand evolution as a familial pedigree, where parental species give evolutionary existence to daughter species. In hindsight, this might have been too anthropo- or animal centric. In recent years, scholars have been trying to map non-genealogical evolutionary relations between various organisms and between various species. The acquisition of foreign DNA through lateral gene transfer undoes the rigid demand to think about evolution in terms of common descent with modification. Instead, modification of species, and exchange of foreign DNA, can occur between distantly and non-related species. Exchange can even occur between living organisms and viral genetic agents which have traditionally been conceived as non-living beings.

Evolutionary theory no longer exclusively encompasses the study of biogenesis, or how existing life brings forth new life. The recent pleas to include the abiotic environment into evolutionary iconography, and to cartography the various metabolic pathways that have evolved, equally open up new and inspiring ways to think about how evolution brings forth patterns that extend the "germ line."

The biggest "take home" message of HGT research, however, is that genetic material is anything but a "frozen accident." After the double-helix was discovered, Crick (1968), and following physicists such as Erwin Schrödinger, wrote a very famous paper wherein he characterized DNA as such. DNA was "frozen" because it was assumed to be a very rigid structure that only occasionally underwent genetic modification by mutations which were merely understood as random copying errors. The specific structures of the DNA base pairs and the rigid translation machinery whereby codons are translated into amino acids made scholars assume that the genetic code was a "frozen" structure. The idea that the specific structure of DNA was an "accident" resulted from theories as they were introduced by Manfred Eigen, who characterized the formation of the double helix as the result of hypercyclic organization which he characterized as a "once forever" event (for a discussion, see Gontier 2005).

Besides genes, the genome was furthermore argued to consist of round and about 75 % of "Junk DNA" that served no purpose. Rather it was just the "fallout," as it were, the remains of the formation of the "frozen accident" that, because of the presumed rigid DNA-structure, could not become eradicated. At best, they served "selfish DNA" theory, where the Junk DNA was argued to have no other purpose than to selfishly replicate itself inside the vehicles it rides. When Dawkins (1976) introduced his selfish gene theory, he was battling Ernst Mayr's idea of the "unity of the genotype." He was a true visionary when he, as one of the first, started to defragment the genome into multiple replicators that each can be studied in and of itself. However, such defragmentation has brought to light that even the smallest DNA fragments are anything but passive elements waiting to become replicated.

Instead, our genome consists of numerous genetic elements that portray an enormous flexibility and mobility. Today, we know that these regions, formerly designated as Junk, contain these fascinating mobile genetic elements that include the viral particles, transposons and retrotransposons, integrons, etc. The means by which genes can be moved and the means whereby they cut and copy themselves within the genome and across genomes demonstrate a remarkable form of flexibility. How much of this flexibility is non-accidental remains an open question. But it is becoming more and more evident that DNA not merely "moves" out of selfish "needs" for propagation. Many mutualistic benefits can be identified to result from HGT, including DNA repair, genome growth, and acquisition of novel functions.

This in turn raises interesting questions on biochemical communication. The mobile elements possess the locks and keys to the cells and genomes they exit or enter, and our bodies equally possess our immunizing recognition systems that demonstrate an intricate coevolutionary and symbiotically evolved biochemical communication system.

HGT also demonstrates that such genetic flexibility comes at an incredibly fast pace. When antibiotics became administered, in less then 10 years did the first instances of resistance and horizontal spread of resistance become reported. Chromosomal breakages, epigenetic changes, and gene loss or acquisition due to transposition can at once, and even within an individual's life span, alter their genetic endowment.

Many of these ontogenetically acquired changes can furthermore be passed on to future generations, and besides natural selection, we need to bring in symbiogenesis theory, evolutionary developmental theory, and punctuated equilibria theory to try and make sense of it all. Gould has long critiqued the Modern Synthesis for its adherence to only one evolutionary mechanism. Darwin was a pluralist, and punctuated equilibria, for example, already demonstrated that abiotic factors can have a causal influence on the further evolution of life, as does development. Gould therefore pleaded for process pluralism, and in recent years, Bapteste and Doolittle have added the requirement of pattern pluralism. The study of reticulate evolution brings to light a new evolutionary pattern whereby life evolves, a pattern that takes on the shape of a web or net of life. To make sense of both this pattern and process pluralism, however, we also need epistemic pluralism. We need to find a way to extend the evolutionary synthesis into an evolutionary framework that enables us to conceptualize the numerous evolutionary mechanisms whereby life evolves.

Acknowledgments This work was written with the support of the Portuguese Fund for Scientific Research (grant ID SFRH/BPD/89195/2012 and project number UID/FIL/00678/2013) and the John Templeton Foundation (grant ID 36288).

Timeline

1717	Mary Wortley Montagu introduces "variolation," an immunizing technique against smallpox (Variola)
1798	Edward Jenner injects cowpox as an immunizing technique against smallpox (Variola). His work avalanches a series of inoculation experiments that underlie vaccination therapy
1809	Jean Baptiste Chevalier de Lamarck publishes his *Philosophie Zoologique*
1817–1828	The fields of embryology, epigenetics, and evo-devo take off with the works of Heinz Christian Pander, Karl Ernst Ritter von Baer, and Edler von Huthorn
1837/1838	Darwin draws the "I think" diagram in his Notebook B
1838	Matthias Schleiden contends that all plants are made up of cells
1839	Theodor Schwan declares that all animals are made up of cells
1848	Wilhelm Hofmeister describes mitosis
1855	Rudolf Virchow declares that all cells come from pre-existing cells

1859	Darwin publishes his *Origin of Species* and uses the "tree of life" metaphor in Chap. 4. The book contains a hypothetical branching diagram that illustrates how species gradually diverge through time by means of natural selection
1861	Heinrich Anton de Bary identifies microorganisms as the cause of plant diseases, and he later introduces the concept of symbiosis
1866	Gregor Mendel introduces his laws of inheritance
1866	Ernst Haeckel introduces the first non-hypothetical "Tree of Life"
1868	Johannes Friedrich Miesher names a substance inside the nucleus of cells "nuclein" (DNA)
1870	Thomas Henry Huxley distinguishes biogenesis from abiogenesis and denies abiogenesis (alternatively known as spontaneous generation)
1873/1875	Pierre Joseph van Beneden identifies parasitic microorganisms as the cause of animal diseases and distinguishes between commensalism, parasitism, and mutualism
1875	Ferdinand Cohn introduces a first classification of bacteria
1876	Robert Koch identifies the *Anthrax bacillus* responsible for "Milzbrand-Krankheit" (anthrax disease) and proves earlier theoretical versions of the germ theory of disease to be true
1877	Paul Ehrlich starts his career by developing new techniques to color bacteria. These techniques will enable him to specify the various types of blood cells there exist, research that will found the study of both serology and immunology
1878	Louis Pasteur's work on the germ theory of disease is read before the French Academy of Sciences
1879	Timothy Lewis identifies microorganisms inside the bloodstream of humans and links them to disease
1880	Charles Louis Alphonse Laveran identifies flagella-like motile unicellular organisms that he identifies as causal agents of malaria
1882	Ilya Ilyich Mechnikov observes what he later calls phagocytosis: cell eating. Phagocytosis is crucial to understand immunity as well as primary, secondary, and tertiary symbiosis
1884	Robert Koch publishes his etiology of tuberculosis that proves that the *tubercle bacillus* is the disease-causing agent of tuberculosis
1884	Hans Christian Joachim Gram develops the Gram stain technique that enables to differentiate between "gram-negative" and "gram-positive" bacteria
1885	Auguste Weismann develops his "transmutation hypothesis." The work is foundational for the "Weismann barrier" that puts a halt to (neo-)Lamarckian theories
1886	Theodor Escherich identifies a "bacterium coli commune" that resides in the human gut (*E. coli*)
1886	D.E. Salmon and Theobald Smith improve vaccination therapies by injecting whole heat-killed cells of virulent strains
1886	Adolf Mayer describes the tobacco mosaic disease

1888	The Pasteur Institute is founded in France
1890	The Cold Spring Harbor Laboratory is founded in Brooklyn, New York
1891	The Robert Koch Institute is founded in Germany
1891	Paul Ehrlich discovers antibodies
1892	Dmitri Iwanowski demonstrates that the tobacco mosaic disease is caused by a non-bacterial infectious agent
1898	Martinus Beijerinck defines the agent responsible for the tobacco mosaic disease as a virus which he characterizes as a "living and fluid infectious agent"
1900	Mendel's hereditary laws are (re)discovered by Hugo de Vries, Carl Correns, and Erich von Tschermak
1902/1910	Fred Neufeld classifies *Pneumococci* into three different types
1905	Constantin Mehrezkowsky introduces a double-origin theory of life, a view he illustrates with a reticulate "tree of life" in 1910
1909	Theodor Boveri and Walter Sutton introduce the chromosome theory
1909	Wilhelm Johannsen distinguishes between the genotype and phenotype
1912	Friedrich Karl von Faber introduces the notion of "*erbliche Zusammenleben*" (hereditary symbiosis)
1915	Hermann Reinheimer introduces the metaphor of the "web of life"
1915	Frederick Twort discovers bacterial lysis and assumes it is induced by viral agents that infect bacteria
1917	Félix d'Herelle cultures viruses that infect bacteria and calls them "bacteriophages"
1928	Frederick Griffith reports on bacterial transformation
1929	Alexander Fleming reports that the mold *Penicillium notatum* undertakes "antibacterial action" against gram-positive microorganisms
1931	Ernst Ruska and Max Knoll build the first electron microscopes
1932	Julius Petrie introduces serological typing
1938	Warren Weaver coins the term "molecular biology"
1941	George Beadle and Edward Tatum demonstrate that protein synthesis as well as the function of enzymes is controlled by genes and they introduce the "one gene–one enzyme theory" (a term coined by Norman Horowitz)
1942	Conrad Waddington coins the term "epigenetics"
1942	Julian Huxley characterizes the late nineteenth century as the "eclipse of Darwinism"
1943	Salvador Luria and Max Delbrück demonstrate that bacteria evolve according to Darwinian principles (they "mutate" randomly)
1944	Oswald Avery, Colin MacLeod, and Maclyn McCarthy confirm that bacteria can transform and they identify DNA as the transforming principle
1944	Barbra McClintock discovers "jumping genes," what we now call "transposons" or "mobile genetic elements" in maize

1944	Albert Schatz isolates the antibiotic streptomycin from *Streptomyces griseus* at Selman Waksman's laboratory which becomes administered against tuberculosis
1946	Joshua Lederberg and Edward Tatum report on bacterial conjugation in the *E. coli* K-12 strain
1950/1953	André Lwoff and Antoinette Gutmann distinguish between the lysogenic and lytic phases of bacteriophages and introduce the concept of "prophage"
1950	Antibiotics such as streptomycin, penicillin, and chloramphenicol are massively produced and administered
1951	Victor Freeman reports on HGT from a bacteriophage to *C. diphtheria*
1951	Esther Lederberg discovers that *E. coli* can become infected with a bacteriophage that she calls *lambda*
1952	Joshua Lederberg, Luigi-Luca Cavalli Sforza, and Esther Lederberg, and independently William Hayes, report on the Fertility factor in *E. coli* that enables bacterial conjugation
1952	Norton Zinder and Joshua Lederberg report on phage-mediated bacterial transduction
1952	Joshua Lederberg introduces the plasmid concept to designate all extrachromosomal DNA by which he intends to include mitochondrial and chloroplast DNA (still a theoretical notion) and viral prophages, and he applies the notion of "hereditary symbiosis" as well as "infective heredity" to the phenomena of bacterial transformation, phage-mediated transduction, and bacterial conjugation
1952	Alfred Hershey and Martha Chase perform the Hershey–Chase experiments with bacteriophage T2 and the *E. coli* bacterium and confirm that DNA, and not proteins, carries hereditary information
1953	X-ray crystallography of DNA performed by Rosalind Franklin leads Francis Crick, Maurice Wilkins, and James Watson to describe the double-helical structure of DNA
1955	Norton Zinder demonstrates transduction of antibiotic resistance genes
1959–1963	Japanese scholars Kunitaro Ochia, Tomoichiro Akiba, and Tsutomu Watanabe report on bacteria that have acquired resistance genes against antibiotics in natural settings and identify bacterial conjugation as the likely mode of transfer
1959	Arthur Pardee, François Jacob, and Jacques Monod (1959) publish the "PaJaMo" paper that demonstrates protein regulation of gene expression, or gene regulatory networks. They base their work on their studies of galactosidase fermentation in *E. coli*
1963	Linus Pauling and Émile Zuckerkandl map the changes in hemoglobin polypeptide chains of different mammalian species and find

	that "semantides": "DNA, RNA, and polypeptides" lend themselves for comparative bimolecular analysis
1969/1970	Stanley Cohen, Annie Chang, and Leslie Hsu demonstrate that *E. coli* can take up plasmids carrying antibiotic resistance genes (R factors)
1970	Howard Temin and S. Mizutani, and independently David Baltimore, discover reverse transcriptase
1972	Susumo Ohno introduces the concept of "Junk DNA"
1976	Richard Dawkins introduces the "selfish gene" theory
1977	Frederic Sanger sequences the first entire genome of a bacteriophage
1977	Carl Woese and George Fox divide prokaryotes into Archaebacteria and Eubacteria which they define as "urkingdoms" or "primary kingdoms"
1977	Bukharo, Shapiro, and Adhya identify insertion sequences
1977	Shine and colleagues identify long terminal repeats (LTRs) in the genome of the avian sarcoma virus
1977	Allan Campbell and colleagues provide a first nomenclature of transposable elements in prokaryotes
1978	Whittaker and Margulis introduce a 5-kingdom classification of life that understands symbiogenesis as the defining mechanism that separates prokaryotes (Monera) from all 4 eukaryotic kingdoms, and in subsequent years, Margulis introduces new, reticulate evolutionary iconography
1979	Variola is declared eradicated
1979	James Shapiro describes the RNA intermediate stage of retrotransposons
1979	Yen, Hu, and Marrs (1979, Marrs 1974) report on "nucleoprotein particles that act as vectors of genetic exchange," i.e., GTAs in *Rhodopseudomonas capsulata* (today called *Rhodobacter capsulatus*)
1980	Alex Champion uses the concepts of "HGT" and "reticulate evolution" to understand the evolution of *Pseudomonas fluorescens*
1981	Joachim Messing develops the shotgun DNA sequencing technique which enables the sequencing of longer stretches of DNA up to whole genomes
1981	Esther Lederberg provides a classification system for insertion sequences
1982	Maxine Singer distinguishes between Short Interspersed Nuclear Elements (SINEs) and Long Interspersed Nuclear Elements (LINEs)
1983	Kary Mullis introduces the polymerase chain reaction (PCR) technique
1983	James Shapiro edits a first anthology on "mobile genetic elements"
1984	Michael Syvänen introduces the notion of "cross-species gene exchange"
1989	Peter Gogarten and colleagues introduce ATPase-based phylogenetic reconstructions of the roots of the tree of life and suggest that these genes were acquired by "HGT"

1989	Jack Heinemann and George Sprague demonstrate that "Bacterial conjugative plasmids mobilize DNA transfer between bacteria and yeast (*Saccharomyces cerevisiae*)"
1989	The American Society for Microbiology publishes their first anthology on mobile DNA, the work is edited by Douglas Berg and Martha Howe
1989	Douglas Berg and Martha Howe differentiate between compositional and non-compositional insertion sequences
1989	Stokes and Hall differentiate integrons as "a novel family of potentially mobile DNA elements encoding site-specific gene integration functions"
1990	Hacker and coworkers introduce the concept of pathogenicity islands to designate specific regions in the genome of bacterial pathogens that are absent in non-pathogenic bacteria
1990	Woese, Kandler, and Wheelis introduce the three-domain classification of life (Archaea, Bacteria, and Eukaryota)
1994	The Tree of Life Web Project (ToL) commences and goes online in 1996
1995	The complete genome of *Haemophilus influenza* is sequenced by Craig Venter, Hamilton O. Smith, and Claire Fraser at the Institute for Genomic Research
1998	Didier Mazel and colleagues discover superintegrons in *Vibrio cholerae* bacteria
1998	Jo Handelsman and colleagues introduce the term "metagenomics" to designate biochemical techniques used to identify the genetic constitution of unidentified soil bacteria
1999	Ford W. Doolittle introduces a hypothetical reticulate image and the metaphor of a "web of life" to visualize and conceptualize the massive HGT that occurs across all three domains of life
1999	Eisterling and colleagues report on a "bacteriophage-like particle," the "voltae transfer agent," of *Methanococcus voltae* PS. This GTA enables transductions between members of the bacterial strain
2000	Andrew S. Lang and J.T. Beatty report on a GTA in the purple nonsulfur bacterium *Rhodobacter capsulatus*
2000	Peter Gogarten introduces the metaphor of a "net" and "network" of life
2001/2002	The American National Science Foundation launches the AToL—Assembling the Tree of Life Project
2002	The American Society for Microbiology publishes their second anthology on mobile DNA, which is edited by Nancy L. Craig, Robert Craigie, Martin Gellert, and Alan M. Lambowitz
2003	The barcoding technique is introduced by Paul Hebert and colleagues
2004	Maria Rivera and James Lake introduce the "ring of life"
2004	The *American Journal of Botany* dedicated a special issue to the tree of life of plants

2005	In an article for *Scientific American*, Ford Doolittle expands his reticulate evolutionary image in order to include the symbiogenetic acquisition of chloroplasts and mitochondria
2005	Fan Ge, Li-San Wang, and Junhyong Kim introduce the metaphor of a "cobweb" of life
2005	The Tree Thinking Group goes online
2006	Bork's team publishes their circular tree of life (Ciccarelli et al. 2006) in *Science* and launches the online iTOL project
2009	*The New Scientist* features a reticulate tree of life image on their January 21st cover and titles it "Darwin was wrong: Cutting down the tree of life." Daniel Dennett, Jerry Coyne, Richard Dawkins, and Paul Meyers argue that the cover feeds into creationism
2009	Tal Dagan and William Martin introduce networks that depict actual horizontal as well as vertical exchange between distinct microbial lineages
2009	*The Philosophical Transactions of the Royal Society, B: Biological Sciences* features a theme issue on "The network of life: genome beginnings and evolution"
2010	Luis Villarreal and Günter Witzany introduce a hypothetical diagram that illustrates the viral origin of life as well as the colonization of all three domains of life by viral agents
2010	The journal *Biology and Philosophy* features a special issue on the tree of life
2011	The journal *Research in Microbiology* dedicates a special issue to "Archaea and the tree of life"
2011	*Biology Direct* publishes an issue titled "Beyond the Tree of Life"
2014	The American National Science Foundation launches the GoLife (Genealogy of Life) project
2014	The *Journal of Ecology* features a special issue on "The tree of life in ecosystems: evolution of plant effects on carbon and nutrient cycling"

Note: This timeline is based upon the timeline provided by the American Society for Microbiology (ASM) available at http://www.asm.org/index.php/choma3/71-membership/archives/7852-significant-events-in-microbiology-since-1861; the Genome News Network site at http://www.genomenewsnetwork.org/resources/timeline; as well as own work.

References

Adams JW, Kaufman RE, Kretschmer PJ, Harrison M, Nienhuis AW (1980) A family of long reiterated DNA sequences, one copy of which is next to the human beta globin gene. Nucleic Acids Res 8(24):6113. doi:10.1093/nar/8.24.6113

Akamatsu T, Taguchi H (2001) Incorporation of the whole chromosomal DNA in protoplast lysates into competent cells of *Bacillus subtilis*. Biosci Biotechnol Biochem 65(4):823–829. doi:10.1271/bbb.65.823

Akiba T, Koyama K, Ishiki Y, Kimura S, Fukushima T (1960) On the mechanism of the development of multiple-drug-resistant clones of *Shigella*. Jpn J Microbiol 4(2):219–227. doi:10.1111/j.1348-0421.1960.tb00170.x

Ambler P, Meyer T, Kamen MD (1979) Anomalies in amino acid sequences of small cytochromes c and cytochromes c′ from two species of purple photosynthetic bacteria. Nature 278:661–662

Andam CP, Williams D, Gogarten JP (2010) Natural taxonomy in light of horizontal gene transfer. Biol Phil 25(4):589–602

Anderson ES (1968) The ecology of transferable drug resistance in the enterobacteria. Annu Rev Microbiol 22:131–180

Andersson DI, Hughes D (2010) Antibiotic resistance and its cost: is it possible to reverse resistance? Nat Rev Microbiol 8:260–271. doi:10.1038/nrmicro2319

Arnold ML (2008) Reticulate evolution and humans: origins and ecology. Oxford University Press, New York

Avery OT, Macleod CM, McCarty M (1944) Studies on the chemical nature of the substance inducing transformation of pneumonococcal types. Induction of transformation by a deoxyribo-nucleic acid fraction isolated from pnuemococcus type III. J Exp Med 79:137–157

Bachmann BJ (1972) Pedigrees of some mutant strains of *Escherichia coli* K-12. Bacteriological Rev 36(4):525–557

Bapteste E, Susko E, Leigh J, MacLeod D, Charlebois RL, Doolittle WF (2005) Do orthologous gene phylogenies really support tree-thinking? Evol Biol 5:33

Bapteste E et al (2009) Prokaryotic evolution and the tree of life are two different things. Biology Direct 4:34

Bardaji L, Añorga M, Jackson RW, Martínez-Bilbao A, Yanguas N, Murillo J (2011) Miniature transposable sequences are frequently mobilized in the bacterial plant pathogen *Pseudomonas syringae*. PLoS ONE 6(10):e25773

Barker A, Clark CA, Manning PA (1994) Identification of VCR, a repeated sequence associated with a locus encoding a hemag glutinin in *Vibrio cholerae* O1. J Bacteriol 176:5450–5458

Behring E, Kitasato S (1890) Ueber das Zustandekommen der Diphtherie-Immunitat und der Tetanus-Immunitat bei Thieren. Deutsche medizinsche Wochenschrift 16:1113–1114

Beijerinck MW (1898) Über ein Contagium vivum fluidum als Ursache der Fleckenkrankheit der Tabaksblätter. Verhandelingen der Koninklijke academie van Wetenschappen te Amsterdam 65:1–22

Belshaw R, Pereira V, Katzourakis A, Talbot G, Paces J, Burt A, Tristem M (2004) Long-term reinfection of the human genome by endogenous retroviruses. PNAS 101(14):4894–4899. doi:10.1073/pnas.0307800101

Berg DE, Howe MM (eds) (1989) Mobile DNA. American Society for Microbiology, Washington DC

Boeke JD (2003) The unusual phylogenetic distribution of retrotransposons: A hypothesis. Genome Res 13:1975–1983

Buchner P (1921) Tier und Pflanze in intrazellularer Symbiose. Berlin

Bukhari AI, Shapiro JA, Adhya SL (1977) DNA insertion elements, plasmids, and episomes. Cold Spring Harbor Laboratory: Cold Spring, Harbor NY

Burnet FM (1934) The bacteriophages. Biol Revs Cambridge Phil Soc 9:332–350

Busslinger M, Rusconi S, Birnstiel ML (1982) An unusual evolutionary behavior of a sea urchin histone gene cluster. EMBO J 1(1):27

Case CL, Chung K (1997) Montagu and Jenner: the campaign against smallpox. SIM News 47(2):58–60

Cavalli Sforza LL (1950) La sessualita new batteri. Boll Ist Sierotera Milano 29:281–289

Champion AB et al (1980) Evolution in *Pseudomonas fluorescens*. J Gen Microbiol 120(2):485–511

Chen I, Dubnau D (2004) DNA uptake during bacterial transformation. Nat Rev Microbiol 2(3):241–249. doi:10.1038/nrmicro844

Chen K, Pachter L (2005) Bioinformatics for whole-genome shotgun sequencing of microbial communities. PLoS Comput Biol 1(2):e24. doi:10.1371/journal.pcbi.0010024

Chiura HX, Kogure K, Hagemann S, Ellinger A, Velimirov B (2011) Evidence for particle-induced horizontal gene transfer and serial transduction between bacteria. FEMS Microbiol Ecol 76:576–591

Ciccarelli FD, Doerks T, von Mering C, Creevey CJ, Snel B, Bork P (2006) Toward automatic reconstruction of a highly resolved tree of life. Science 311:1283–1287

Cohen SN (1976) Transposable genetic elements and plasmid evolution. Nature 263:731–738. doi:10.1038/263731a0

Cohen SN, Miller CA (1969) Multiple molecular species of circular R-factor DNA isolated from *Escherichia coli*. Nature 224:1273–1277

Cohen SN, Miller CA (1970) Molecular nature of R-factors isolated from *Proteus mirabilis* and *Escherichia coli*. J Mol Biol 50(3):671–687

Cohen S, Chang A, Hsu L (1972) Nonchromosomal antibiotic resistance in bacteria: genetic transformation of *Escherichia coli* by R-factor DNA. PNAS 69(8):2110–2114. doi:10.1073/pnas.69.8.2110

Cohn FJ (1875) Untersuchungen über Bakterien. Beitraege zur Biologie der Planzen 1:127–222

Campbell A et al (1977) Nomenclature of transposable elements in prokaryotes, DNA insertion elements, plasmids, and episomes. Cold Spring Harbor Laboratories, New York pp 15–22; also, 1979, Gene, 5, pp 197–206

Cornelissen JH, Cornwell WK (2014) The tree of life in ecosystems: evolution of plant effects on carbon and nutrient cycling. J Ecol 269–274 doi:10.1111/1365-2745.12217

Cotton J (2001) Retroviruses from retrotransposons. Genome Biol 2(2):6

Coughter JP, Stewart GJ (1989) Genetic exchange in the environment. Antonie Van Leeuwenhoek 55(1):15–22

Cowles HC (1915) Hereditary symbiosis. Bot Gaz 59(1):61–63

Craig NL, Craigie R, Gellert M, Lambowitz AM (eds) (2002) Mobile DNA II. ASM Press, Washington

Crick F (1968) The origin of the genetic code. J Mol Biol 38:367–379

Dagan T, Martin W (2006) The tree of one percent. Genome Biol 7:118. doi:10.1186/gb-2006-7-10-118

Dagan T, Martin W (2009) Getting a better picture of microbial evolution en route to a network of genomes. Phil Trans R Soc B 364:2187–2196

Daniels SP et al (1990) Evidence for horizontal transmission of the P transposable element between *Drosophila* species. Genetics 124:339–335

Darwin C (1837–1838) Notebook B: [Transmutation of species (1837–1838)]. CUL-DAR121—transcribed by Kees Rookmaaker. Darwin Online. http://darwin-online.org.uk

Darwin C (1859) On the origin of species. Murray, London

Dawkins R (1976) The selfish gene. Oxford Univ Press, London

De Bary HA (1861) Die gegenwärtig herrschende Kartoffelkrankheit, ihre Ursache und ihre Verhütung: Eine pflanzenphysiologische Untersuchung in allgemein verständlicher Form dargestellt. Leipzig: A. Förstersche Buchhandlung

de Koning AP, Gu W, Castoe TA, Batzer MA, Pollock DD (2011) Repetitive elements may comprise over two-thirds of the human genome. PLoS Genet 7:e1002384. doi:10.1371/journal.pgen.1002384

de Magalhães J, Finch C, Janssens G (2010) Next-generation sequencing in aging research: emerging applications, problems, pitfalls and possible solutions. Ageing Res Rev 9(3):315–323. doi:10.1016/j.arr.2009.10.006

Deininger PL, Batzer MA (2002) Mammalian retroelements. Genome Res 12(10):1455–1465. doi:10.1101/gr.282402

Dennett D, Coyne J, Dawkins R, Myers P (2009) Darwin was right. New Sci 201(2696):25

D'Herelle F (1917) Sur un microbe invisible antagoniste des bacilles dysentériques. CR Acad Sci Paris 165:373–375. doi:10.1098/rstb.2009.0040

Doolittle WF (1999) Phylogenetic classification and the universal tree. Science 284:2124–2128. doi:10.1126/science.284.5423.2124

Doolittle FW (2005) If the tree of life fell, would we recognize the sound? In: Sapp J (ed) Microbial phylogeny and evolution: concepts and controversies. Oxford University Press, New York, pp 119–133
Doolittle WF (2009) The practice of classification and the theory of evolution, and what the demise of Charles Darwin's tree of life hypothesis means for both of them. Proc Natl Acad Sci USA. doi:10.1098/rstb.2009.0032
Doolittle WF, Bapteste E (2007) Pattern pluralism and the tree of life hypothesis. Proc Natl Acad Sci USA 104:2043–2049. doi:10.1073/pnas.0610699104
Doolittle WF, Sapienza C (1980) Selfish genes, the phenotype paradigm and genome evolution. Nature 284(5757):601–603. doi:10.1038/284601a0
Doolittle WF et al (1990) A naturally occurring horizontal gene transfer from a eukaryote to a prokaryote. J Mol Evol 31:383–388
Downie AW (1972) Pneumococcal transformation—a backward view: fourth Griffith memorial lecture. J Gen Microbiol 73(1):1–11. doi:10.1099/00221287-73-1-1
Dubey GP, Ben-Yehuda S (2011) Intercellular nanotubes mediate bacterial communication. Cell 144:590–600
Dubnau D (1999) DNA uptake in bacteria. Annu Rev Microbiol 53(1):217–244. doi:10.1146/annurev.micro.53.1.217
Ehrlich P (1877) Beiträge zur Kenntniss der Anilinfärbungen und ihren Verwendung in der mikroskopischen Technik.In: Schultze M (ed) Archiv für Mikroskopische Anatomie Bd 13. Valetta St. George, Bonn & W. Waldeyer, Strassburg, pp 263–277
Ehrlich P (1879a) Beiträge zur Kenntniss der granulirten Bindegewebszellen und der eosinophilen Leukocythen. Archiv fuer Anatomie und Physiologie, Physiologische Abteilung, pp 166–169
Ehrlich P (1879b) Ueber die specifischen Granulationen des Blutes. Archiv fuer Anatomie und Physiologie, Physiologische Abteilung, pp 571–579
Ehrlich P (1892a) Bemerkungen über die Immunität durch Vererbung und Säugung. Dtsch Med Wochenschr 18:511
Ehrlich P (1892b) Ueber Immunität durch Vererbung und Säugung. Zeitschrift für Hygiene und Infektionskrankheiten, medizinische Mikrobiologie, Immunologie und Virologie 12:183–203
Ehrlich P (1898) Ueber den Zusammenhang von chemischer Constitution und Wirkung. Münchener medizinische Wochenschrift 1654–1655
Ehrlich P (1900) Cellularbiologische Betrachtungen über Immunität. Bericht der Senckenbergischen Naturforschenden Gesellschaft in Frankfurt am Main 147–150
Ehrlich P, Morgenroth J (1902) Die Seitenkettentheorie der Immunität. Anleitung zu hygienischen Untersuchungen: nach den im Hygienischen Institut der königl. Ludwig-Maximilians-Universität zu München üblichen Methoden zusammengestellt 3. Aufl 381–394
Eisen JA (2007) Environmental shotgun sequencing: its potential and challenges for studying the hidden world of microbes. PLoS Biol 5(3):e82. doi:10.1371/journal.pbio.0050082
Eldredge N, Gould SJ (1972) Punctuated equilibria: an alternative to phyletic gradualism. In Schopf TJM (ed) Models in paleobiology. Freeman. Cooper and Co, New York, pp 82–115
Engelmoer DJ, Rozen DE (2011) Competence increases survival during stress in *Streptococcus pneumoniae*. Evolution 65(12):3475–3485. doi:10.1111/j.1558-5646.2011.01402.x
Finkel SE, Kolter R (2001) DNA as a nutrient: novel role for bacterial competence gene homologs. J Bacteriol 183(21):6288–6293. doi:10.1128/JB.183.21.6288-6293.2001
Finnegan DJ (1989) Eukaryotic transposable elements and genome evolution. Trends Genet 5:103–107
Fire A, Xu S, Montgomery MK, Kostas SA, Driver SE, Mello CC (1998) Potent and specific genetic interference by double-stranded RNA in *Caenorhabditis elegans*. Nature 391(6669):806–811
Flavell AJ (1981) Did retroviruses evolve from transposable elements? Nature 289:10–11
Fournier GP, Dick AA, Williams D, Gogarten JP (2011) Evolution of the Archaea: emerging views on origins and phylogeny. Res Microbiol 162(1):92–98

Fox GCA et al (1980) The phylogeny of prokaryotes. Science 209(4455):457–463. doi:10.1126/science.6771870

Freeman VJ (1951) Studies on the virulence of bacteriophage-infected strains of *Corynebacterium diphtheriae*. J Bacter 61(6):675–688

Galun E 2003 Transposable elements: a guide to the perplexed and the novice. With appendices on RNAi, chromatin remodeling and gene tagging. Kluwer Academic Publishers, Dordrecht NL

Ge F, Wang LS, Kim J (2005) The cobweb of life revealed by genome scale estimates of horizontal gene transfer. PLoS Biol 3:e316. doi:10.1371/journal.pbio.0030316

Gil R, Amparo L (2012) Factors behind JUNK DNA in Bacteria. Genes 3:634–650. doi:10.3390/genes3040634

Gogarten JP, Kibak H, Dittrich P et al (1989) Evolution of the vacuolar H + -ATPase: implications for the origin of eukaryotes. Proc Natl Acad Sci USA 86(17):6661–6665

Goldenfeld N, Woese C (2007) Biology's next revolution. Nature 445:369

Golding G, Gupta RS (1995) Protein based phylogenies support a chimeric origin of the eukaryotic genome. Mol Biol Evol 12:1–6

Goldschmitdt R (1940) The material basis of evolution. Yale University Press, New Haven

Gontier N (2007) Universal symbiogenesis: a genuine alternative to universal selectionist accounts. Symbiosis 44:167–181

Gontier N (2011) Depicting the tree of life: the philosophical and historical roots of evolutionary tree diagrams. Evol Educ Outreach 4(3):515–538

Gray M et al (1989) On the evolutionary origin of the plant mitochondrion and its genome. PNAS 86:2267–2271

Gregory RT 2007 A word about 'Junk DNA'. Evolver Zone Genomicron http://www.genomicron.evolverzone.com/2007/04/word-about-junk-dna/

Gribaldo S, Forterre P, Brochier-Armanet C (2011) Editorial: Archaea and the tree of life. Res Microbiol 162:11–14

Griffith F (1928) The significance of pneumococcal types. J Hyg 27:113–159

Griffiths AJF, Miller JH, Suzuki DT, et al (2000) Bacterial conjugation and bacterial transduction. In: An introduction to genetic analysis, 7th edn. WH Freeman, New York. http://www.ncbi.nlm.nih.gov/books/

Grohmann E, Muth G, Espinosa M (2003) Conjugative plasmid transfer in gram-positive bacteria. Microbiol Mol Biol Reviews 67(2):277–301. doi:10.1128/MMBR.67.2.277-301.2003

Gupta RS, Singh B (1994) Phylogenetic analysis of 70kD heat-shock protein sequences suggest a chimeric origin for the eukaryotic nucleus. Curr Biol 4:1104–1114

Guttman B, Griffiths A, Suzuki D, Cullis T (2002) Genetics. Oneworld, Oxford

Haeckel E (1866) Generelle Morphologie der Organismen: allgemeine Grundzüge der organischen Formen-Wissenschaft, mechanisch begründet durch die von Charles Darwin reformirte Descendenz-Theorie. Fisher, Jena

Hall RM, Stokes HW (2004) Integrons or super integrons? Microbiology 150(Pt 1):3–4. doi:10.1099/mic.0.26854-0

Handelsman J, Rondon MR, Brady SF, Clardy J, Goodman RM (1998) Molecular biological access to the chemistry of unknown soil microbes: a new frontier for natural products. Chem Biol 5(10):R245–R249. doi:10.1016/S1074-5521(98)90108-9

Hartley RW (1980) Homology between prokaryotic and eukaryotic ribonucleases. J Mol Evol 15(4):355–358

Hayes W (1952) Recombination in *Bact. coli* K 12: unidirectional transfer of genetic material. Nature 169:118–119

Hayes W (1953) Observations on a transmissible agent determining sexual differentiation in *Bacterium coli*. J Gen Microbiol 8:72–88

Hebert P et al (2003) Biological identifications through DNA barcodes. Proc R Soc B 270:313–321. doi:10.1098/rspb.2002.2218

Heinemann JA, Sprague GF Jr (1989) Bacterial conjugative plasmids mobilize DNA transfer between bacteria and yeast. Nature 340:205–209. doi:10.1038/340205a0

Hershey AD, Chase M (1952) Independent functions of viral protein and nucleic acid in growth of bacteriophage. J Gen Physiol 36(1):39–56

Hilario E, Gogarten JP (1993) Horizontal transfer of ATPase genes—the tree of life becomes a net of life. Biosystems 31:111–119. doi:10.1016/0303-2647(93)90038-E

Hoelzer MA, Michod RE (1991) DNA repair and the evolution of transformation in *Bacillus subtilis*. III. Sex with damaged DNA. Genetics 128(2):215–223

Holloway B, Broda P (1996) William Hayes 1918–1994. Hist Rec Aust Sci 11(2):213–228

Hong D-Y, Chen Z-D, Qiu Y-L, Donoghue MJ (2008) Patterns of evolution and thc trcc of life (a symposium volume). J Syst Evol 46(3)

Hugenholz P, Goebel BM, Pace NR (1998) Impact of culture-independent studies on the emerging phylogenetic view of bacterial diversity. J Bacteriol 180(18):4765–4774

Huxley J (1942) Evolution: the Modern Synthesis. Allen & Unwin, London

Iwanowski D (1892) Über die Mosaikkrankheit der Tabakspflanze. Bulletin Scientifique publié par l'Académie Impériale des Sciences de Saint-Pétersbourg/Nouvelle Serie III (St. Petersburg) 35:67–70

Iwasaki W, Takagi T (2009) Rapid pathway evolution facilitated by horizontal gene transfers across prokaryotic lineages. PLoS Genet 5(3):e1000402

Jacob F (1955) Transduction of lysogeny in *Eschercia coli*. Virology 1:207–220

Jacob F, Wollman EL (1958) Les épisomes, elements génétiques ajoutés. Comptes Rendus des Académie des Sciences, Paris 247:154–156

Jacob F, Schaeffer P, Wollman EL (1960) Microbial genetics. Symposium Soc Gen Microbial 10(67):352

Jenner E (1798) An inquiry into the causes and effects of the Vaccinæ, Or Cow-Pox. The Harvard Classics

Johnsborg O, Eldholm V, Håvarstein LS (2007) Natural genetic transformation: prevalence, mechanisms and function. Res Microbiol 158(10):767–778. doi:10.1016/j.resmic.2007.09.004

Jones D, Sneath HA (1970) Genetic transfer and bacterial taxonomy. Bacteriological Rev 34:40431

Karnovsky ML (1981) Metchnikoff in Messina: a century of studies on phagocytosis. N Engl J Med 304(19):1178–1180. doi:10.1056/NEJM198105073041923

Keeling PJ, Palmer JD (2008) Horizontal gene transfer in eukaryotic evolution. Nat Rev Gen 9:605–618. doi:10.1038/nrg2386

Kellis M et al (2014) Defining functional DNA elements in the human genome. PNAS 111(17):6131–6138. doi:10.1073/pnas.1318948111

Khodosevich K, Lebedev L, Sverdolv E (2002) Endogenous retroviruses and human evolution. Comp Funct Genomics 3:494–498. doi:10.1002/cfg.216

Kidwell MG, Novy JB (1979) Hybrid dysgenesis in *Drosophila melanogaster*: sterility resulting from gonodal dysgenesis in the P-M system. Genetics 92:1127–1140

Koch R (1876) Untersuchungen ueber Bakterien V. Die Aetiologie der Milzbrand-Krankheit, begruendent auf die Entwicklungsgeschichte des Bacillus Anthracis. Beitr z Biol D Pflanzen 2:277–310

Koch R (1882) Die Aetiologie der Tuberculose. Berliner Klinische Wochenschrift 19:221–230

Kovalevskaia NP (2002) Mobile gene cassettes and DNA integration elements. Mol Biol 36(2):261–267

Kunin V, Goldovsky L, Darzentas N, Ouzounis CA (2005) The net of life: reconstructing the microbial phylogenetic network. Genome Res 15:954–959. doi:10.1101/gr.3666505

Lang AS, Beatty JT (2000) Genetic analysis of a bacterial genetic exchange element: the Gene Transfer Agent of *Rhodobacter capsulatus*. PNAS 97(2):859–864. doi:10.1073/pnas.97.2.859

Lang AS, Zhaxybayeva O, Beatty TJ (2012) Gene transfer agents: phage-like elements of genetic exchange. Nature Rev Mircobiol 10:472–482

Laveran A (1880) A new parasite found in the blood of malarial patients: parasitic origin of malarial attacks. Bull mem soc med hosp Paris 17:158–164

Lawton G (2009) Axing Darwin's tree: the tree of life is an iconic image, but it could be time to fell it. New Scientist 201(2692):34–39
Lederberg J (1952) Cell genetics and hereditary symbiosis. Physiol Rev 32(4):403–430
Lederberg J (1955) Recombination mechanisms in bacteria. J Cell Comp Physiol 45(Suppl 2):75
Lederberg J (1956) Genetic transduction. Am Sci 14(3):264–280
Lederberg EM (1981) Plasmid reference center registry of transposon (Tn) allocations through. Gene 16:59–61
Lederberg J (2003) Infectious history. Science 288(5464):27
Lederberg EM, Lederberg J (1953) Genetic studies of lysogenicity in *Escherchia coli*. Genetics 38:51–64
Lederberg J, Tatum EL (1946) Gene recombination in *E coli*. Nature 158(4016):558. doi:10.1038/158558a0
Lederberg J, Lederberg EM, Zinder ND, Lively ER (1951) Recombination analysis of bacterial heredity. Cold Spring Harbor Symp Quant Biol 16:413–441
Lederberg J, Cavalli LL, Lederberg EM (1952) Sex compatibility in *Escherichia coli*. Genetics 37(6):720–730
Lennox ES (1955) Transduction of linked genetic characters of the host by bacteriophage P1. Virology 1:190–206
Lodish H, Berk A, Zipursky SL et al (2000) Molecular cell biology, 4th edn. Freeman, New York
Lopez P, Bapteste E (2009) Molecular phylogeny: reconstructing the forest. Curr Biol 332:171
Lwoff A (1953) Lysogeny. Bact Rev 17:269–337
Lwoff A (1965) Nobel lecture: interaction among virus, cell, and organism. Nobelprizeorg Nobel Media AB http://www.nobelprize.org/nobel_prizes/medicine/laureates/1965/lwoff-lecture.html
Lwoff A, Gutmann A (1950) Recherches sur un bacillus megatherium lysogène. Ann Inst Pasteur (Paris) 78(6):711–739
Mahillon J, Chandler M (1998) Insertion sequences. Microbiol Mol Biol Rev 62(3):725–774
Mandel M, Higa A (1970) Calcium-dependent bacteriophage DNA infection. J Mol Biol 53(1):159–162. doi:10.1016/0022-2836(70)90051-3
Margulis L (1970) Origin of eukaryotic cells. Yale University Press, New Haven
Margulis L (ed) (1991) Symbiosis as a source of evolutionary innovation: speciation and morphogenesis. The MIT Press, Boston
Margulis L (1998) The symbiotic planet: a new look at evolution. Weidenfeld & Nicolson, London
Margulis L, Schwartz KV (1997) Five kingdoms: an illustrated guide to the phyla of life on earth. W.H. Freeman & Company
Marrs BL (1974) Genetic recombination in *Rhodopseudomonas capsulata*. Proc Natl Acad Sci USA 71:971–973
Martin W (1999) Mosaic bacterial chromosomes: a challenge en route to a tree of genomes. BioEssays 21:99–104
Martinez E, de la Cruz F (1988) Transposon Tn21 encodes a RecA-independent site-specific integration system. Mol Gen Genet 211:320–325
Mashburn-Warren LM, Whiteley M (2006) Special delivery: vesicle trafficking in prokaryotes. Mol Microbiol 61:839–846
Mattick JS (2002) Type IV pili and twitching motility. Annu Rev Microbiol 56(1):289–314. doi:10.1146/annurev.micro.56.012302.160938
Mayer A (1886) Über die Mosaikkrankheit des Tabaks. Die Landwirtschaftliche Versuchsstationen 32:451–467
Mazel D (2006) Integrons: agents of bacterial evolution. Nature Rev Microbiol 4:608–620. doi:10.1038/nrmicro1462
Mazel D, Dychinco B, Webb VA, Davies J (1998) A distinctive class of integron in the *Vibrio cholerae* genome. Science 280(5363):605–608. doi:10.1126/science.280.5363.605
McClintock B (1941) The stability of broken ends of chromosomes in Zea Mays. Genetics 26(2):234–282

McClintock B (1950) The origin and behavior of mutable loci in maize. PNAS 36(6):344–355
McClintock B (1953) Induction of instability at selected loci in maize. Genetics 38(6):579–599
Merezhkowsky C (1905) Über natur und ursprung der chromatophoren im pflanzenreiche. Biol Centralbl 25(593–604):689–691
Merezhkowsky C (1910) Theorie der zwei Plasmaarten als Grundlage der Symbiogenese, einer neuen Lehre von der Entstehung der Organismen, Biologisches Centralblatt 30: 278–288, 289–303, 321–347, 353–367
Messing J, Crea R, Seeburg PH (1981) A system for shotgun DNA sequencing. Nucleic Acids Res 9(2):309–321. doi:10.1093/nar/9.2.309
Mindell DP, Villarreal LP (2003) Don't forget about viruses. Science 5:1677
Mitsuhashi S, Harada K, Hashimoto H, Egawa R (1961) On the drug-resistance of enteric bacteria. Jpn J Exp Med 31:47–52
Moran NA, Jarvik T (2010) Lateral transfer of genes from fungi underlies carotenoid production in Aphids. Science 328:624–627
Morange M (2000) A history of molecular biology. Harvard University Press, Cambridge (New Edited edition)
Morgan GJ (1998) Emile Zuckerkandl, Linus Pauling, and the molecular evolutionary clock, 1959–1965. J Hist Biol 31(2):155–178. doi:10.1023/A:1004394418084
Morse ML, Lederberg EM, Lederberg J (1956) Transduction in *Eschercia coli* K-12. Genetics 41:142–156
Nelson PN, Hooley P, Molecular Immunology Research Group (2004) Human endogenous retroviruses: transposable elements with potential? Clin Exp Immunol 138(1):1–9 doi:10.1111/j.1365-2249.2004.02592.x
Neufeld F (1902) Über die agglutination der pneumokokken und über die theorieen der agglutination. Z Hyg Infektionskr 40:54–72
Neufeld F, Händel L (1910) Weitere Untersuchungen über Pneumokokken Heilsera. III Mitteilung. Über Vorkommen und Bedeutung atypischer Varietäten des Pneumokokkus. Arbeit a.d. Kaiserlichen Gesundheitsamte 34:293–304
O'Malley M, Martin W, Dupré J (2010) The tree of life: introduction to an evolutionary debate. Biol Philos 25(4):441–453
Ochia K, Yamanaka T, Kimura K, Sawada O (1959) Inheritance of drug resistance (and its transfer) between *Shigella* strains and between *Shigella* and *E. coli* strains. Nihon Iji Shimpo 1861:34
Ohno S (1972) So much junk DNA in our genome. In: Smith HH (ed) Evolution of genetic systems. Gordon and Breach, New York, pp 366–370
Ohshima K, Okada N (2005) SINEs and LINEs: symbionts of eukaryotic genomes with a common tail. Cytogenet Genome Res 110(1–4):475–490. doi:10.1159/000084981
O'Malley MA, Koonin EV (2011) How stands the tree of life a century and a half after the origin? Biology Direct 6:32
Palmer JD, Soltis DE, Chase MW (2004) The plant tree of life: an overview of some points of view. Am J Bot 91(10):1437–1445
Pardee AB, Jacob F, Monod J (1959) The genetic control and cytoplasmic expression of inducibility in the synthesis of beta-galactosidase by *E. coli*. J Mol Biol 1:165–178
Pasteur L (1880) De l'attenuation du virus cholera des poules. CR Acad Sci 91:673–680
Pettersson E, Lundeberg J, Ahmadian A (2009) Generations of sequencing technologies. Genomics 93(2):105–111. doi:10.1016/j.ygeno.2008.10.003
Proft T, Baker EN (2009) Pili in gram-negative and gram-positive bacteria—structure, assembly and their role in disease. Cell Mol Life Sci 66(4):613–635. doi:10.1007/s00018-008-8477-4
Ragan MA, McInerney JO, Lake J (2009) The network of life: genome beginnings and evolution. Phil Trans R Soc B 364(1527):2169–2175
Redfield R, Schrag M, Dead A (1997) The evolution of bacterial transformation: sex with poor relations. Genetics 146(1):27–38
Reinheimer H (1915) Symbiogenesis: the universal law of progressive evolution. Knapp, Drewett and Sons Ltd, Westminster

Rivera MC, Lake JA (2004) The ring of life provides evidence for a genome fusion origin of eukaryotes. Nature 431(2005):152–155. doi:10.1038/nature02848
Ryan F (2004) Human endogenous retroviruses in health and disease: a symbiotic perspective. J Roy Soc Med 97:560–565
Ryan F (2009) Virolution. Harper Collins, London
Sakai T, Iseki S (1954) Transduction of flagellar antigen in *Salmonella E* group. Gunma Jour Med Sco 3:195–199
Salmon DE, Smith T (1886) On a new method of producing immunity from contagious diseases. Proc Biol Soc 3:29–33
San Mauro D, Agorreta A (2010) Molecular systematics: a synthesis of the common methods and the state of knowledge. Cell Mol Biol Lett 15(2):311–341. doi:10.2478/s11658-010-0010-8
Sanger F, Coulson AR (1975) A rapid method for determining sequences in DNA by primed synthesis with DNA polymerase. J Mol Biol 94(3):441–448. doi:10.1016/0022-2836(75)90213-2
Sapp J (1994) Evolution by association: a history of symbiosis. Oxford University Press, New York
Sapp J (2003) Genesis: the evolution of biology. Oxford University Press, New York
Sapp J (2009) The new foundations of evolution: on the tree of life. Oxford Univ. Press, New York
Schuster S (2008) Next-generation sequencing transforms today's biology. Nat Methods 5(1):16–8 doi:10.1038/nmeth1156 (PMID 18165802)
Scott JR, Zähner D (2006) Pili with strong attachments: gram-positive bacteria do it differently. Mol Microbiol 62(2):320–330
Serrelli E, Gontier N (eds) (2015) Macroevolution: explanation, interpretation and evidence. Springer, Dordrecht
Shapiro JA (1979) Molecular model for the transposition and replication of bacteriophage *Mu* and other transposable elements. Proc Natl Acad Sci USA 76(4):1933–1937. doi:10.1073/pnas.76.4.1933
Shapiro JA (ed) (1983) Mobile genetic elements. Academic Press, Waltham
Shine J, Czernilofsky AP, Friedrich R, Bishop JM, Goodman HM (1977) Nucleotide sequence at the 5′ terminus of the avian sarcoma virus genome. Proc Natl Acad Sci USA 74(4):1473–1477. doi:10.1073/pnas.74.4.1473
Singer MF (1982) SINEs and LINEs: highly repeated short and long interspersed sequences in mammalian genomes. Cell 28(3):433–434. doi:10.1016/0092-8674(82)90194-5
Sisco KL, Smith HO (1979) Sequence-specific DNA uptake *in Haemophilus* transformation. PNAS 76(2):972–976. doi:10.1073/pnas.76.2.972
Spradling AC, Rubin GM (1982) Transposition of cloned P elements into *Drosophila* germ line chromosomes. Science 218(4570):341–347. doi:10.1126/science.6289435
Staden R (1979) A strategy of DNA sequencing employing computer programs. Nucleic Acids Res 6(7):2601–2610. doi:10.1093/nar/6.7.2601
Stanto TB (2007) Prophage-like gene transfer agents-novel mechanisms of gene exchange for *Methanococcus, Desulfovibrio, Brachyspira,* and *Rhodobacter* species. Anaerobe 13(2):43–49
Stokes HW, Hall RM (1989) A novel family of potentially mobile DNA elements encoding site-specific gene-integration functions: integrons. Mol Microbiol 3:1669–1683
Summers WC (2006) Phage and the early development of molecular biology. In: Bacteriophages The (ed) Calendar R. Oxford Univ Press, New York
Syvänen M (1984a) Conserved regions in mammalian ß-globins: could they arise by cross-species gene exchange? J Theor Biol 107:685–696
Syvänen M (1984b) The evolutionary implications of mobile genetic elements. Annu Rev Genet 18:271–293
Syvänen M (1985) Cross-species gene transfer, implications for a new theory of evolution. J Theor Biol 112(2):333–343. doi:10.1016/S0022-5193(85)80291-5

Syvänen M (1986) Cross-species gene transfer: a major factor in evolution? Trends Genetic 4:1–4

Syvanen M (1987) Molecular clocks and evolutionary relationships: possible distortions due to horizontal gene flow. J Mol Evol 26(1–2):16–23

Syvanen M, Kado CI (eds) (1998) Horizontal gene transfer. Chapman & Hall, London

Tan SY, Dee MK (2009) Elie Metchnikoff (1845–1916): discoverer of phagocytosis. Signapore Med L 50(5):456

Tauber AI (2003) Metchnikoff and the phagocytosis theory. Nat Rev Mol Cell Biol 4:897–901

Temin HM (1980) Origin of retroviruses from cellular movable genetic elements. Cell 21:599–600

The ENCODE Project Consortium et al (2012) An integrated encyclopedia of DNA elements in the human genome. Nature 489(7414):57–74. doi:10.1038/nature11247

Thomas CM (2000) Horizontal gene pool: bacterial plasmids and gene spread. CRC Press, Boca Raton

Trevors JT, Barkay T, Bourquin AW (1987) Gene transfer among bacteria in soil and aquatic environments: a review. Canad J Microbiol 33(3):191–198

Twort F (1915) An investigation on the nature of ultra-microscopic viruses. Lancet 2:1241–1243

Van Beneden PJ (1873) Un mot sur la vie sociale des animaux inférieurs. Bull Acad R Belgique série 2(36):779–796

Van Beneden PJ (1875) Les comensaux et les parasites dans le règne animal. Biblio Sci. Int, Paris

Villareal LP, Defilipps V (2000) A hypothesis for DNA viruses as the origin of eukaryotic replication proteins. J Virol 74(15):7079–7084

Villarreal LP, Witzany G (2010) Viruses are essential agents within the roots and stem of the tree of life. J Theor Biol 262(4):698–710

Von Faber FC (1912) Das erbliche zusammenleben von bacterien und tropischen pflanzen. Jahrb Wiss Bot 51:285–375

Wallin IE (1927) Symbionticism and the origin of species. Williams and Wilkins company, Baltimore

Watanabe T (1971) The problems of drug-resistant pathogenic bacteria: the origin of R factors. Ann NY Acad Sci 182:126–140

Weickert MJ, Adhya S (1993) The galactose regulon of *Escherichia coli*. Mol Microbio 10(2):245–251

Weiner AM (2002) SINEs ans LINEs: the art of biting the hand that feeds you. Curr Opin Cell Biol 14:343–350

Weismann A (1885) Die Continuität des Keimplasma's als Grundlage einer Theorie der Vererbung. Fischer, Jena

Whittaker RH, Margulis L (1978) Protist classification and the kingdoms of organisms. Biosystems 10:3–18

Wicker T, Sabot F, Hua-Van A et al (2007) A unified classification system for eukaryotic transposable elements. Nat Rev Genet 8(12):973–982. doi:10.1038/nrg2165

Williams D, Fournier GP, Lapierre P et al (2011) A rooted net of life. Biology Direct 6:45

Woese CR (1967) The genetic code: the molecular basis for genetic expression. Harper & Row

Woese CR (1998) The universal ancestor. PNAS 95(12):6854–6859

Woese CR, Fox GE (1977) Phylogenetic structure of the prokaryotic domain: the primary kingdoms. PNAS 74(11):5088–5090. doi:10.1073/pnas.74.11.5088

Woese CR, Kandler O, Wheelis ML (1990) Towards a natural system of organisms: proposal for the domains Archaea, Bacteria, and Eucarya. Proc Natl Acad Sci USA 87:4576–4579. doi: 10.1073/pnas.87.12.4576

Wolf K, Delgiudice L (1987) Horizontal gene transfer between mitochondrial genomes. Endocytobiosis Cell Res 4(2):103–120

Wollman EL, Jacob F (1955) Sur le mécanisme du transfer de matériel génétique au cours de la recombination chez *E. coli* K12. Compt Rend Acad Sci 240:2449–2451

Yen HC, Hu NT, Marrs BL (1979) Characterization of the gene transfer agent made by an overproducer mutant of *Rhodopseudomonas capsulata*. J Mol Biol 131:157–168

Zhang Q, Arbuckle J, Wessler SR (2000) Recent, extensive, and preferential insertion of members of the miniature inverted-repeat transposable element family *Heartbreaker* into genic regions of maize. Proc Natl Acad Sci USA 97:1160–1165

Zhaxybayeva O, Doolittle WF (2011) Lateral gene transfer. Curr Biol 21(7):R242–R246

Zinder ND (1955) Bacterial transduction. J Comp Physiol 45(Suppl 2):23–49

Zinder ND (1992) Forty years ago: the discovery of bacterial transduction. Genetics 132(2):291–294

Zinder ND, Lederberg J (1952) Genetic exchange in *Salmonella*. J Bact 64:679–699

Zuckerkandl E, Pauling L (1962) Molecular disease, evolution, and genic heterogeneity. In: Kasha M, Pullman B (eds) Horizons in biochemistry. Academic Press, New York, pp 189–225

Zuckerkandl E, Pauling L (1965a) Evolutionary divergence and convergence in proteins. In: Bryson V, Vogel HJ (eds) Evolving genes and proteins. Academic Press, New York, pp 97–166

Zuckerkandl E, Pauling L (1965b) Molecules as documents of evolutionary history. J Theor Biol 8(2):357–366

Plasmids: Histories of a Concept

William C. Summers

Abstract The plasmid concept is rooted in the notion of particulate determinants of inheritance and the chromosome theory of heredity, but some biologists saw genes as determinants of the way an organism developed from a fertilized ovum into a mature adult; some of these determinants seem to be passed on through cytoplasmic transfer. In a 1952 review, J. Lederberg proposed that all "extrachromosomal hereditary determinants" be designated "plasmids." In 1958, Jacob and Wollman suggested that genetic elements which were optionally associated with the chromosomes, such as the F-factor, the colicinogenic factor, and bacteriophage lambda, be termed "episomes." Allan Campbell (Adv Genetics 11:101–145, 1962) proposed a beautifully simple solution to the problem of how episomes could be associated with the chromosome when he suggested the recombinational interaction of one circular molecule with another. The key to the modern concept of the plasmid was the confirmation that DNA molecules can, and often do, exist as circular structures. Many observations (mainly on yeast and protozoans) suggested that nonchromosomal heredity exists in eucaryotes as well, and eventually, cytochemical, electron microscopic, and biochemical evidence established the existence of cytoplasmic genes in eucaryotes. By the end of the 1960s, both the genetic and physical understanding of plasmids and cytoplasmic heredity had reached a level of detail to allow exploitation of these genetic elements as tools to manipulate and study cell genetics by various techniques of lateral gene transfer.

Keywords Cytoplasmic heredity · Episome · Lederberg · Plasmid

W.C. Summers (✉)
Program on History of Science and Medicine, Yale University, 450 Saint Ronan St, New Haven, CT 06511, USA
e-mail: William.summers@yale.edu

N. Gontier (ed.), *Reticulate Evolution*, Interdisciplinary Evolution Research 3,
DOI 10.1007/978-3-319-16345-1_6

1 Particulate Heredity

The early history of the plasmid concept is rooted in the notion of particulate determinants of inheritance. In the first decade of the twentieth century, the chromosome theory was developed and two key papers are usually cited: Sutton (1903) and Boveri (1904). These workers argued from diverse observations that the cytologically observable structures in the cell nucleus are the physical units that determine the Mendelian characters. Of course, it was very unclear just what a "Mendelian character" was, but in the second decade of the twentieth century, Thomas Hunt Morgan and his group, in experiments with fruit flies (*Drosophila melanogaster*), presented evidence for the formal agreement of the behavior of several "Mendelian factors" and the behavior of the physical structures known as chromosomes. Morgan and his school generalized these results into a broad "Theory of the Gene" which held that the Mendelian factors (genes) were arranged linearly on the visible structures (chromosomes) that resided in the nucleus of every cell and which were duplicated and partitioned equally to the daughter cells during cell division (Morgan 1926). Thus was solved (at one level, at least) the age-old problem of "how like begets like." Many biologists took up this approach and gathered much evidence to supports its validity and universality.

2 Cytoplasmic Contributions

At the same time, other biologists, working on problems of embryology and morphology, saw genes as determinants of the way an organism developed from a fertilized ovum into a mature adult. For them, genetics was not about transmission of characters across the generations, but about how gene action worked to make the organism a nearly exact copy of its parents, that is, a different version of the age-old problem of "how like begets like." For many of these biologists, the determinants of the characters involved in development and differentiation seemed to be neither obviously nuclear, nor chromosomal. Some of these genes seem to be passed on through cytoplasmic transfer. For example, the eminent biologist Ross Harrison wrote:

> The prestige of success enjoyed by the gene theory might easily become a hindrance to the understanding of development by directing our attention solely to the genome, whereas cell movements, differentiation and in fact all developmental processes are actually effected by the cytoplasm. Already we have theories that refer the process of development to genic action and regard the whole performance as no more that the realization of the potencies of the genes. Such theories are altogether too one-sided. (Harrison 1937, p 372)

By the mid-1930s, these cytoplasmic determinants came to be known as "plasmagenes." Plasmagenes, however, were often invoked to explain the possible mechanisms of "inheritance of acquired traits" and played directly into the schemes of the Michurinist/Lysenkoist genetics in the Soviet Union. At the time, then, plasmagenes acquired the extra baggage of Cold War ideology (Sapp 1987).

3 Genes in Bacteria

The existence of genes in bacteria was much debated in the first half of the twentieth century. Without a visible nucleus, without visible chromosomes, without a known dimorphic sexual phase, and without many distinguishing characteristics, it was easy to believe that bacteria were altogether different from organisms which reproduced sexually. In 1942, the famous British biologist Julian Huxley wrote:

> Bacteria (and *a fortiori* viruses, if they can be considered to be true organisms), in spite of occasional reports of a sexual cycle, appear to be not only wholly asexual but pre-mitotic. Their hereditary constitution is not differentiated into special parts with different functions. They have no genes in the sense of accurately quantized portions of hereditary substance; and therefore they have no need for the accurate division of the genetic system which is accomplished by mitosis.... We must, in fact, expect that the process of variation, heredity, and evolution in bacteria are quite different from the corresponding processes in multicellular organisms. But their secret has not yet been unraveled. (Huxley 1942, pp 131–132)

By 1946, however, the experiments of Lederberg and Tatum (1946) challenged and clarified the understanding of genes in bacteria. Without dealing with the physical nature of the genetic structures in bacteria (there was considerable debate about the existence of a bacterial nucleus), they obtained clear support for a mating system in a bacterium (*Escherichia coli*, strain K-12) and subsequently, Lederberg (1947) employed Sturtevant's paradigm of genetic linkage (Sturtevant 1913), to establish a genetic map in *E. coli* based on the frequency of recombination of genetic determinants observed in standardized "matings." At this time, the dominant model was based on the sexual processes in higher cells: cell fusion with zygote formation, recombination, and marker segregation and cell division.

In 1949, Cavalli and Heslot (1949) and, in 1951, Lederberg (1951) surveyed other strains of *E. coli* for their ability to mate with Lederberg's strain K-12 and found that only 9 of 140 isolates could mate with K-12. Thus, there seemed to be something peculiar about mating in *E. coli*. In London, William Hayes started to study the kinetics of the mating process in 1950 and at the suggestion of Denis Mitchison, conceived of bacterial mating as an asymmetric process involving a gene donor and a gene acceptor. This model for bacterial mating fitted the data Hayes was obtaining in various bacterial matings much better than a classical cell fusion model, and in 1952, he published this work (Hayes 1952) and presented it at meetings in the summer of 1952. As James D. Watson described the event:

> Bill's appearance was the sleeper of the three day gathering; before his talk no one except Cavalli-Sforza knew he existed. As soon as he had finished his unassuming report, however, everyone in the audience knew that a bombshell had exploded in the world of Joshua Lederberg! (Watson 1968, pp 141–142)

4 The F-Factor

The directionality and polarity of the bacterial mating process, first suggested by Hayes, greatly clarified the understanding of bacterial genetics as studied by mating experiments. The problem of sexual compatibility, however, remained. The rather rare property of a given *E. coli* strain to mate was a puzzle. In 1952, Lederberg et al. (1952) and, a year later, Hayes (1953) independently reported that the ability to act as a donor in a bacterial mating was a property controlled by a "factor" designated "F" (fertility) that seemed to behave as "an infectious particle."

5 Lambda Bacteriophage and Colicines

In the mid-1950s, two other anomalous hereditary "factors" were discovered to behave as "infectious particles" as well. One was the bacteriophage lambda, a lysogenic phage found in *E. coli* K-12 by Lederberg (1950), and the other was the factor determining the production of colicine, a killer substance, produced by some strains of *E. coli,* originally discovered by Gratia (1925), and studied extensively by Fredericq (1963).

6 Plasmids and Episomes

In a broad review, Lederberg (1952) proposed that all "extrachromosomal hereditary determinants" be subsumed under the designation "plasmid." He did not distinguish nuclear or cytoplasmic location, nor the possibility of association of such determinants with the chromosome on some occasions. In a more limited review of bacterial genetic systems, Jacob and Wollman (1958) suggested that genetic elements which were optionally associated with the chromosomes of the cell be termed "episomes." They used the F-factor, the colicinogenic factor, and bacteriophage lambda as prototypic episomes. By this time, it was known that the F-factor could become associated with the bacterial chromosome and result in the transfer of chromosomal genes with high frequency in mating experiments (Hfr strains).

By the end of the 1950s, the recognition of genetic determinants which were not able to be located on the genetic map in standard crosses established the concept of "plasmid" (episome was used rather interchangeably with plasmid by some, but William Hayes, for one, calling the F-factor "a small, supernumerary chromosome," stated:

> We think the word "episome," although an excellent substitute for "plasmid," has become a source of confusion because the existence of alternative chromosomal and cytoplasmic states was central to its original usage. … It now seems to us that the most meaningful biological distinction is between plasmids which promote conjugation, which we will classify as sex factors, and those which do not. (Hayes 1968, pp 747–748)

7 Chromosomal Associations

The understanding of the possibility of the attachment (by some unknown mechanism, often diagramed as a "bump" on a linear diagram of a chromosome) of the F-factor to the chromosome in Hfr strains probably helped the understanding of the linage of the fertility property and the genetic determinant for lactose fermentation (lac) in the work of Jacob and Adelberg (1959). They concluded that the F-factor could become associated with cell genes which then became part of the "infectious hereditary particle" that was the F-factor. Soon, these "augmented" F-factors became known as "F-prime" factors. Soon, many variant F-primes were found and it was realized that F-prime plasmids carrying any desired part of the bacterial chromosome could be constructed. Elie Wollman recalled the history of F-prime factors:

> Adelberg had brought back to Berkeley some of our Hfr strains. I spent the year 1958–59 in Berkeley – finishing the writing of our book [Wollman and Jacob 1959]. Once Ed Adelberg came to me telling me that one of the Hfr strains had changed: the frequency of recombinants was less than the expected, but all were donors of intermediate frequency. I suggested that, by comparison with HFT phage the sex factor had left its site accompanied by neighboring genetic fragments. This was verified experimentally. Lwoff, who had come to visit, brought the news back to Francois Jacob who immediately used it for making partial Lac diploids. This is the history of F-prime factors. (quoted in Brock 1990, p 104)

8 The Physical Nature of Plasmids and Episomes: DNA

By 1960, it was clear, of course, that "the genetic material is DNA," but the identification of cytoplasmic DNA was still questionable. Likewise, the structure of DNA in genes and in chromosomes was debated. The sizes of DNA "molecules" seemed to increase each year as the methods of preparation improved and as the techniques for study of large, linear polymers got better. The recognition that shear forces could easily break large DNA molecules was especially important. Since bacteriophage was believed to be simple models for the genetic material of the cell, the nature of the DNA in phage was thought to be relevant. The sizes of the DNA in phages were rather ingeniously and indirectly determined by a technique known as "stargazing" (Levinthal 1956). This method compared the amount of DNA radioactivity (^{32}P) in a single phage particle, with the amount of radioactivity in the isolated DNA molecules released from the same phages under very gentle conditions. The radioactivity was detected by counting (under a microscope) the tracks in photographic emulsion in which the phages and the DNA were embedded. Each phage particle and each DNA molecule formed a "star" of such tracks. Since the number of tracks was the same for the intact phage particle and released DNA molecule, it was concluded that the DNA was present in the phage particle as one (possibly two) long piece (perhaps held together by some non-DNA linkages). From the chemical composition of the phage and the bulk specific activity of the DNA, it was possible to calculate the molecular weight of the phage chromosome.

That plasmids are DNA was rather conclusively demonstrated in physical experiments, first reported by Marmur et al. (1961) who used the CsCl buoyant density separation of DNA based on nucleotide base composition to show that "light density *E. coli*-like DNA" appeared in *Serratia marcescens* (which has a somewhat denser DNA) after transfer of the F-factor to *Serratia*. Silver and Ozeki (1962) provided evidence for the same conclusion based on labeled DNA transfer of the colicine factor. By the late 1960s, detailed studies of such transfer allowed Rupp and Ihler (1968) to demonstrate that episome transfer is driven by a strand-specific DNA replication of the plasmid and thus explained the unidirectional transfer of the genetic markers observed much earlier.

9 The "Campbell Model"

Most experiments on the chemistry of DNA confirmed that DNA molecules were very long, linear, non-branched structures. How, then, to envision the attachment of episomes to the chromosome? In a 1962 review on episomes, Campbell (1962) proposed a beautifully simple solution to this problem: the recombinational interaction of one circular molecule with another. "The Campbell Model," as it came to be known, explained the reversible association of some episomes with the chromosome, the inversion of the genetic map of lambda bacteriophage upon lysogeny as recently reported by Calef and Licciardello (1960), and the formation of double lysogens and defective heterogenotes in lambda phage (Whitfield and Appleyard 1958). Campbell's crucial insight was that the episome must exist as a physically circular DNA structure. Interestingly, he reasoned from the circular genetic map of phage T4 (there were, of course, no physical structures established for genomes).

> Detailed linkage studies lead to the conclusion that the genome of one phage (T4) is indeed circular (Streisinger, Edgar and Harrar, quoted by Stahl 1961). If circularity is a property of phages in general, the equivalent of the insertion hypothesis is to the one circle out of two.[…] If the phage genome [now referring to lambda] is circular rather than linear, the lambda chromosome need not be split into parts [to account for map inversion in lysogens] but rather could be cut at a specific point on the circle when it lysogenizes. It is actually very simple (on paper) to insert a circular phage chromosome into a linear bacterial chromosome by reciprocal crossing over (Fig. 2). (Campbell 1962, p 112)

It is, of course, interesting to note that while Campbell based his argument on the T4 genome, which turns out to be linear although it has a circular map, he applied it to lambda which turns out to have a linear map, but a circular intracellular form.

10 Circular DNA

While the genetics of plasmids pointed the way to circular forms, the chemistry of DNA was just becoming clearer. The key step in the modern concept of the plasmid was the confirmation that DNA molecules can and, often do, exist as circular structures. The first confirmation of this fact came again, from the study of phage biology. In an attempt to study the smallest life-form, biologists had been studying bacteriophages, and the smallest known phages were two related phages φX174 and S13. Sinsheimer (1959) had shown that φX174 was unusual in that it contained the single-stranded form of DNA rather than the double-helical DNA of the Watson–Crick model. Using the recently characterized nucleases with specificity for exonucleolytic attack coupled with hydrodynamic studies, Fiers and Sinsheimer (1962) asserted that φX174 DNA was in the form of a small circular, single-stranded DNA molecule.

This precedent for circular DNA molecules was soon followed by the discovery in 1963 of:

1. the cohesive ends of the DNA of bacteriophage lambda and its ability to form circles (called "folded molecules" at the time) (Hershey et al. 1963);
2. the circular structure of the *E. coli* genome by autoradiography (Cairns 1963); and
3. the evidence that the DNA from polyomavirus is circular (Dulbecco and Vogt 1963; Weil and Vinograd 1963).

Even though these studies with phage and viral DNAs provided the methods and concepts to characterize circular DNAs, the study of the physical nature of most plasmids was complicated by the difficulty in separation of the plasmid DNA from the mass of chromosomal DNA. This problem was solved in 1967 by the introduction of the dye-buoyant density method (Radloff et al. 1967). This method depended on the restriction on binding of a DNA-intercalating dye such as ethidium bromide by covalently closed circular DNA molecules in comparison with the linear and nicked circular molecules. These dye-DNA complexes could be separated in density gradients of the dense salt CsCl formed in the ultracentrifuge. Plasmid DNAs were easily isolated by this method for detailed characterization, and Bazaral and Helinski (1968) applied this method to colicine factors E1, E2, and E3 and showed that these factors were circular DNA molecules of homogeneous molecular weights. Beginning in 1959, electron microscopic visualization began to be applied to DNA molecules spread in protein films (Kleinschmidt and Zahn 1959), and this technique soon allowed "direct" visualization of both phage and plasmid DNAs and provided dramatic confirmation of the circular nature of plasmid DNAs.

11 R-Factors

Another important class of plasmids which were discovered in relation to their pathogenesis is the R-factor. In the early 1950s, it was observed in Japan that multiple antibiotic resistance was developing in a single step in patients with enteric infections. Akiba et al. (1960) described this phenomenon, and in 1961, Watanabe and Fukasawa (1961) reported that this multiple drug resistance was being transferred by a plasmid (? an episome) which they called a resistance transfer factor (RTF or R-factor).

12 Organelle Genetics

As difficult as it was to elaborate an understanding of the genetic and physical basis for nonchromosomal heredity in bacteria, the parallel history of eucaryotic cells is even more tortuous. While many observations in eucaryotes (mainly yeast and protozoans, single-cell organisms more amenable to genetic analysis than many multi-cellular organisms) suggested that nonchromosomal, especially cytoplasmic, heredity exists, the acceptance of this conclusion and evidence for its physical basis was long in coming.

Mitochondrial genetics, pioneered in the 1950s by Ephrussi and Słonimski (1955) in their studies of the respiratory-deficient petite mutants of yeast (such mutants grow slowly, depending as they do on glycolysis, and give small colonies, hence the designation, petite), became well established only in the late 1960s when many additional mutants were identified that were associated with the mitochondria. Also, as early as 1954, some mutations in *Chlamydomonas* were found by Sager (1954) to behave in non-Mendelian fashion and were attributed to mutations in the chloroplasts. Finding the physical basis of organelle heredity (that is, the DNA in these structures) proved difficult as well. Cytochemical, electron microscopic, and biochemical evidence were offered (Rhoades 1955; Ris and Plaut 1962), but until the techniques for study of DNA based on sequence comparisons (first nucleic acid hybridization and more recently direct nucleotide sequence analysis), the existence of cytoplasmic genes in eucaryotes was controversial. As Ruth Sager noted in 1972:

> The pendulum of opinion had swung from one extreme – cytoplasmic genes do not exist because we do not see cytoplasmic chromosomes to the other extreme – cytoplasmic DNA's exist, and therefore there must be cytoplasmic genes. (Sager 1972, p 2)

13 The "Modern Period" of Plasmid Research

By the end of the 1960s, then, both the genetic and physical understanding of plasmids and cytoplasmic heredity had reached a level of detail which allowed the subsequent massive exploitation of these genetic elements as tools to study key cellular processes such as DNA replication as well as to manipulate and engineer the genetic contents of cells at will by means of the newly devised methods of in vitro recombinant DNA chemistry (see Clowes 1972).

14 Plasmid Early History Timeline

1903	Walter S. Sutton and Theodor Boveri independently developed the hypothesis that the units of heredity are physically located on chromosomes, thus giving a physical location for heredity
1910	Thomas Hunt Morgan described the association of heritable properties in *Drosophila* with a specific chromosome and began the analysis of genes in the nucleus
1920s–1940	Embryological observations suggested that there are hereditary determinants in the cytoplasm
1946	Joshua Lederberg and Edward Tatum reported strong evidence for a sexual phase in *E. coli* K-12
1949–1951	J. Lederberg and L Luca Cavalli and Henri Heslot found that most strains of *E. coli* will not mate with K-12
1950	André Lwoff and Antoinette Gutmann clarified the nature of phage lysogeny
1951	Esther Lederberg discovered the lysogenic bacteriophage lambda in *E. coli* K-12
1950s	Respiratory deficient mutants in yeast (petites) are studied by Piotr Słonimski and Boris Ephrussi and are attributed to cytoplasmic hereditary units in the mitochondria. Mutations in *Chlamydomonas* are attributed to hereditary units in the chloroplasts by Ruth Sager

1950–1952	William Hayes suggested that mating in *E. coli* is an asymmetric (unidirectional) process rather than one analogous to cell fusion and zygote formation in higher organisms
1952	J. Lederberg reviewed the literature on cell heredity and suggested the term "plasmid" for all extrachromosomal hereditary determinants
1952–1953	Hayes, and J. Lederberg, Cavalli, and E. Lederberg reported that the ability to mate is controlled by a factor (F) that seems to be an infectious particle not associated with the chromosome
1954	Pierre Fredericq and colleagues showed that colicines behave as genetic factors independent of the chromosome
1958	François Jacob and Elie Wollman proposed the term "Episome" to describe genetic elements such as F, colicine, and phage lambda which can exist both in association with the chromosome and independent of it
1959	Jacob and Edward Adelberg found that the F-factor can associate with cell genes and identify F-prime factors
1959	Albrecht Kleinschmidt and Rudolf Zahn showed that DNA molecules can be studied in the EM by spreading the DNA in protein films on the surface of water
1960–1961	Tomoichiro Akiba, Kotaro Koyama, Yoshito Isshiki, Sadao Kimura, and Toshio Fukushima, and Tsutomu Watanabe and Toshio Fukusawa described multiple drug resistance transferred by an episome designated the R-factor
1961	Physical experiments involving DNA labeling (either by density [Julius Marmur et al.] or radioactivity [Simon Silver and Haruo Ozeki]) show that mating in bacteria is accompanied by transfer of DNA from the donor to the recipient
1962	In a review on episomes, Allan Campbell proposed the reciprocal recombination of circular episome DNA molecules with the chromosomal DNA as a way to physically insert the episome DNA linearly into the chromosome
1962	Circular DNA is found to actually exist by Walter Fiers and Robert Sinsheimer in the genome of the small phage φX174
1963	Alfred Hershey showed that bacteriophage lambda can form circles in vitro by virtue of its "cohesive ends." Other circular DNAs are also reported: the *E. coli* genome by John Cairns; and polyomavirus DNA by Renato Dulbecco and Marguerite Vogt, and by Roger Weil and Jerome Vinograd
1967	Roger Radloff, William Bauer, and Jerome Vinograd described the dye-buoyant density method to separate closed circular DNA from open circles and linear DNA, thus facilitating the physical study of plasmids
1968	W. Dean Rupp and Garret Ihler showed that episome transfer involves only one of the two strands being transferred by a explicative mechanism
1969	Michael Bazaral and Donald R. Helinski showed that several colicine factors are homogeneous circular DNA molecules

References

Akiba T, Koyama T, Isshiki Y, Kimura S, Fukushima T (1960) On the mechanisms of the development of multiple drug-resistant clones of *Shigella*. Jap J Microbiol 4:219–227
Bazaral M, Helinski DR (1968) Circular DNA forms of colicinogenic factors E1, E2 and E3 from *Escherichia coli*. J Mol Biol 36:185–194
Boveri T (1904) Ergebnisse über die Konstitution der chromatischen Substanz des Zellkerns. Fischer, Jena
Brock T (1990) The emergence of bacterial genetics. Cold Spring Harbor Press, Cold Spring Harbor, p 104
Cairns J (1963) The bacterial chromosome and its manner of replication as seen by autoradiography. J Mol Biol 6:208–213
Calef E, Licciardello G (1960) Recombination experiments on prophage host relationships. Virology 12:81–103
Campbell A (1962) Episomes. Adv Genetics 11:101–145
Cavalli LL, Heslot H (1949) Recombination in bacteria: outcrossing *Escherichia coli* K 12. Nature 164:1057–1058
Clowes RC (1972) Molecular structure of bacterial plasmids. Bacteriol Rev 36:361–405
Dulbecco R, Vogt M (1963) Evidence for a ring structure of polyoma virus DNA. Proc Natl Acad Sci USA 50:236–243
Ephrussi B, Słonimski PP (1955) Yeast mitochondria: subcellular units involved in the synthesis of respiratory enzymes in yeast. Nature 176:1207–1208
Fiers W, Sinsheimer RL (1962) The structure of the DNA of bacteriophage φX174: I. The action of exopolynucleotidases. J Mol Biol 5:408–434
Fredericq P (1963) On the nature of colicinogenic factors: a review. J Theor Biol 4:159–165
Gratia A (1925) Sur un remarquable exemple d'antagonisme entre deux souches de colibacille. Compt Rend Soc Biol 93:1040–1041
Harrison R (1937) Embryology and its relations. Science 85:369–374
Hayes W (1952) Recombination in *Bact. coli* K12: unidirectional transfer of genetic material. Nature 169:118–119
Hayes W (1953) Observations on a transmissible agent determining sexual differentiation in *Bacterium coli*. J Gen Microbiol 8:72–88
Hayes W (1968) The genetics of bacteria and their viruses, 2nd edn. Blackwell, Oxford
Hershey AD, Burgi E, Ingraham L (1963) Cohesion of DNA molecules isolated from phage lambda. Proc Natl Acad Sci USA 49:748–755
Huxley J (1942) Evolution: the modern synthesis. G Allen and Unwin, London
Jacob F, Adelberg EA (1959) Transfert de caracteres genetiques par incorporation au facteur sexuel d'*Escherichia-coli*. Compt Rend Acad Sci 249:189–191
Jacob F, Wollman E (1958) Les épisomes, éléments génétiques ajoutés. Compt Rend Acad Sci 247:154–156
Kleinschmidt AK, Zahn RK (1959) Deoxyribonucleic acid molecules in protein-mixed films. Z Naturforsch 146:770–779
Lederberg J (1947) Gene recombination and linked segregations in *Escherichia coli*. Genetics 32:505
Lederberg EM (1950) Lysogenicity in *Escherichia coli* strain K-12. Microb Gen Bull 1:5–8
Lederberg J (1951) Prevalence of *Escherichia coli* strains exhibiting genetic recombination. Science 114:68–69
Lederberg J (1952) Cell genetics and hereditary symbiosis. Physiol Rev 32:403–430
Lederberg J, Tatum EL (1946) Gene recombination in *Escherichia coli*. Nature 158:558
Lederberg J, Cavalli LL, Lederberg EM (1952) Sex compatibility in *Escherichia coli*. Genetics 37:720–730
Levinthal C (1956) The mechanism of DNA replication and genetic recombination in phage. Proc Natl Acad Sci USA 42:394–404

Marmur J, Rownd R, Falkow S, Baron LS, Schildkraut C, Doty P (1961) The nature of intergeneric episomal infection. Proc Natl Acad Sci USA 47:972–979

Morgan TH (1926) Theory of the gene. Yale Univ Press, New Haven

Radloff R, Bauer W, Vinograd J (1967) A dye-buoyant-density method for the detection and isolation of closed circular duplex DNA: the closed circular DNA in HeLa cells. Proc Natl Acad Sci USA 57:1514–1521

Rhoades MM (1955) Interaction of genic and non-genic hereditary units and the physiology of non-genic function. In: Ruhland W (ed) Encyclopedia of Plant Physiology, vol 1. Springer, Berlin

Ris H, Plaut W (1962) Ultrastructure of DNA-containing areas in the chloroplast of *Chlamydomonas*. J Cell Biol 13:383–391

Rupp WD, Ihler G (1968) Strand selection during bacterial mating. Cold Spring Harb Symp Quant Biol 33:647–650

Sager R (1954) Mendelian and non-Mendelian inheritance of streptomycin resistance in *Chlamydomonas reinhardi*. Proc Natl Acad Sci USA 40:356–363

Sager R (1972) Cytoplasmic genes and organelles. Academic Press, New York

Sapp J (1987) The cold war in genetics. In: Beyond the gene. Oxford Univ Press, Oxford

Silver S, Ozeki H (1962) Transfer of deoxyribonucleic acid accompanying the transmission of colicinogenic properties by cell mating. Nature 195:875–876

Sinsheimer RL (1959) A single-stranded deoxyribonucleic acid from bacteriophage φX174. J Mol Biol 1:43–53

Sturtevant AH (1913) The linear arrangement of six sex-linked factors in *Drosophila*, as shown by their mode of association. J Exptl Zool 14:43–59

Sutton WS (1903) The chromosomes in heredity. Biol Bull 4:231–251

Watanabe T, Fukasawa T (1961) Episome-mediated transfer of drug resistance in Enterobacteriaceae ii.: elimination of resistance factors with acridine dyes. J Bact 81:679–683

Watson JD (1968) The double helix. Atheneum, New York, pp 141–142

Weil R, Vinograd J (1963) The cyclic helix and cyclic coil forms of polyoma viral DNA. Proc Natl Acad Sci USA 50:730–738

Whitfield JF, Appleyard RK (1958) Recombination and phenotypic mixing during phage growth in strains of *Escherichia coli* doubly lysogenic for coliphage lambda. Virology 5:275–290

Wollman E, Jacob F (1959) La sexualité des bactéries. Masson, Paris

Symbiosis Between Non-Transferable Plasmids and Prokaryotic Cells

Francisco Dionisio, João Alves Gama and André F.P. Carvalho

Abstract Plasmids are common in the prokaryotic world, both in bacteria and archaea. Most of these extrachromosomal DNA molecules do not code for essential genes. One may expect that the replication of plasmids and the expression of plasmidic genes impose a fitness cost to their host. Given this cost, and given that plasmid-free cells often arise, it is striking that so many non-transferable plasmids are able to maintain themselves inside prokaryotic cells without being counter-selected in favor of plasmid-free cells. A solution to this paradox would be the evolution of controlling mechanisms to regulate rivalry between plasmids for the stability of these symbiotic relationships. In this chapter, we discuss the evolutionary selective conditions for such mechanisms to evolve.

Keywords Plasmid · Competition · Conflict · Rivalry · Relatedness · Virulence · Mutualism · Commensalism

F. Dionisio · J.A. Gama · A.F.P. Carvalho
Centre for Environmental Biology (CBA), University of Lisbon,
Edifício C2.5.46 Campo Grande, 1749-016 Lisbon, Portugal

F. Dionisio · J.A. Gama · A.F.P. Carvalho
Gulbenkian Institute for Science (IGC), Rua da Quinta Grande, 6, 2780-156 Oeiras, Portugal

F. Dionisio · A.F.P. Carvalho
Centre for Ecology, Evolution and Environmental Change (CE3C), University of Lisbon,
Edifício C2.5.46 Campo Grande, 1749-016 Lisbon, Portugal

F. Dionisio (✉)
Department of Plant Biology, Faculty of Science, University of Lisbon,
Edifício C2.2.24 Campo Grande, 1749-016 Lisbon, Portugal
e-mail: dionisio@fc.ul.pt; francisco.dionisio@gmail.com

J.A. Gama
Centre for Ecology, Evolution and Environmental Change (CE3C), Faculty of Science,
University of Lisbon, Edifício C2.5.46 Campo Grande, 1749-016 Lisbon, Portugal

N. Gontier (ed.), *Reticulate Evolution*, Interdisciplinary Evolution Research 3,
DOI 10.1007/978-3-319-16345-1_7

1 Plasmids

Plasmids are double-stranded DNA molecules that have the ability of self-replication inside prokaryotic cells. Plasmids are generally able to cope with cell division and ensure that at least one plasmid copy stays in each daughter cell. Like viruses, plasmids lack a proper metabolism, but, unlike them, plasmids cannot persist freely outside the host because they lack a protective structure (such as viral capsides) against environmental stresses.

Some plasmids are able to transfer horizontally to other cells, and there are two major groups. Conjugative plasmids are those harboring all the necessary genes for horizontal transfer, including (i) the mobility genes and (ii) a membrane-associated mating pair formation complex. Mobilizable plasmids lack the genes for the mating pair formation complex; hence, they are unable to transfer horizontally to another cell. This inability can be circumvented if the cell also harbors a conjugative plasmid encoding these genes. However, there is a third major group of plasmids henceforth named non-transferable plasmids, that cannot transfer horizontally to other cells, even if they coexist with plasmids encoding the transfer apparatus. A recent study has shown that 48 % of proteobacterial plasmids are of this latter kind: neither conjugative nor mobilizable (Smillie et al. 2010).

2 Evolution of Virulence of Non-Transferable Plasmids

One may expect that the replication of plasmids and the expression of plasmidic genes implicate a fitness cost to their host, specially when the cell population just received a plasmid. This fact has been observed very often with several types of plasmids. Figure 1 shows the fitness of *Escherichia coli* cells shortly after receiving a plasmid relative to the fitness before receiving the plasmid.

At first glance, and according to the trade-off hypothesis for the evolution of virulence, dyads of prokaryotic cells and non-transferable plasmids should evolve toward commensalism or mutualism (Bull 1994; Levin 1996; Messenger et al. 1999). The trade-off hypothesis proposes that there is a compromise between the fitness cost (virulence) imposed by the parasite and the rate of horizontal transmission of the parasite. Therefore, to maximize the spread of the parasite, evolution leads to a balance between virulence and transmission mutualism (Bull 1994; Levin 1996). This hypothesis has been corroborated by several authors (see, e.g., Messenger et al. 1999; Turner et al. 1998). The inability to transfer horizontally would lead non-transferable plasmids evolve towards commensalism or even mutualism because plasmids' success is tightly associated with the host's success. However, non-transferable plasmids are an interesting case where the null rate of horizontal transmission does not necessarily maximize genetic relatedness (Frank 1996; Chao et al. 2000). The reason for this is very simple: despite their inability to transfer to other cells, their host may sometimes have to compete with incoming transferable plasmids. This possibility eventually forces the resident

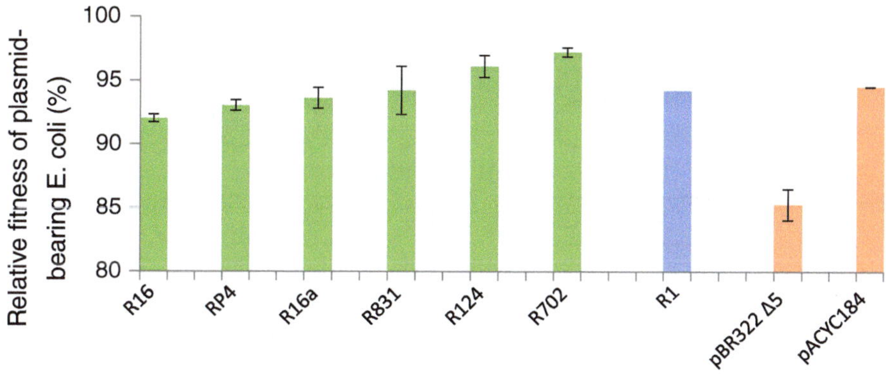

Fig. 1 Fitness of plasmid-bearing *E. coli* cells relative to the fitness of plamid-free cells. Plasmids R16, RP4, R16a, R831, R124, R702, and R1 are conjugative. Plasmids pBR322Δ5 and pACYC184 are non-transferable. The plasmid pBR322Δ5 is artificial and commonly used in molecular biology

(non-transferable) plasmid to have mechanisms to compete with other plasmids, hence imposing a cost on the host. To ensure vertical transmission, a conceivable mechanism available to the non-transmissible resident plasmid is to increase copy-number in the cell. The drawback of such strategy is that it increases the cost imposed to the host.

Unsurprisingly, plasmids encode mechanisms for copy-number control, typically by repressing plasmid replication when copy-number achieves a certain value (for a review, see Paulsson and Ehrenberg 2001). For example, the well-studied conjugative plasmid R1 controls its copy-number in two steps. Firstly, R1 plasmids encode a replication inhibitor with an ingenious mechanism to "count" the number of plasmid copies—the inhibitor molecule is unstable which implies that the concentration of plasmid copies is proportional to the replication inhibitor molecules. Secondly, the inhibitor is the antisense RNA of the mRNA responsible for initiation of plasmid replication—hence blocking translation of the mRNA and the replication of the plasmid. In this way, the number of inhibitors increases with the number of plasmid copies (for a review, Nordstrom 2006).

Something remarkable happens when two plasmids use very similar regulation mechanisms, that is, when they react to similar replication repressors: In a growing population of host cells, the two plasmids cannot stably coexist in the same cells. Having similar replication control systems, they tend to respond to each other's replication activators and repressors. In other words, the two plasmids are incompatible (for a review, see Novick 1987). As an example, suppose that, when isolated in their own host cell, plasmid A and plasmid B have similar copy-numbers $c_A = c_B = 2$ a few moments before cell division. Now, further suppose that, for some reason, they find themselves in the same cell. Given that they count the other plasmid as one of itself, neither plasmid replicates. Each plasmid acts as if the other plasmid is already a copy of itself. Thus, after cell division, the two plasmids A and B go to different daughter cells (Fig. 2).

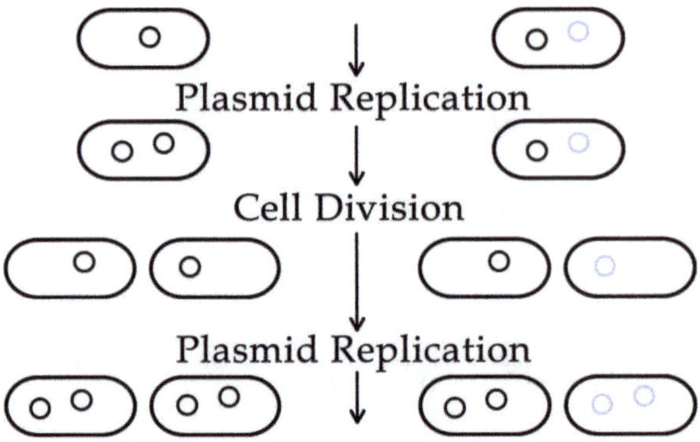

Fig. 2 Incompatibility between similar plasmids. Plasmids are represented as *small circles* inside bacterial cells—represented with an *ovoid shape*. Chromosomes are not represented. Two plasmids with similar replication control systems (here represented by *black* and *gray circles*) tend to respond to each other's replication activators and repressors. When this happens, the two different plasmids cannot stably maintain in bacterial cells. In other words, these plasmids are said to be *incompatible*. Being similar, each plasmid *counts* the other plasmid as one of itself, i.e., each plasmid acts as if the other plasmid is already a copy of itself. Ultimately, two plasmids find themselves in *different* daughter cells

Precisely because of this phenomenon of incompatibility, a good strategy for one of the plasmids to compete with the other plasmid would be to evolve toward a higher copy-number—hence ensuring vertical transmission to both daughter cells. Such a process would be advantageous to the winning plasmid, but this victory implies the production of extra copies, which, most likely, increased the cost imposed on the host. Competition between hosts imposes a drawback to the cells with higher copy-numbers. This is a clear example of multilevel selection: the levels of selection are not aligned, implying that an advantage at a given level (e.g., at the level of the plasmids, increasing copy-number) may imply a disadvantage at another level (cell level, suffering a burden from the presence of the high plasmid copy-number) (Paulsson 2002).

The emergence of plasmid incompatibility poses a remarkable paradox. Apparently, regulation mechanisms evolved to control for copy-number. Yet, such mechanisms often lead to incompatibility between similar plasmids. However, when there is incompatibility, increasing copy-number is a winning strategy.

The copy-number control mechanism of most plasmids (e.g., the R1 plasmid mentioned above) has an interesting characteristic. While the molecules responsible for the initiation of plasmid replication replicates only the plasmid where they are encoded, the inhibitor always constrains the replication of all plasmid molecules, not only the plasmid molecule where it is encoded. Would it be a good strategy for the "other" plasmids to mutate the receptor of the inhibitor in order to become deaf to it? Kentzoglanakis et al. (2013) analyzed the conflict between obedience and deafness to the control mechanism with computer simulations to show that the evolution of plasmid copy-number control mechanisms improves plasmid stability (Kentzoglanakis et al. 2013).

Before focusing on non-transferable plasmids, it is instructive to start by discussing the conditions for the maintenance of transmissible plasmids. Such conditions are very stringent, mainly because the transfer rate of many plasmids seems to be too low to explain their maintenance (for a review, see Dionisio et al. 2012).

3 What Do We Know About Maintenance Conditions of Conjugative Plasmids?

Stewart and Levin (1977) designed a theoretical model of a chemostat with plasmid-free cells and plasmid-bearing cells to analyze the conditions for plasmid maintenance. A chemostat is a bioreactor with continuous influx of fresh culture medium and efflux of culture, at a constant rate. By changing the rate with which fresh medium is added to the bioreactor, the growth rate of the microorganism can be controlled (Novick and Szilard 1950). Stewart and Levin (1977) found that the transfer rate of conjugative plasmids has to be high enough to compensate for the following: (i) plasmid fitness cost, (ii) turnover of the chemostat, and (iii) segregation rate (i.e., rate at which plasmid-bearing cells loose the plasmid upon cell division) (Stewart and Levin 1977). However, not all conjugative plasmids seem to transfer to other cells with enough speed (Dionisio et al. 2002; Gordon 1992). This creates a difficulty to explain the existence of transferable plasmids (let alone non-transferable plasmids): If conjugative and mobilizable plasmids impose a cost to their hosts and their transfer rate is low, why do so many prokaryotic cells contain these plasmids? Recently, it has been observed that, under specific conditions, cells may take up plasmids from the surrounding environment, but this phenomenon seems to be too rare to explain the maintenance of plasmids in nature (Maeda et al. 2006; Kurono et al. 2012; Matsuda et al. 2012; Perez-Mendoza and de la Cruz 2009). Of course, one can hypothesize that plasmids encode some genes that are advantageous in other conditions—other than laboratory conditions of a chemostat at stable temperature, aeration, and nutrition (Stewart and Levin 1977). This solves the puzzle, but generates a paradox.

One may assume that, to maintain plasmids, cells have to spend resources to at least express plasmidic genes and replicate plasmidic DNA (Bentley et al. 1990; Diaz Ricci and Hernandez 2000; Harrison et al. 2012; Lenski 1997). From the point of view of the chromosome, the best solution would be recruiting the advantageous genes and discard the rest of the plasmid (Levin and Bergstrom 2000). Therefore, plasmids should not exist even if they encode certain genes useful for the host—these genes should be present in the chromosome.

Smith (2001) proposed an interesting hypothesis to explain the maintenance of certain genes on mobile genetic elements rather than on the chromosome (Smith 2001). This hypothesis focuses on secreted factors that play their role outside cells, that is, metabolites that constitute *public goods*—the term used in micro-economy to name goods (here, the secreted metabolites) that are accessible to all individuals (non-excludability as defined in micro-economy), and such that, the use of the good by one individual does not reduce availability of the good to other individuals (non-rivalry as defined in micro-economy). For an account on the relation between public goods, the tragedy of the commons and the prisoner's dilemma in economical sciences and in evolutionary biology, see (Dionisio and Gordo 2006).

As a public good, the metabolite is accessible, not only to producer cells, but also to non-producer cells. In principle, one may assume that the production of these metabolites impose a fitness cost to producers. Given this cost, non-producers will increase in frequency. If the genes encoding for the production of secreted metabolites are on transferable plasmids, plasmid-bearing cells may force non-producers to produce the metabolite (Smith 2001). With this process, producers avoid the cost of competition with cheaters and, at the same time, gain a few more collaborators to produce the metabolite (Smith 2001).

Mc Ginty et al. (2011) pinpoint a flaw in this reasoning. It involves surface and entry exclusion, very often encoded by plasmids. With any of these mechanisms, the resident plasmid strongly decreases the probability that similar plasmids enter into its host (for a review, see Garcillan-Barcia and de la Cruz 2008). For example, the transfer rate of the F plasmid is about 100-fold higher if the recipient cell has no F plasmid than if the recipient cell already harbors the F plasmid (Achtman et al. 1977).

Mc Ginty et al. (2011) posed the following question: What happens if some of these plasmids loose the gene(s) coding for the public good? Cells containing these defective plasmids do not pay the cost for metabolite production, although they have the cost of bearing the plasmid. However, because these cells carry the (mutated) plasmid, both surface and entry exclusion prevent the entry of plasmids that code for the public good. In homogenous populations, this effect will be responsible for the displacement of cells with cooperating plasmids in favor of cells carrying cheater plasmids (Mc Ginty et al. 2011). That is, without plasmids, competition between chromosomes favors non-producers of the public good. When plasmids rather than chromosomes code for the public good, competition occurs at the level of cells: between cells that code and those that do not code for the public good (Mc Ginty et al. 2011). However, in structured populations, one rescues Smith's hypothesis. In a non-homogeneous population, horizontal gene transfer benefits plasmid-carried public goods through the twofold effect of increasing local relatedness and through the effects of transmission (Mc Ginty et al. 2013). This is an important point to stress because actual prokaryote populations are not homogeneous.

Bioinformatic analysis of plasmidic and chromosomal sequences gave strong support to Smith's hypothesis. Indeed, many of the genes carried by mobile elements code for metabolites that constitute public goods. These metabolites are either released into the extracellular environment or exhibited at the cell surface (Mc Ginty et al. 2011; Nogueira et al. 2009). These proteins may have diverse functions, from foraging, to shelter, or even for microbial virulence. Moreover, these proteins use amino acids with low biosynthetic cost [i.e., cheaper than the other proteins, even highly expressed proteins (Nogueira et al. 2009)].

Smith's hypothesis explains the presence of certain genes in plasmids, rather than in chromosomes, with the fact that bacterial cells often have to interact with competitor bacterial cells. In this sense, one may classify this hypothesis as *social*. As such, competition, cooperation, altruism, or spite are social behaviors. Consider, for example, a (new) gene (or a new gene version—mutation) that enables a bacterium to replicate faster by consuming a certain sugar though in an inefficient way; as such, less sugar is left to other cells, which means that this is a social interaction. In contrast, if a certain bacterium has a mutation that turns it more resistant to, say, detergents, then its expected reproductive success increases by its own—without directly influencing the fate of other cells.

Other explanations for plasmid maintenance based on social interactions have been proposed and are reviewed elsewhere (Dionisio et al. 2012).

In conclusion, the point we want to stress is that, to assure maintenance among prokaryote populations, either plasmid-borne genes are involved in host social behavior (Mc Ginty et al. 2013; Nogueira et al. 2009; Smith 2001), or plasmids have a very high transfer rate to compensate for plasmid segregation and fitness cost imposed on the host (Stewart and Levin 1977).

Both possibilities involve the transfer ability of plasmids. So, how should one explain the maintenance of non-transmissible plasmids among prokaryote cells?

4 What Do We Know About Maintenance Conditions of Non-Transferable Plasmids?

Consider a population of bacterial cells that just received a plasmid. Usually, the growth rate of this population declines a few percentages, typically not more than 10 % (Dionisio et al. 2012). Now, if one selects for the plasmid (hence imposing plasmid maintenance) for a few hundred generations, the cost often disappears. That is, populations of evolved plasmid-bearing cells now grow as fast or almost as fast as populations of plasmid-free cells. Several authors have seen this outcome with different plasmid-bacterium dyads and in different conditions. Details on where mutations occur (plasmid or chromosome) differ from dyad to dyad and from experiment to experiment, but they all have in common the fact that a population of the evolved dyad regains the growth rate of a population of plasmid-free cells (Bouma and Lenski 1988; Dahlberg and Chao 2003; Dionisio et al. 2005; Modi and Adams 1991). Why is that so? Moreover, why is this observed both in transferable and non-transferable plasmids?

When a transferable plasmid "moves" to another cell, it leaves a copy in the original cell. This is rather uncommon among other types of prokaryote parasites. For example, viruses typically kill their host to transfer to another host. Most bacterial viruses are either lysogenic or lytic viruses. Lytic viruses invade the host cell, takeover cell metabolism to replicate themselves several times, and, more often than not, kill the host, releasing tens, hundreds, or more viruses (Campbell 1996). Lysogenic viruses, however, have two possibilities: (1) either they follow the lytic cycle as just described for lytic viruses or, instead, (2) they integrate into the bacterial chromosome. In the latter case, the virus genome becomes part of the bacterial chromosome over several cell generations. Some conditions (e.g., UV radiation) trigger the viral lytic cycle, producing and releasing viral progeny and thus killing the host. The newly released lysogenic viruses are now ready to infect new cells, again with the two options described above. Other bacterial viruses are neither lytic nor lysogenic. Consider, for example, the M13 "chronic" virus that, after infecting an *E. coli* cell, the cell does not replicate anymore because it is too busy replicating the virus and releasing around 200 new M13 viruses per hour. These hundreds of viruses are now ready to infect hundreds of *E. coli* cells. The point here is that, in contrast to what happens with plasmids, newly produced viruses ultimately leave the host. Therefore, the fitness of any plasmid (conjugative, mobilizable, or non-transferable) has a direct relationship with the reproductive success of the host, which is not the case for bacterial viruses.

5 Problems that Plasmids Have to Solve to Avoid Heavy Costs

Despite using the replication machinery of the cell, plasmid replication is autonomous—in the sense that it occurs independently of chromosomal or cell duplication. In the process of cell division, plasmid loss may occur if, by chance, one of thc two daughter cells receives all the plasmid copies. Generally, this will not happen with the chromosome because the cell uses a partition system to assure that each daughter cell inherits a chromosomal copy.

In the case of plasmids without a partition system, one expects binomial partitioning (Novick 1987; Paulsson 2002). Assuming that the probability of ending up in one or the other daughter cell is ½, the probability that all of the *n* copies go into the same cell is $2\left(\frac{1}{2}\right)^n$. The factor *2* comes from the fact that the inheritance of all the copies may occur in any of the two cells: the other becomes plasmid-free. If, for a given plasmid type, there are two plasmid copies at the moment of cell replication (n = 2), the probability that the two plasmid copies end up in the same daughter cell is 50 % if there is no partition system. This probability of plasmid-free cells to arise is too high because bacterial populations are often composed of million or billion of cells and such a value would allow the emergence of fast-growing plasmid-free cells.

Suppose that, in the moment of cell division, the plasmid assures a plasmid copy-number of $n = 20$; now, the probability that one daughter cell is plasmid-free is $2\left(\frac{1}{2}\right)^{20} \approx 2 \times 10^{-6}$. Is this probability low enough? Perhaps not and the reason is the following. Consider a plasmid-bearing bacterial cell (e.g., *E. coli*) that divides every 20 min. In the laboratory environment, at 37 °C and with good aeration, the population grows from one cell to 10^{10} cells in 11 h only. To reach this cell number, they divided almost 10^{10} times and, every time a cell divides, there is a probability of 2×10^{-6} that a plasmid-free cell arises. Therefore, it is almost certain that at least one plasmid-free cell arises in this process, probably hundreds of them. Having no cost of bearing a plasmid, plasmid-free cells may outcompete plasmid-bearing cells.

Sometimes plasmids recombine on each other. For example, two plasmids join to form a single plasmid with all genetic information duplicated. These recombination events may involve more than two plasmids. The formation of these circular molecules composed of tandem repeats of the plasmid—called multimers—may strongly increase plasmid loss upon cell divisions. If, among the n plasmids, k plasmids form a multimer, the probability for the creation of a plasmid-free cell rises 2^{k-1}-fold. For example, if two plasmids recombine on each other to form a multimer, the probability of plasmid-free cells to arise is doubled in comparison with before the recombination event.

Plasmids contain different types of genes to solve these problems, such as those encoding mechanisms for partition (Austin and Abeles 1983a, b; Nordstrom and Austin 1989), for multimer resolution (Summers 1994; Summers et al. 1993; Summers and Sherratt 1984), as well as for the tight control of plasmid copy-number (Lestas et al. 2010; Paulsson and Ehrenberg 2001). We are not going to review these mechanisms here. We rather discuss the following: how should one explain that most or all plasmids, transferable or not, evolve toward very low or null cost for its host?

6 Selective Conditions for the Evolution of Control Mechanisms of Rivalry in Non-Transmissible Plasmids

Consider a non-transmissible plasmid without partition systems, that is, without mechanisms to ensure transmission to both cells upon cell division. Assuming that this plasmid (and its copies) is the only one inside a prokaryotic cell, an efficient method to assure plasmid vertical transmission is by achieving a high copy-number inside cells. Indeed, and as we saw above, with very high numbers of plasmid copies, the probability that plasmid-free cells arise at cell division becomes very low. In other words, the individual reproductive success of a plasmid is higher for higher values of copy-number. However, a cell containing "greedy" plasmids (replicating themselves too many times) would decline in frequency when growing in competition with other cells with low copy-number plasmids.

As mentioned above, plasmids encode for mechanisms to regulate copy-number, which may lead to incompatibility when two similar plasmids find themselves in the same cell. We also mentioned that a good strategy for a given focal plasmid to compete with other plasmids would be to evolve toward higher copy-number—higher than the *other* plasmids. In other words, relative copy-number may be a crucial factor, not only the absolute copy-number (Dionisio, in preparation).

For different plasmids, the relevance of absolute copy-number or relative copy-number inside host cells may change. Suppose that only relative copy-number is important for the individual component of plasmid fitness. For this case (though in a different context), Frank (1995) has shown that a policing mechanism to control rivalry would be selected for low values of relatedness and/or low values of costs of the policing mechanism. The term *policing* or *policing mechanism* is used here as a mechanism that regulates a certain behavior to prevent individual deviations or selfish behavior. For high values of relatedness, self-restraint yields superior success than policing. If relatedness is high, there is no point in competing too much; hence, there is no reason to spend resources (which have a cost) to control for plasmid rivalry.

Should non-transferable plasmids invest in such a mechanism to control rivalry among plasmids? Given that they do not transfer to other cells, one might expect that the genetic similarity between plasmids within each cell is very high (high relatedness). However, cells containing non-transferable plasmids may receive transferable plasmids from other cells and both may have to compete with each other particulary if they belong to the same incompatibility group. For non-transferable plasmids that do not rely on partition systems, the absolute copy-number seems to be more important for fitness than the relative copy-number (Dionisio, in preparation).

According to (Dionisio, in preparation), the selective conditions for the evolution of a control mechanism to regulate rivalry among plasmids are wider for low values of relatedness between plasmids and for low-cost mechanisms (as seen above, Frank 1995). Moreover, such mechanisms are also more prone to be selected if the importance of relative copy-number for plasmid fitness is low. Indeed, if the rivalry term is of low importance for plasmid fitness, it always pays to control for rivalry.

Non-transferable plasmids have to deal with rivalry only if one of the plasmid copies mutates or if another plasmid comes in; in other words, when plasmid diversity within the bacterial cell increases. This is in contrast to the case of conjugative plasmids: By transferring to several cells, the probability of having to deal with rivalry is much higher than if it was non-transferable. Therefore, non-transferable plasmids are more prone to evolve a policing mechanism to control for rivalry than mobilizable or conjugative plasmids. At first, may be surprising that plasmids less involved in rivalry are the ones more disposed to control for it. The explanation for this is that plasmids for which the rivalry term is less important are also those plasmids with less to lose if the rivalry term decreases by the action of the policing mechanism.

7 Epistatic Interactions Between Plasmids

Consider a given pair of plasmid types, say A and B. When each plasmid type is alone in their host cell, they are all equal to each other within each cell and no rivalry is expected (Frank 1995). Therefore, there is no need for a policing mechanism to diminish rivalry. When the two plasmid types are together in the same cell, relatedness decreases ($r < 1$), and policing mechanisms encoded by both plasmids A and B may control for rivalry between them. Therefore, the fitness cost of the two plasmids together is expected to be lower than the sum of the fitness cost of each plasmid isolated—the case of positive epistasis.

However, if two plasmids spend resources to control for rivalry between them but without success, one expects negative epistasis because the cost of both plasmids together is higher than the sum of the fitness costs when plasmids infect distinct cells.

This could explain the results obtained by Silva et al. (2011) where four out of nine pairs of conjugative plasmids showed positive epistasis and three out of nine showed positive epistasis (i.e., 7/9 = 78 % cases of non-zero epistasis). In contrast, the authors show in the same study (with the same plasmids) that the proportion of non-zero epistasis between single plasmids and chromosomal mutations is 20/50 = 40 % (Silva et al. 2011). Of course, these results involving epistasis between plasmids and the above prediction require further testing and do not explain why most natural plasmids evolved toward commensalism—they just suggest an explanation for a low cost.

8 Conclusion

Understanding the existence of transferable plasmids is not easy (reviewed by Dionisio et al. (2012). The transfer rate of many conjugative plasmids is not high enough to explain their existence as parasites only, and it is insufficient to explain plasmid existence on the grounds of their "beneficial" genes because it does not explain why they are carried in plasmids. Under what conditions should cells maintain certain genes in mobile elements instead of recruiting them into the chromosome? These questions gave rise to the hypothesis that these specific genes placed in plasmids have social functions, that is, functions that influence the fitness of other cells: By being transferable to other cells, plasmid donors avoid exploitation by cheater cells and force others to cooperate to produce public goods.

The work by Nogueira et al. (2009) discussed above corroborates this hypothesis. Interestingly, however, Nogueira et al. (2009) does not distinguish between fully transferable plasmids and non-transferable plasmids. Indeed, the authors placed all types of plasmids and other mobile elements on one side and the core genome of the chromosome on the other and showed that the genes coding for secreted molecules predominate in the first set (Nogueira et al. 2009). Given that non-transferable plasmids have genes coding for secreted molecules, they are probably transient states between a conjugative plasmid and plasmid extinction (Mc Ginty et al. 2011; Smith 2001; Nogueira et al. 2009; Mc Ginty et al. 2013).

According to Smith (2001) and Mc Ginty et al. (2013), if a given gene has its product secreted, it makes sense to place it in a conjugative plasmid (Mc Ginty et al. 2013; Smith 2001). However, if the plasmid became non-transferable, it should be advantageous for the cell to recruit the gene into the chromosome (and discard the rest of the non-transferable plasmid), or, even better, place that gene in a transferable plasmid (and discard the non-transferable plasmid). If this hypothesis were true, it would imply that there is a continuous appearance of non-transferable plasmids by mutation of transferable plasmids and of plasmid-free cells from cells plasmid-bearing cells, and, somehow, continuous formation of new conjugative and/or mobilizable plasmids in nature.

Interestingly (Dionisio, in preparation), theoretical models suggest a more complex solution to this problem. As argued above, non-transferable plasmids without partition systems (hence strongly investing on high copy-numbers) are more prone to be involved in policing for rivalry than conjugative plasmids. With very low values of rivalry, stable coexistence with conjugative and mobilizable plasmids is more likely. This creates good conditions for the non-transferable plasmid to become, again, a transferable plasmid: non-transferable plasmids just have to regain the missing genes (mobility genes or genes involved in mating pair formation complex) from coexisting transferable plasmids. According to this prediction, plasmids may lose and regain these genes several times and, each time, from different transferable plasmids.

The main objective of this chapter is to answer the question: How should one explain that most or all plasmids, transferable or not, evolve toward very low or null cost for its host? If plasmids code for useful genes, i.e., useful to the bacterial cell, it appears that the chromosome *should* recruit the useful genes and discard the *backbone* of the plasmid—e.g., genes for plasmid replication and other genes apparently useful only to plasmids themselves. Probably this is indeed occurring in nature all the time, but one still finds plasmids inside most of natural bacterial strains and one has to be able to explain this. Here, we present some explanatory hypothesis: plasmids as carriers of genes that code for public goods; plasmids avoid entrance of other plasmids; plasmids repress rivalry. What these hypotheses have in common is that they all have a social dimension. That is, plasmids are important to cope with other, less domesticated, incoming plasmids. As argued, none of these hypotheses alone seems to be sufficient to explain plasmid existence and their low or null cost among most of natural bacterial cells. Future work, mainly experiments and computer simulations, should address this hypothesis.

Glossary

Antisense RNA Regulatory RNA molecule complementary to the target RNA (Nordström et al. 1996)

Chemostat Bioreactor with continuous influx of fresh culture medium and efflux of culture, at a constant rate

Symbiosis (commensalism and mutualism) A prolonged relationship between organisms (mutualism if both parts have a benefit or commensalism if the benefit is unilateral)

Conjugation (bacterial conjugation) Transfer mechanism of plasmids (and conjugative transposons) requiring contact between donor and recipient cells (Pinto et al. 2012)

Copy-number See Plasmid copy-number

Epistasis Interaction between genes (Phillips 2008)

Excludability A good is excludable if it is possible to prevent non-contributing individuals from having access to it. Cable TV is excludable, but public TV broadcasts is non-excludable. A metabolite used only by the producer cell (probably inside the cell) is excludable; a metabolite that has its function outside cells are non-excludable (Dionisio and Gordo 2006)

Partition system System regulating the distribution of plasmids between two daughter cells (Pinto et al. 2012)

Plasmid copy-number Average number of copies of a plasmid per cell (Pinto et al. 2012)

Policing mechanism A mechanism that regulates a certain behavior to prevent individual deviations or selfish behavior (Frank 1995)

Prokaryote Organism without a defined cellular nucleus. Both bacteria and archaea are prokaryotes

Proteobacteria A group (phylum) of gram-negative bacteria

Public good A resource available to all interacting individuals (Dionisio and Gordo 2006)

Rivalry Competition for an exhaustible resource: consumption or use by one individual does not reduce the amount available for others (Dionisio and Gordo 2006)

References

Achtman M, Kennedy N, Skurray R (1977) Cell–cell interactions in conjugating *Escherichia coli*: role of traT protein in surface exclusion. Proc Natl Acad Sci USA 74:5104–5108

Austin S, Abeles A (1983a) Partition of unit-copy miniplasmids to daughter cells I: P1 and F miniplasmids contain discrete, interchangeable sequences sufficient to promote equipartition. J Mol Biol 169:353–372

Austin S, Abeles A (1983b) Partition of unit-copy miniplasmids to daughter cells II: the partition region of miniplasmid P1 encodes an essential protein and a centromere-like site at which it acts. J Mol Biol 169:373–387

Bentley WE, Mirjalili N, Andersen DC, Davis RH, Kompala DS (1990) Plasmid-encoded protein: the principal factor in the metabolic burden associated with recombinant bacteria. Biotechnol Bioeng 35:668–681

Bouma JE, Lenski RE (1988) Evolution of a bacteria plasmid association. Nature 335:351–352

Bull JJ (1994) Perspective—virulence. Evolution 48:1423–1437

Campbell AM (1996) Bacteriophages. In: Neidhart FC et al. (eds) *Escherichia coli* and Salmonella—cellular and molecular biology. American Society for Microbiology, Washinghton

Chao L,Hanley KA, Burch CL, Dahlberg C, Turner PE (2000) Kin selection and parasite evolution: higher and lower virulence with hard and soft selection. Q Rev Biol 75:261-275

Dahlberg C, Chao L (2003) Amelioration of the cost of conjugative plasmid carriage in *Eschericha coli* K12. Genetics 165:1641–1649

Diaz Ricci JC, Hernandez ME (2000) Plasmid effects on *Escherichia coli* metabolism. Crit Rev Biotechnol 20:79–108

Dionisio F, Gordo I (2006) The tragedy of the commons, the public goods dilemma, and the meaning of rivalry and excludability in evolutionary biology. Evol Ecol Res 8:321–332

Dionisio F, Matic I, Radman M, Rodrigues OR, Taddei F (2002) Plasmids spread very fast in heterogeneous bacterial communities. Genetics 162:1525–1532

Dionisio F, Conceicao IC, Marques ACR, Fernandes L, Gordo I (2005) The evolution of a conjugative plasmid and its ability to increase bacterial fitness. Biol Lett 1:250–252

Dionisio F, Nogueira T, Carvalho LM, Mendes-Soares H, Mendonça SCM, Domingues I, Moreira B, Reis AM (2012) What maintains plasmids among bacteria? In: Francino MP (ed) Horizontal gene transfer in microorganisms. Caister Academic Press, Norwich, pp 131–154

Frank SA (1995) Mutual policing and repression of competition in the evolution of cooperative groups. Nature 377:520–522

Frank SA (1996) Models of parasite virulence. Q Rev Biol 71:37–78

Garcillan-Barcia MP, de la Cruz F (2008) Why is entry exclusion an essential feature of conjugative plasmids? Plasmid 60:1–18

Gordon DM (1992) Rate of plasmid transfer among *Escherichia-coli* strains isolated from natural-populations. J Gen Microbiol 138:17–21

Harrison E, Koufopanou V, Burt A, MacLean RC (2012) The cost of copy number in a selfish genetic element: the 2-mum plasmid of *Saccharomyces cerevisiae*. J Evol Biol 25:2348–2356

Kentzoglanakis K, Lopez DG, Brown SP, Goldstein RA (2013) the evolution of collective restraint: policing and obedience among non-conjugative plasmids. PLoS Comput Biol 9(4):e1003036

Kurono N, Matsuda A, Etchuya R, Sobue R, Sasaki Y, Ito M, Ando T, Maeda S (2012) Genome-wide screening of *Escherichia coli* genes involved in execution and promotion of cell-to-cell transfer of non-conjugative plasmids: rodZ (yfgA) is essential for plasmid acceptance in recipient cells. Biochem Biophys Res Commun 421:119–123

Lenski RE (1997) The cost of antibiotic resistance—from the perspective of a bacterium. Ciba F Symp 207:131–140

Lestas I, Vinnicombe G, Paulsson J (2010) Fundamental limits on the suppression of molecular fluctuations. Nature 467:174–178

Levin BR (1996) The evolution and maintenance of virulence in microparasites. Emerg Infect Dis 2:93–102

Levin BR, Bergstrom CT (2000) Bacteria are different: observations, interpretations, speculations, and opinions about the mechanisms of adaptive evolution in prokaryotes. Proc Natl Acad Sci USA 97:6981–6985

Maeda S, Ito M, Ando T, Ishimoto Y, Fujisawa Y, Takahashi H, Matsuda A, Sawamura A, Kato S (2006) Horizontal transfer of nonconjugative plasmids in a colony biofilm of *Escherichia coli*. FEMS Microbiol Lett 255:115–120

Matsuda A, Kurono N, Kawano C, Shirota K, Hirabayashi A, Horino M, Etchuya R, Sobue R, Sasaki Y, Miyaue S et al (2012) Genome-wide screen for *Escherichia coli* genes involved in repressing cell-to-cell transfer of non-conjugative plasmids. Biochem Biophys Res Commun 428:445–450

Mc Ginty SE, Rankin DJ, Brown SP (2011) Horizontal gene transfer and the evolution of bacterial cooperation. Evolution 65:21–32

Mc Ginty SE, Lehmann L, Brown SP, Rankin DJ (2013) The interplay between relatedness and horizontal gene transfer drives the evolution of plasmid-carried public goods. Proc R Soc B Biol Sci 280(1761):20130400

Messenger SL, Molineux IJ, Bull JJ (1999) Virulence evolution in a virus obeys a trade-off. Proc R Soc Lond Ser B-Biol Sci 266:397–404

Modi RI, Adams J (1991) Coevolution in bacterial-plasmid populations. Evolution 45:656–667

Nogueira T, Rankin DJ, Touchon M, Taddei F, Brown SP, Rocha EPC (2009) Horizontal gene transfer of the secretome drives the evolution of bacterial cooperation and virulence. Curr Biol 19:1683–1691

Nordstrom K (2006) Plasmid R1—replication and its control. Plasmid 55:1–26

Nordstrom K, Austin SJ (1989) Mechanisms that contribute to the stable segregation of plasmids. Annu Rev Genet 23:37–69

Nordström K, Cohen S, Simons R (1996) Antisense RNA. In: Resnekov O, Gabain A (eds) Post-transcriptional control of gene expression. Springer, Berlin, pp 231–261

Novick RP (1987) Plasmid incompatibility. Microbiol Rev 51:381–395

Novick A, Szilard L (1950) Description of the chemostat. Science 112:715–716

Paulsson J (2002) Multileveled selection on plasmid replication. Genetics 161:1373–1384
Paulsson J, Ehrenberg M (2001) Noise in a minimal regulatory network: plasmid copy number control. Q Rev Biophys 34:1–59
Perez-Mendoza D, de la Cruz F (2009) *Escherichia coli* genes affecting recipient ability in plasmid conjugation: are there any? BMC Genomics 10(1):71
Phillips PC (2008) Epistasis—the essential role of gene interactions in the structure and evolution of genetic systems. Nat Rev Genet 9:855–867
Pinto UM, Pappas KM, Winans SC (2012) The ABCs of plasmid replication and segregation. Nat Rev Microbiol 10:755–765
Silva RF, Mendonca SCM, Carvalho LM, Reis AM, Gordo I, Trindade S, Dionisio F (2011) Pervasive sign epistasis between conjugative plasmids and drug-resistance chromosomal mutations. PLoS Genet 7:e1002181
Smillie C, Garcillan-Barcia MP, Francia MV, Rocha EPC, de la Cruz F (2010) Mobility of plasmids. Microbiol Mol Biol Rev 74(3):434–452
Smith J (2001) The social evolution of bacterial pathogenesis. Proc R Soc B 268:61–69
Stewart FM, Levin BR (1977) Population biology of bacterial plasmids—a priori conditions for existence of conjugationally transmitted factors. Genetics 87:209–228
Summers DK (1994) The origins and consequences of genetic instability in prokaryotes. Dev Biol Stand 83:7–11
Summers DK, Sherratt DJ (1984) Multimerization of high copy number plasmids causes instability: ColE1 encodes a determinant essential for plasmid monomerization and stability. Cell 36:1097–1103
Summers DK, Beton CW, Withers HL (1993) Multicopy plasmid instability: the dimer catastrophe hypothesis. Mol Microbiol 8:1031–1038
Turner PE, Cooper VS, Lenski RE (1998) Tradeoff between horizontal and vertical modes of transmission in bacterial plasmids. Evolution 52:315–329

Host–Symbiont–Pathogen–Host Interactions: Wolbachia, Vector-Transmitted Human Pathogens, and the Importance of Quantitative Models of Multipartite Coevolution

Caetano Souto-Maior

Abstract Infectious disease has been recognized for a long time as an important evolutionary force: It created the need for and shaped the evolution of immune systems and influenced reproduction as well as behavior of many host species. Infectious agents themselves also evolve and must have adapted to host strategies to evade infection, to multiple external and internal environments, and to transmission between hosts. Given the pressure to evolve on both sides, coevolution is expected. Evolution is indeed observed when looking at either host, pathogen, or at other microorganisms directly or indirectly involved and is dependent to some degree on all species interacting. Vector-transmitted diseases with high burden to humans such as malaria and dengue fever are some of many examples where parasites evade the immune system of both mosquito and human hosts, thereby maximizing the vector's transmission and persistence. Arthropod hosts such as mosquitos may also be carriers of vertically transmitted endosymbionts, such as the *Wolbachia* bacterium, that also induce a complex modification of the arthropod's life history traits. This sort of scenario illustrates the need to consider ecological, multipartite, and evolutionary models—the relevance to human health, together with extensive data collection from epidemiological surveillance, provides an opportunity to expand and improve evolutionary theory.

Keywords Infectious diseases · Host–microorganism interactions · Dengue virus · *Wolbachia* · Population biology · Coevolution

C. Souto-Maior (✉)
Gulbenkian Institute for Science (IGC), Apartado 14, 2780-901 Oeiras, Portugal
e-mail: cmmendes@igc.gulbenkian.pt

N. Gontier (ed.), *Reticulate Evolution*, Interdisciplinary Evolution Research 3,
DOI 10.1007/978-3-319-16345-1_8

1 Host–Pathogen, Host–Symbiont, and Symbiont–Pathogen Interactions: The Underlying Concepts

Strict definitions apart, the relationship between hosts and parasites is probably almost as old as life itself. The relevance of disease for human health made it of interest already in ancient societies, much earlier than any scientific methods could be applied or were available to investigate their properties or etiology, which was attributed to spirits or other ethereal entities such as "bad air" ("mal aire," Italian words that originated the name "malaria"). With becoming sick being such a widespread and easily recognizable phenomenon, finding out why and how it happened quickly became an obvious research program, which really gained traction with the postulation of the germ theory of infectious diseases by Pasteur (1878, revisited by Absolon et al. 1970) and Koch's postulates (Koch 1880). These and other observations that disease could be transmitted from person to person, from animals, or from foul stuff such as rotting things essentially established that all microorganisms causing disease are horizontally acquired from pathogens' reservoirs—a view which may be valid to a large extent, but is by no means complete.

Horizontal transmission implies the pathogen or microorganism can be transmitted from any host carrying it to any non-carrier, while vertical transmission is more restrictive, with transmission from parent to offspring establishing a closely related tracing of host and microorganism lineages. Notwithstanding the fact that many microorganisms were known to be transmitted by different routes and could potentially compete for hosts, the modern population biology study of host–microorganism interactions, especially of infectious disease in humans, has nevertheless been generally formulated as that of the horizontal transmission of pathogens to their hosts (Keeling and Rohani 2008). The paradigm is embraced by epidemiology (Anderson and May 1979; May and Anderson 1979), as well as by the research areas concerned with quantitative mechanistic formulations of the biological process of disease transmission in populations—the theory underlying disease ecology also dubbed theoretical epidemiology—and has been an active field of research since the first work in the early twentieth century, today almost a hundred years old (Ross 1916; Smith et al. 2012).

Vertical transmission of microorganisms, on the other hand, has been less studied and formalized under adequate ecological models although symbiosis has been a research topic of interest due to its ubiquitousness (Moran 2006). Theory on vertical transmission was not greatly furthered once some early work suggested only mutualistic associations could be stably maintained (Fine 1978), while parasitic relationships required some degree of horizontal transmission for persistence of the parasite (Lipsitch et al. 1995)—a kind of relationship that could be as easily explained was probably not as interesting. More importantly, the population biology of host–symbiont interactions was considered separate and independent from that of the same host and its pathogens, that is, host–pathogen and host–symbiont ecology and evolution were treated as distinct pairs.

Symbiont–pathogen interactions, the third possible pairwise combination of three-way interaction, however, are plausible enough if one thinks about them under an ecological perspective, as two species occupying the same niche: the host. In that case, the population biology of host, symbiont, and pathogen would all be inextricably linked. A host carrying a microorganism, say a bacterium, may be more affected if infected by a virus—e.g., it has a greater total burden of parasites—or it may be more protected—the bacterium is able to occupy the place and exclude the virus or make infection more difficult. Lively et al. (2005) suggested and formalized mathematically the possibility of a vertically transmitted parasite (VTP) to become an indirect mutualist in the presence of a more virulent horizontally transmitted parasite (HTP); simulations showed that, contrary to what previous work suggested (Fine 1978; Lipsitch et al. 1995), persistence of an otherwise parasitic symbiont that could not be transmitted horizontally was indeed possible. The result blurred even further the already blurry definition of parasite/pathogen and mutualistic symbiont.

Most of these developments do not factor in variation and natural selection, although J.B.S. Haldane has proposed, as early as in the 1940s, that disease was an important evolutionary force (Haldane 1949), and important developments have been made over the last decades, such as the modeling of reciprocal interactions between evolution and ecology (Reznick 2013; Luo and Koelle 2013), the application of models to fast-mutating viruses such as HIV (Perelson 2002), and integration of population genetic frameworks into the study of transmission dynamics (Grenfell et al. 2004; Wakeley 2005; Wakeley and Sargsyan 2009).

2 Wolbachia: Manipulation, Invasion, and Evolution

2.1 Reproductive Manipulation and Invasion

Wolbachia is obligatory intracellular, maternally transmitted symbionts of the α-proteobacteria class that are present in a large number of arthropod and nematode species (insects and worms) (Hilgenboecker et al. 2008). In arthropods, Wolbachia is a facultative symbiont which can have either a parasitic or mutualistic effect on its host (Weeks et al. 2007), although examples can be found that point to a more intimate relationships (Hosokawa et al. 2010); in filarial nematodes (round worms), for example, association is not facultative but obligatory, which happens to make antibiotic treatment effective for filarial worm infections by killing the bacteria (Beeching and Gill 2014). Wolbachia is considered a striking example of manipulation of the host by the symbiont (For a review see Werren et al. 2008); because the symbiont is transmitted maternally, any phenotype manipulation that distorts the male to female ratio such as feminization, male killing, and parthenogenesis favors persistence of *Wolbachia*.

Most notably, in some species of arthropods, *Wolbachia* can induce cytoplasmic incompatibility (CI), a phenotype by which crosses of *Wolbachia*-carrying males with non-carrying females result in little or no offspring, while that of female carriers with male non-carriers have normal viability. Because the former are crosses with non-carrying mothers (which do not result in symbiont-carrying progeny) and the latter results in carriers, CI gives *Wolbachia* carriers a selective advantage, increasing the frequency of the symbiont in the population (Fig. 1).

Cytoplasmic incompatibility is often conceptualized as a modification–rescue, or lock-and-key mechanism (Merçot and Poinsot 2009): Individuals carrying *Wolbachia* are able to induce incompatibility by modifying the sperm's cytoplasm, and, in order to produce viable progeny, the egg's cytoplasm must be able to be rescued—a compatible Wolbachia in the mother can come to the rescue. Modification and rescue can in principle be uncoupled, giving rise to different phenotypes, e.g., suicide symbionts, which are able to induce incompatibility when in the male but not resist it when in a female body. These types and their importance for persistence and evolution of the symbiont are discussed in the following section; for clarity, unless stated otherwise, when CI-inducing *Wolbachia* is mentioned here, it refers to microbes that both induce incompatibility to hosts that are not carriers and that give rise to viable offspring in an otherwise incompatible cross with a carrier.

Evolution of arthropod carriage of *Wolbachia* (at this point ignoring host, symbiont or any other coevolutionary responses to the presence of the symbiont) is conceptually equivalent to that of a mitochondrial gene—i.e., exclusively transmitted by mothers to their offspring. Using basic selection theory, Caspari and Watson (1959) described the dynamics of the frequency of (what was postulated to be) a maternally transmitted unidirectional cytoplasmic sterility factor between the Oggelshausen (*Og*) and Hamburg (*Ha*) strains of *Culex pipiens* mosquitoes: Female *Ha* X male *Og* crosses were viable, but male *Ha* X female *Og* crosses would give no progeny, all offspring dying as embryos. Still under the population genetics framework of basic selection theory, Turelli and Hoffman (1991) described the dynamics of *Wolbachia* based on the observation that in *Drosophila simulans* it induced both unidirectional incompatibility in crosses and a fecundity cost to carriers (Turelli and Hoffman 1995; Carrington et al. 2011).

A large body of theoretical work was developed around the main result that establishment or elimination of *Wolbachia* is depended on the initial frequency of

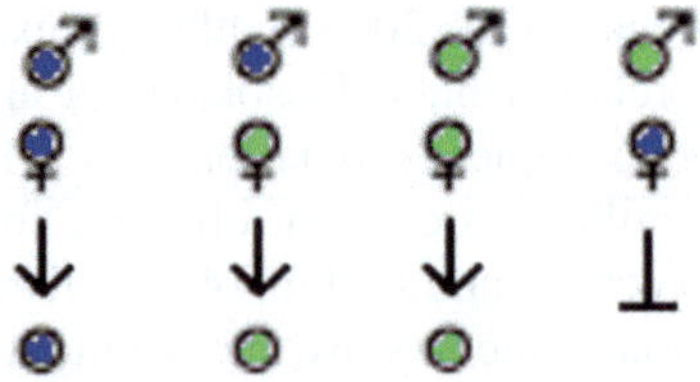

Fig. 1 Cytoplasmic incompatibility. Four crosses are possible between male and female carriers or non-carriers of *Wolbachia*. When the symbiont induces cytoplasmic incompatibility, the cross between a male carrier and a female non-carrier is sterile, or significantly reduced; therefore, carriers of *Wolbachia* have the advantage of having two out of four viable crosses against only one out of four of non-carriers

the symbiont in the population: If the initial frequency was above a given threshold, *Wolbachia* would get fixed; otherwise, it would be eliminated (Hoffman et al. 1986; Jansen et al. 2008; Turelli 2010; Barton and Turelli 2011). Under this model, the invasion threshold was found to be determined by the ratio of fecundity costs to intensity of cytoplasmic incompatibility. Additionally, other costs or *Wolbachia*-associated effects could be present and it does not follow trivially that a simple cost/advantage ratio describes the invasion threshold in those cases. In fact, taking into consideration the specific ecological processes such as birth and death rates, for instance, the threshold can be shown to be dependent on these rates (Souto-Maior et al. 2015) and that the reason they do not appear as such in some of the previous formulations is because the processes are either absent from the model or fall into a special case where they are aggregated and equivalent to fecundity (Hoffmann et al. 1990; Hancock et al. 2011).

One example of a somewhat unexpected effect associated with *Wolbachia* is antiviral protection conferred by naturally occurring strains of *Drosophila melanogaster*, described first in fruit flies challenged with Drosophila C virus and Flockhouse virus (Teixeira et al. 2008; Hedges et al. 2008). The *D. melanogaster*-derived strain *w*MelPop was adapted through passage in mosquito cell lines and transferred to *Aedes aegypti* mosquitoes; it was found to confer protection against RNA viruses causing human disease such as dengue virus (DENV) and chikungunya, as well as against *Plasmodium galinaceum*, a malaria parasite that does not infect humans (Moreira et al. 2009). *Wolbachia* of the *w*MelPop strain transinfected to mosquitoes imposed both fecundity and longevity costs, while on the other hand, protecting mosquitoes against mosquito pathogens, the dynamics of symbiont carriage is more complicated in that case, and population genetic models fall short of describing it.

Picking up on the tripartite interaction framework (Lively et al. 2005; Jones et al. 2007), ecological models were used to describe the effect of protection in *Wolbachia* invasion (Fenton et al. 2011), showing that protection against a virulent pathogen combined to CI facilitated invasion of Wolbachia imposing a fecundity cost and having imperfect transmission from mother to offspring. Simulations that factored in age structure showed that timing of introductions affected the probability of invasion, so planned releases should consider timing (single versus multiple releases), because the threshold alone no longer predicted success of the invasion (Hancock et al. 2011). Further theoretical work also demonstrated that the threshold of invasion could be analytically calculated in the presence of fecundity, longevity costs, and protection against pathogens, that each *Wolbachia*-associated effect had a weight in the threshold proportional to importance of the ecological process affected, and that heterogeneity in these effects could impact invasion (Souto-Maior et al. 2015). The usual culprits of dismantling predictions of simple, deterministic models, spatial structure and stochasticity can also affect the probability of invasion (Hancock and Godfray 2012; Barton and Turelli 2011; Jansen et al. 2008). Therefore, unlike some early results suggested, not all costs affect invasion in the same way; ecology and age structure affect establishment of *Wolbachia*; and spatial spread does not follow trivially from local invasion.

Despite the vast theoretical work developed around a single symbiont, and most notably one particular reproduction manipulation phenotype out of the many observed, obtaining estimates for all known relevant parameters is not trivial, and more elaborate frameworks may be necessary to estimate some of them (Pessoa et al. 2014), or quantify environmental effects such as pathogen burden, which is likely to depend on a multitude of microbial species (Calzolari et al. 2012). Furthermore, the models discussed so far concern only evolution of the host insofar as there is selection to carry or not carry the symbiont—i.e., the symbiont is a fixed set of genes inherited separately from host nuclear genes, and it may be or not be advantageous to have those genes. Host and symbiont genes are of course not fixed, but exhibit heritable variation among individuals. Although not formalized into a quantitative framework, there are arguments that there should be host responses to manipulation by the symbionts (Merçot and Poinsot 2009; Vavre and Charlat 2012; Tortosa et al. 2010). However, the closest to any kind of formalization of the a real-world situation came from the need to assess the impact of *Wolbachia* on DENV transmission—since releases already took place in different parts of the world and more were scheduled to happen in 2014 in more countries—and is in the form of an "evolutionary forecast" (Bull and Turelli 2013). Likely scenarios were forwarded with preliminary predictions that need to be further studied to better understand the consequences of such a provoked human intervention in a complex environment.

2.2 Symbiont Evolution

Besides carrying or not carrying *Wolbachia*, evolution of host nuclear genes and of *Wolbachia* itself is expected and, as mentioned above, the host may respond to the effects of the symbiont, which in turn may tweak its effects to persist, all that conditional on variants being present are selected for the advantageous traits.

Turelli (1994) formalized the evolution of incompatibility inducing microbes into a mathematical model and concluded that after being driven in by cytoplasmic incompatibility, there would no longer be any selection for the trait (as long as any microbe variants present were mutually compatible) and that rather, selection would tend to attenuate any costly effects of these symbionts. Merçot and Poinsot (2009) reasoned that, in that case, and considering cytoplasmic incompatibility as a modification–rescue phenotype and with no pressure to maintain CI, selection or random genetic drift could cause variants not inducing the phenotype to be fixed in the population instead. As soon as there were no CI-inducing individuals, a rescue-only variant would not have any pressure to maintain that trait either, and the result could be a symbiont carrier neither inducing nor resisting CI. Although this is conceptual and not quantitative reasoning, the expected intermediate variants of *Wolbachia* have been observed in nature (Merçot and Poinsot 1998; Bourtzis et al. 1998; Charlat et al. 2003; Zabalou et al. 2004).

Additional population structure could favor some selection for CI persistence, although it is thought not to be as strong as selection for increased fecundity (i.e., decreased fecundity cost), so persistence of the CI phenotype in many species is not entirely accounted for (Haygood and Turelli 2009). *Wolbachia* in those cases would be "passengers" that were able to squeeze in through manipulation, differently from mutualists, which always have selection favor their persistence, or their status in nematodes, where the association is obligatory for host survival; in the latter cases, the symbiont would actually be a "resident" (Merçot and Poinsot, 2009), propagating and persisting together with its host as one unit.

Uni- and bidirectional cytoplasmic incompatibility are striking, conspicuous manipulative phenotypes; explaining its emergence and persistence is of interest from the point of view of the evolution of complex traits, of manipulation in symbioses, and their possible role in speciation (Faria and Sucena 2014). In a much more straightforward prediction, if fecundity is heritable and variants with high fecundity are present, the trait is always expected to evolve, because by definition, they have increased fitness: They leave more offspring and increase in frequency—as discussed above, they may even offset stronger or more striking phenotypes such as CI in the long run (Haygood and Turelli 2009). In *D. melanogaster*, the intensity of CI (percentage of offspring killed in incompatible crosses) is quite weak (Hoffmann et al. 1990); in the sister species *D. simulans*, CI is stronger, and still the effect of *Wolbachia* on fecundity was observed to evolve from costly to beneficial: The association of *Wolbachia* and *Drosophila* evolved from a parasitic to a mutualistic one (Weeks et al. 2007).

Less obvious is the impact of other traits induced by symbionts, such as reduction in life span (or longevity cost) or protection against pathogens; even though both are supposed to have a positive effect, these depend on exactly how they increase fitness (i.e., how they translate into a selection coefficient), e.g., a longevity cost may have no impact if it causes individuals to die only after laying all eggs it could. Protection against pathogens depends on a burden of pathogens actually being present. Also, trade-offs between traits may cause selection of one but not other trait, i.e., individuals with higher resistance against pathogens may have lower fecundity. There is evidence, for instance, of a global replacement of more protecting strains by others that would protect less, but have increased longevity (Riegler et al. 2005; Chrostek et al. 2013).

Extending what earlier population genetics purported to do into the evolution of a symbiont inducing multiple changes in the host life history is a challenge, especially when the host is likely to coevolve in response to the new association: an arthropod host with a specific genetic background will evolve either while carrying *Wolbachia* or without it. The genes can arrive at a host through various combinations of crosses: A mother carrying *Wolbachia* may give rise to some offspring without *Wolbachia* (due to imperfect transmission), and a father without *Wolbachia* in a cross with a female carrier will see its offspring emerge as carriers. Evolution of *Wolbachia* will only happen inside its host, but will also be conditioned on host genetic background and host–symbiont feedbacks. Disentangling this sort of confusion is necessary to minimally describe coevolution.

Fuzzy as it may seem, though, this still only describes host–symbiont coevolution, ignoring other microorganisms and longer term genome interactions that may occur over greater evolutionary timescales. Despite not being entirely within the scope of theoretical population genetics, some observations of natural populations can hint at the evolution of host and symbionts in the presence of other microorganisms.

3 Host–Microorganism Genome Interactions and Long-Term Relationships

The origin of organelles such as mitochondria is generally accepted to be of endosymbiotic nature. Many variations of the same organelle theme suggest that the host nuclear genome incorporated many of the genes originally in these free-living organisms turned resident machinery. Although it is different, transfer of genetic material is postulated to have happened from *Wolbachia* to its hosts (Nikoh et al. 2008).

In *D. melanogaster*, the association with *w*Mel was estimated to have the symbiont's most recent common ancestor at around 8000 years ago, with strong association between the phylogenetic relationship of the symbiont and mitochondria, therefore forwarding the idea that the currently observed patterns are best explained by a single acquisition and subsequent loss of *Wolbachia* in some lineages afterward (Richardson et al. 2012). Indeed, large chunks of *Wolbachia* genes have been found to be incorporated into the nuclear genome of some insect and nematode species and are found to be transcribed (Hotopp et al. 2007; Nikoh et al. 2008); as with mitochondria, it is unclear to which extent nuclear genes are functional and could take over the functions performed by the symbiont—thereby rendering it useless—but any such events could greatly affect coevolution of host and symbiont, possibly shifting the balance in any existing marriage conflicts. Lateral gene transfers and host jumps (symbiont horizontal transmission) are believed to be rare and are therefore not expected to be observed in shorter timescales; nevertheless, artificially introducing a *D. melanogaster*-derived strain of *Wolbachia* *w*Mel into a new species, such as *A. aegypti*, may see many features of a novel association, which may be compared to what is observed in older couples.

Some statements about old and new associations need to be reconciled, but it is not always easy to do so; one of the most difficult to make sense of is "*Wolbachia* confers antiviral protection to its host." Once more strict assertions are made, what can really be said is something less general: *w*Mel, the *D. melanogaster*-derived *Wolbachia* strain, confers antiviral protection to its *Drosophila* species (Teixeira et al. 2008; Martinez et al. 2014), and indeed, the same strain (and the related *w*MelPop) confers protection against DENV in mosquitoes *A. aegypti* and *Aedes albopictus* (Moreira et al. 2009; Walker et al. 2011; Blagrove et al. 2012). However, *A. albopictus* is naturally a carrier of not only one, but two different strains of Wolbachia (*w*AlbA and *w*AlbB), and is a competent vector of DENV:

Although usually considered an inferior vector than *A. aegypti*, in the absence of the sister species, it can efficiently transmit disease (Lambrechts et al. 2010). So the endogenous strain in one old host, *D. melanogaster*, confers protection, but in another, *A. albopictus*, it does not; while it can be argued that these are strain-specific properties or that the density of symbiont can explain those properties (Lu et al. 2012), the possibility of these observations being the result of a coevolutionary process involving both host and *Wolbachia* cannot be discarded. This is a concern particular for *Wolbachia*-based interventions aimed at human health, like that of provoked invasions of *Wolbachia*-carrying *A. aegypti* to block DENV transmission.

For a disease control intervention based on *Wolbachia*, the ideal scenario is that of a successful introduction of a reproductively costly, life-shortening, DENV-blocking strain that could overall reduce transmission to levels below the epidemic threshold, i.e., no sustained transmission could be maintained. This could also be achieved with a DENV-blocking strain with high protection (measuring absolute protection is tricky, but see Gomes et al. 2014; Pessoa et al. 2014). If no evolution was expected, continuous DENV transmission could be eliminated in areas where *A. aegypti* is the only or main vector, which is the case in urban areas; because the *w*Mel–*Aedes* association is not a natural, and therefore unlikely to be an evolutionary stable one, there is a reason to expect immediate response of the host, if it is evolutionarily perceived as an aggression and if there is variation in the population that can be selected to counter these effects. Some speculation could be done on how the mosquitoes could be unhappy at their new guests, but actually identifying the tools to respond is more complicated, and experiments rely on artificially selecting insects for many generations (Martins et al. 2013).

Additional insight can be gained by looking at different species with different kinds and times of associations; *w*Mel in *A. aegypti*, for instance, is seen to increase expression of immune genes (Bian et al. 2010). Even though this is not observed in its original *Drosophila* host even with foreign symbionts (Chrostek et al. 2014), it raises the question of whether the antiviral protection in mosquitoes is a result of priming the immune system (Ye et al. 2013) with a new, strange thing and that the immune upregulation could go away once the host got adapted to its new passenger—having over-activated immunity could also explain some fitness costs in terms of collateral damage inflicted by immune cells that normally would not be active, but are out to kill something (Schneider 2011).

That would also be consistent with *A. albopictus* not having noticeable protection despite naturally carrying two *Wolbachia* strains, but being more protected when carrying the newly transinfected *w*Mel. Punctual observations such as these comparative analyses, however, can suggest some hypothesis and rule out a few predictions, but cannot explain or predict a quantitative outcome such as elimination of endemic transmission, and stability of the strategy in the long run.

Existing long-term relationships of arthropods and *Wolbachia* can suggest possible outcomes of new introductions of the symbiont into a new host; however, because host jumps occur, naturally occurring associations are not guaranteed to be old, and estimates of the time since the symbiosis exists are necessary to assess whether enough time has elapsed for any kind of equilibrium to be attained (and even so there is still no guarantee that it has). Ancient symbioses may have a more complicated interpretation if the partners exchanged genes. Many of those aspects are either beyond the scope of population genetics or cannot be analyzed under classic population genetics, or are too hard to formalize at this point, and even if it was done, measuring and characterizing the existing variation requires a heroic effort to measure a single (albeit possibly one that could be very important) parameter. Like the more general retrospective phylogenetic analyses, prediction in this situation is more akin to some sort of weather forecast than the rigorous hypothesis population genetics usually forwards.

4 Dengue Virus: The Case for Evolutionary Medicine and Evolving Vaccines

DENV belongs to the flaviviruses, a single-stranded RNA virus family; it can alternatively be more loosely classified as an arbovirus, or viruses that are borne by arthropods. It comprises four related serotypes that are moderately close (uncreatively named DENV-1 through 4), sharing a little over 65 % of genetic sequence similarity (Guzman et al. 2010). The viral structural and non-structural genes, general characteristics, as well as the clinical disease caused by any of the serotypes are similar, which includes undifferentiated fever, joint pain, strong headaches, rash, as well as usually mild bleeding manifestations; unique characteristics of any one serotype are unknown or unresolved (Halstead 2007a). The disease is transmitted to humans by female mosquitoes of the genus *Aedes*, the transmission cycle being maintained by continual transmission from a human to an arthropod host, but a sylvatic transmission cycle can be maintained between monkey and other mosquito species normally not interacting with humans (Vasilakis et al. 2011).

Almost 3 billion people live in areas with the risk of DENV transmission, and it is estimated that there are some 300 million cases every year (Bhatt et al. 2013); there is no specific treatment for the disease, and management consists of supportive treatment of the symptoms. Uncomplicated cases resolve themselves, but more severe manifestations of dengue hemorrhagic fever or dengue shock syndrome (DHF/DSS) can have a case fatality rate of 50 % if untreated due to hemorrhagic manifestations. DHF/DSS cases are associated with secondary DENV infections, but have been observed in primary infections, particularly in children (Halstead 2007a, b), and where factors affecting coagulation are involved—which is why the use of salicylic acid or other drugs affecting blood clotting is not recommended to relieve symptoms. It has been argued that because of the increased risk associated with secondary infections, an effective vaccine should confer high protection against all four serotypes; clinical trials of a vaccine have recently been conducted, with mixed results (Costa et al. 2014). Given the difficulties, most control measures focus on the management of the vector population.

After an infectious mosquito bite, there is an incubation period which, although normally not possible to measure directly or with high precision, is generally accepted to be somewhere between the 3–14 days interval (WHO); if infection is successful, disease follows and is normally self-limiting—at least in uncomplicated cases—after about 7 days, period during which the human host is most likely to transmit to any uninfected mosquitoes that feed on his or her blood (Nguyet et al. 2013). After an infectious blood meal, the female mosquito may develop disease and become infectious after about 10–14 days (Halstead 2007b); infected mosquitoes are believed not to recover from DENV, and the disease halves the life span of the mosquito (Maciel-de-Freitas 2011).

The DENV serotypes are antigenically distinct—infection with one serotype confers lifelong protection against a second infection with the same serotype, but infection by any of the other three serotypes is still possible—it is still a matter of discussion whether secondary infection is less, equally, or more likely that primary infection, or if instead a temporary cross-immunity, plays any role in DENV epidemiology (Johansson et al. 2011). The chain of transmission is shown in Fig. 2.

The processes just described can be formalized into a mathematical model (or alternative models) of DENV transmission (Nishiura 2006), with the observed waiting times of each process being converted into rates or probabilities that each event happens (Johansson et al. 2011), i.e., the model parameters. Although some of these parameters can only be measured imprecisely, indirectly, or not at

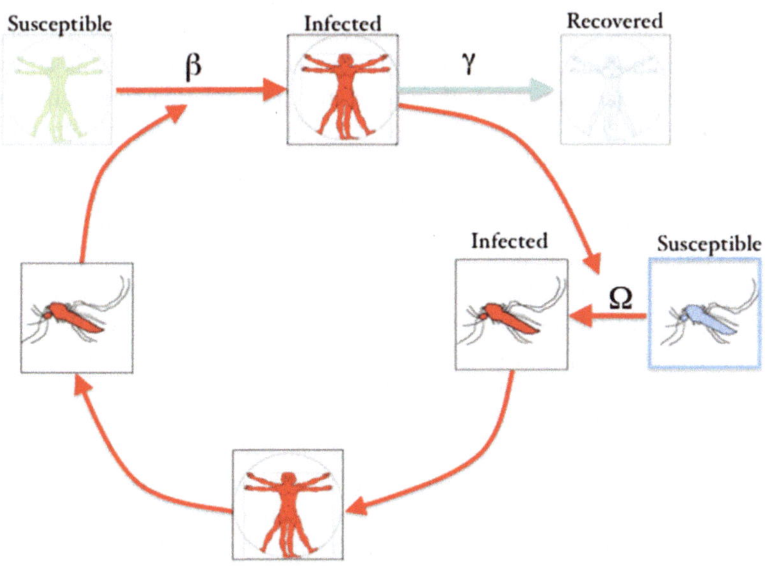

Fig. 2 Dengue virus transmission cycle. Infected mosquitoes bite and infect susceptible humans with probability β, which later recover at rate γ. Infected humans bitten by a susceptible mosquito transmit disease with probability Ω. The continual cycle maintains endemic transmission if the transmission rates are high enough

all, building such a model allows for simulating disease transmission within what is thought to be a reasonable range of parameters; the patterns can then be analyzed and compared to the observed disease incidence (Keeling and Rohani 2008). Despite the series of assumptions and simplifications that are needed to come up with what is already a quite complex model (Gunawardena 2014), some general features of interest arise, like the sustained multiannual oscillations in the number of cases that are a result of the interaction of two or more serotypes (Esteva and Vargas 1999; Luz et al. 2003; Wearing and Rohani 2006; Nagao and Koelle 2008; WHO-VMI Dengue Vaccine Modeling Group 2012).

Multiannual cycles are observed in epidemics of DENV in places where more than one serotype was known to have circulated; nevertheless, whether the cycles are a result of heterologous serotype temporary cross-immunity, permanent cross-protection or cross-enhancement cannot be straightforwardly resolved by simulation alone (Adams et al. 2006; Cummings et al. 2005), since the model outputs are similar for either mechanism. Simulation can therefore only find whether the proposed mechanism can broadly generate the observed behavior. Despite all the complexity involved, it is worth reminding that these models assume all population traits are fixed, i.e., there is no heritable phenotypic variation in any of the populations and therefore no evolution.

On the evolution side, *Aedes* mosquitoes have been studied using the modern synthesis–era theory of population genetics, with measures of expected genetic differentiation of species such as F_{IS} and F_{ST}, (e.g., Lourenço-de-Oliveira et al. 2004, Bracco et al. 2007), which do not consider any specific models of mosquito population dynamics and do not explicitly account for any selective pressures arising from pathogens or symbionts. The question answered by this approach is mainly "is the genetic structure of two (or more) populations different from one another?" but not "how much does having (or not having) the presence of pathogens or symbionts (possibly at varying levels) affect genetic structure?" These measures are also not easily connected to the selection of any specific traits or population processes and only indirectly answer questions about the processes that generated the current diversity.

Evolution of DENV, in turn, is studied mostly via phylogenetics (Miagostovich et al. 2006; de Castro et al. 2013; Faria et al. 2013; de Araújo et al. 2012), trying to make sense of clade replacements in terms of selection in favor of a specific genotype; increased incidence of an outbreak is associated with a genotype being used as a proxy for greater fitness. Despite having a mutation model implicit in the phylogenetic clustering algorithms, results are essentially decoupled from any mechanistic model of disease transmission, and in most cases, results cannot be used to interpret the patterns of disease transmission quantitatively (but see Mondini et al. 2009; Rasmussen et al. 2014 for work that begins pushing in that direction in the context of neutral evolution).

Observations such as the low in vitro replication rate (which could be a component of virus fitness) of the American DENV-2 are nevertheless inconsistent with the large outbreaks observed for the strain, for which newly arisen virulence mutation markers could not be identified (Halstead 2007a). Some research has also attempted to make sense of increased fitness of a clade in the context of immunity conferred by other serotypes, where a reasonably large amount of serological a genetic data was available from patients (OhAinle et al. 2011); however, the number of hypotheses, correlations, and comparisons is quite large—interpretation of the results can be cumbersome, and statistical power to identify large significant effects is probably low.

Describing phenotypic evolution of DENV in terms of mechanistic processes in disease transmission requires the heritable genetic variation in the population of viruses (and, if coevolution is to be modeled, also in mosquitoes and humans) to be quantified; it is not trivial to find whether variation is present for, for instance, insect resistance against pathogens (Martins et al. 2013) and much less for all relevant parameters in the transmission model, so a mechanistic description of phenotypic evolution that is realistic would require a large amount of prior information. Furthermore, selective pressures acting on the virus are likely to be very different whether inside the human rather than the mosquito host, and evolution then depends on a balance of what is selected in each case, as well as on the rate of transmission between hosts, which is also closely related to viral fitness.

Neutral molecular evolution, on the other hand, has recently had reasonable success in being incorporated into parametric models (Volz 2012; Volz et al. 2013); as usual, several assumptions go into making that possible, such as assuming infection consists of a single viral sequence, and not a dynamic, mutating population of viruses—so even the observation that infection consists of a polymorphic viral population has yet to be developed (but see Gordo and Campos 2007; Gordo et al. 2009).

Whether evolution of the human population feeds back (interacts reciprocally) into evolution of DENV is not clear, since the disease does not seem to impose a high mortality burden; for mosquitoes, despite the high virulence (Maciel-de-Freitas et al. 2011), it is believed that less than 1 % of mosquitoes would be infected with DENV at any one time, so that the virus could not affect much evolution of the vector—quantification of these predictions is not straightforward.

Models were built to predict the evolution of the virus under interventions aimed at controlling the disease (Medlock et al. 2009); the use of *Wolbachia* to introduce resistance into and manipulate the life history of the mosquito population has to be analyzed under a similar light, although the interactions are much more complex in the latter case. As mentioned above, the tripartite interaction could induce feedback in the evolution of host, symbiont, and pathogen and affect the immunity profile of the human population if transmission was halted or reduced to low levels. While difficult to formalize the multiple interactions, mathematical models can help analyze possible outcomes, when intuition alone is at loss given the number of subpopulations and parameters.

5 Using Human Disease as a Model to Study Evolution: Difficulties in the Description and Prediction of Host–Microorganism Associations and "Evolutionary Forecasts"

Evolutionary theory has come a long way from the Darwinian proposal of natural selection; theoretical population genetics was in the forefront of formalizing its concepts, thus consolidating the modern synthesis, proposing new ones such as neutral evolution, and providing testable quantitative hypotheses. Mathematical modeling of ecological and epidemiological dynamics, as first introduced by Ross (1916), has also come a long way, and computer simulation of huge, complex systems that are mathematically intractable has more or less recently become possible. Statistical inference is still a somewhat limiting factor, although Bayesian methods started to be more widely applied and have recently used the power of computers to push likelihood-based methods beyond what the more traditional least-squares fitting methods can achieve (Myung 2003; Lavine 1999).

More limiting is our ability to develop models that, at the same time, capture enough detail of the processes of interest and are tractable enough to be analyzed and explored and amenable to statistical inference. Currently, a great challenge is to use the mathematical and statistical tools to expand the description of biological systems and obtain high-resolution information about evolutionary processes; in the processes, some classic results may be improved, while others will be refuted, which will help overturn current consensus and dogmas that may be in the way of scientific progress.

Given the relevance of human pathogens, exploring and trying to predict the consequences of medical interventions is indeed a matter of life and death. Some of the theory has been put in place to develop and further the field of evolutionary medicine and epidemiology (Price 1970; Gandon and Day 2007), and there have been calls to make good use of the models (Vavre and Charlat 2012), but a full-fledged framework is still lacking that is applicable to real epidemics and interventions such as vaccination programs and more experimental strategies such as the use of *Wolbachia* (which conceptually is equivalent to vaccinating mosquitoes).

Vavre and Charlat (2012) argue for extending the study of the dynamics of the host–*Wolbachia*–pathogen "*menage a trois,*" and the coevolution of pathogen virulence and prevalence together with symbiont-mediated protection—they also note that the presence of CI may release selective pressure for high protection—an important, unexplored question. Not mentioned is that selection for CI can be relaxed in the absence of non-carriers, as discussed above.

Bull and Turelli (2013) propose and discuss plausible scenarios for the post-intervention of at least three strategies involving the release of *Wolbachia*-carrying mosquitoes: life-shortening *Wolbachia*, CI-based population suppression, and DENV-blocking *Wolbachia*. They argue that in the case of life-shortening, there would be intense selection for decreasing the effect in both host and symbiont, since life-shortening decreases fitness of both, while the virus would see more (but may be not a lot more, since there it should already be strong) selection for a reduced incubation period that allowed it to be transmitted sooner. For suppression, there would be strong selection to escape CI or avoid mating with carriers; in *D. simulans,* there seems to be little or no genetic variation for these traits, and if it is also true for *A. aegypti,* short-term success is likely, and over the long run, selection should overturn the results. In the case of blocking, the situation is the most complicated; there is a conflict among the partners, with the host enjoying antiviral protection, no direct selection on *Wolbachia*, and the virus being selected to escape. Information on standing genetic variation for those traits is essentially unknown. The authors also acknowledge that extending the reasoning to a model-based inference is unreliable due to the multitude of parameters and potential trade-offs.

The difficulties that arise in trying to predict evolution stem from the fact that it requires formalizing a model and its parameters, specifying the heritable variation for each of them (or at least the important ones or those expected to evolve), as well as characterizing trade-offs between any two or more of them—not a straightforward task at all. More worrying, a model-based framework to estimate pre- and post-intervention parameters is essentially absent; evaluation of interventions is done through randomized trials, and general conclusions cannot be extracted from them beyond whether the intervention has any success in that particular place and time (but see Gomes et al. 2014). Research on this interface could go many ways, with each specific research area informing the others: evolution, ecology, medicine, epidemiology, and modeling of complex systems. A summary of the interactions and illustrative evolutionary outcomes of host, a pathogen, and a symbiont is shown in Fig. 3.

5.1 Multipartite Interactions

Tripartite and multipartite interactions are, however, not an exclusivity of host–microorganism interactions; one can picture that a carnivorous animal, an herbivorous one, and one (or more) species of plant are coevolving in response to forces exerted directly or indirectly by each of the players that may decrease the fitness (most easily depicted by the carnivore killing and eating the herbivore, thereby reducing the fitness of the latter to zero after the event).

Nevertheless, Wright–Fisher population models and simple extensions thereof (models with constant population size or constant intervals) have been used for a long time to obtain estimates of population parameters of animal species (Rosenberg and Nordborg 2002; Quéméré et al. 2012; Heller et al. 2013); there is no a priori reason why host–microorganism interactions should be any different, and yet epidemiologists are usually not satisfied with the direct application of theoretical population genetics to disease transmission and host–microorganism interactions in general (Volz et al. 2013). There are a few reasons why that is so, and these probably offer new opportunities to population genetics and possibly to evolutionary theory alike.

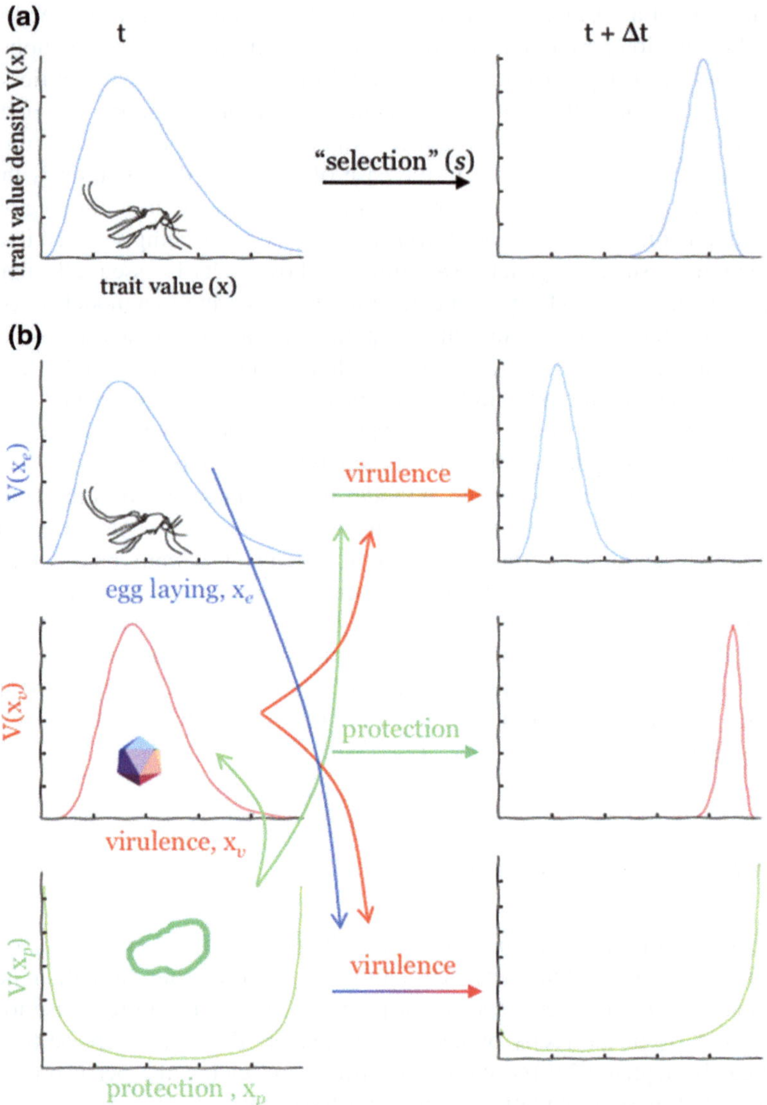

Fig. 3 In more traditional population genetics models, a trait may be thought to be under a selective pressure, broadly speaking, that would, for instance, select for higher values of the trait. The distribution of the trait after a time Δt is the function of the phenotypic variation, $V(x)$, and the selective pressures, s (A). In models of interactions of host, pathogen, and symbiont, the selective pressures arise from the interaction of the different species on each other, and therefore, all of the trait distributions after a time Δt are functions of all original distributions ($V(x_e)$, $V(x_v)$, $V(x_p)$) as well as of the different selective pressures (virulence, protection, etc.), which may all vary nonlinearly over time (as opposed to being a constant "s" coefficient)

First, the theoretical framework to describe the population dynamics of pathogen infections represents an entire new field of research, and theoretical epidemiologists are probably not ready to abandon it entirely in favor of simpler and even naïve models, even if consolidated as useful tools in population genetics. Second, the timescale and resolution investigated for animal systems are much larger, and the fine-grained population dynamics are secondary, with major features being more important because they are able to give insight into unobservable processes removed so far back in the past that any information is very valuable nonetheless. Third, and putting the second and first together, disease ecology and epidemiology aim to answer specific questions concerning the distribution and risk of disease, which are not necessarily the same questions of interest from animals or plants, e.g., for conservation purposes, it is of interest to know whether an animal population had its diversity affected by a bottleneck or fragmentation of its habitat (Heller et al. 2013), while for disease bottlenecks are part of everyday life, and their high mutation rates help them make up for the diversity lost after a population crash (also known as elimination or disease control). Additionally, important classic metrics of population genetics often do not have an obvious meaning for epidemiologists; effective number of infections (Frost and Volz 2010), the epidemiological equivalent to the effective population size, is not immediately interpretable or useful for medical purposes.

On the flip side, disease ecology allows observation of some (in animal ecology) unobservables. While the number of animals is generally unavailable, or requiring extensive efforts to obtain a vague idea of its census population, a reasonable-though-imperfect count of the number of infections—the size of the pathogen population—is not breaking news. The number of disease cases is often public and obtained through health surveillance systems designed specifically to record not only incidence/prevalence of disease, but increasingly also patient serological information and pathogen strain/genotype (SINAN). Therefore, inference from sequence data can in principle be directly compared to the observed population dynamics, i.e., estimates such as effective population size can be compared to actual population size to see whether the quantity actually gives information about the real population. Because of the short timescale and high mutation rates, this kind of inference can be made for many "replicates" of the natural experiment. Also, because of the simplicity of viral pathogens, arguably with little or no recombination and strictly asexual reproduction, they best approximate the common assumptions of population genetics and their genomes are closest to genetic recording machines (although for many reasons not nearly as much as researchers think or would like them to).

6 Concluding Remarks

While it requires working on the interface between ecology; theoretical population genetics; medicine and epidemiology; and biomathematics and statistics, studying disease as an evolving system can help answer interesting questions to the different research areas. The difficulties in mathematical and statistical modeling apply throughout; however, the ecological framework is in place, researchers start to pick up where classical population genetics left off, and furthermore, epidemiological data are available, but yet to be fully incorporated to the analyses. A new movement to synthesize these parallel efforts, as well as hard work on the methods, can help extend evolutionary theory further.

Acknowledgments I would like to thank Elvira Lafuente, Ricardo Cavalcanti, Vitor Faria, Rafael Maciel-de-Freitas, Luis Teixeira, and Gabriela Gomes for critical reading, comments, and references.

The author is the beneficiary of a doctoral grant from the AXA Research Fund.

Glossary

Artificial selection Consists of creating conditions (usually in a laboratory setting) that favor certain variants provided they exist in the previous populations—to be more represented in the following generations, e.g., infecting flies with a virulent pathogen will cause resistant flies to increase in frequency. Artificial selection is useful for detecting and quantifying variation for certain traits

Bayesian statistics One of the two main statistical philosophies, as opposed to frequentist statistics, based on the original work of Bayes, requires using (or assuming to the best of knowledge lack of) prior knowledge as input and dealing with uncertainty in the parameter estimates

Cross-protection or cross-enhancement Phenomenon by which infection with a pathogen (or serotype) either makes a secondary infection less (protection) or more (enhancement) likely

Deterministic models Models that do not take chance events into account and approximate the occurrence of events by the average or expected rate

Disease ecology The study of disease as species that colonize niches (usually hosts, but also intermediate environmental stages or secondary hosts)

Endemic transmission Continual transmission of disease, which may still vary for reasons such as seasons or immunization, but does not depend on external introduction of the pathogen

Evolutionary stable (strategy) An ESS is a trait value or a combination of values that cannot be beaten by any other, and therefore, once achieved, it stays the same in the population (provided the environment does not change). If more than one organism is evolving in response to the others, a coevolutionary stable strategy, CSS, can be achieved, so that individuals of any species with different trait values cannot succeed in the population better than the ones that have achieved it

F_{IS}, F_{ST} Measures of genetic differentiation. F_{IS} is the inbreeding coefficient, which is a measure of consanguinity, or relationship. F_{ST} is a measure of variation within a subpopulation compared to the variation in the total population

Fitness The capacity of an individual or population to propagate and persist in the population fitness has many components such as fecundity (the more offspring a variant has, the more successful it will be), survival (the better it survives, the more it will be present), and many others

Genetic background The genetic sequence of an individual. Clones of a laboratory animal share a same genetic background

Heritable genetic variation The distribution in a population of variants of any genes, i.e., variants that can be transmitted genetically from parent to offspring

Incidence (of disease) The number of new infections in a given time period, e.g., the weekly incidence of dengue fever is the number of new cases of the disease in a particular week (also usually defined for a geographic region, like a city)

Least-squares fitting Frequentist estimation method that minimizes the square of the distance between the data points and a curve (straight line, function, or model) and should explain the trends observed

Neutral evolution Evolution that does not depend on natural selection, most readily exemplified by sequence variation that has no function and therefore does not affect survival, i.e., is neutral

Phenotypic evolution Evolution of observable traits, usually equated to evolution of traits under selection

Phenotypic variation Distribution of trait values, e.g., height in a human population

Polymorphism Variation. A polymorphic gene is a gene that differs in individuals (or the pair of chromosomes of a single individual)

Prevalence (of disease) The number of disease cases in a certain point in time

Priming (Immune system) Upregulating immune responses that would then respond more readily to an aggression, e.g., exposing the immune system to a bacterium may activate responses that would then help kill viruses introduced afterwards

Random genetic drift Process by which, due to chance in reproduction, some individuals pass on offspring (and therefore genes) to the next generation, while others do not. As a consequence, over some time, there is a finite probability that some genes get lost and others get fixed without there being any selection for them. In Motoo Kimura's words, it is the "survival of the luckiest"

Reciprocal interactions, Feedback Any mechanism that allows a process to self-regulate or self-enhance is termed feedback; ecology affects evolution, and in the case where evolution of traits affects ecology in the short term, it is said that the relationship between ecology and evolution is reciprocal

Selection coefficient The mathematical representation of selection, usually as a single parameter (s)

Selection, selective pressure Any process that favors specific variants in the population, e.g., pesticides favor the survival of resistant pests, and therefore selects them for future generations

Serotype Pathogen type that elicits a specific immune response and therefore is distinguished from similar pathogens based on its antigens and matching antibodies produced against them

Statistical inference, fitting Any method that adjusts free parameters of a curve (function or model) to find the values that best explain the observed data, i.e., that best "fits" the data

Sylvatic transmission cycle Disease transmission that happens among wild animals, independently from humans

Temporary cross-immunity Temporary cross-protection that wanes after some time

Theoretical epidemiology Broadly similar to disease ecology, but associated more with the theoretical aspects of the practice of epidemiology, as opposed to the study of diversity of pathogens

References

Absolon KB, Absolon MJ, Zientek R (1970) From antisepsis to asepsis. Louis Pasteur's publication on the germ theory and its application to medicine and surgery. Rev Surg 27:245–258

Adams B, Holmes EC, Zhang C, Mammen MP, Nimmannitya S, Kalayanarooj S, Boots M (2006) Cross-protective immunity can account for the alternating epidemic pattern of dengue virus serotypes circulating in Bangkok. Proc Natl Acad Sci USA 103:14234–14239. doi:10.1073/pnas.0602768103

Anderson RM, May RM (1979) Population biology of infectious diseases: Part I. Nature 280:361–367

May RM, Anderson RM (1979) Population biology of infectious diseases: part II. Nature 280:455–461

De Araújo JMG, Bello G, Romero H, Nogueira RMR (2012) Origin and evolution of dengue virus type 3 in Brazil. PLoS Negl Trop Dis 6:e1784. doi:10.1371/journal.pntd.0001784

Barton NH, Turelli M (2011) Spatial waves of advance with bistable dynamics: cytoplasmic and genetic analogues of allee effects. Am Nat 178: E48–E75

Beeching N, Gill GV (2014) Chapter 14: Onchocerciasis, filariasis and loiasis. In Lecture notes: tropical medicine, 7th edn Wiley, Hoboken, pp 126

Bhatt S, Gething PW, Brady OJ et al (2013) The global distribution and burden of dengue. Nature 496:504–507. doi:10.1038/nature12060

Bian G, Xu Y, Lu P et al (2010) The endosymbiotic bacterium Wolbachia induces resistance to dengue virus in Aedes aegypti. PLoS Pathog 6:e1000833. doi:10.1371/journal.ppat.1000833

Bourtzis K, Dobson SL, Braig HR, O'Neill SL (1998) Rescuing Wolbachia have been overlooked. Nature 391:852–853

Blagrove M, Arias-Goeta C, AB F, Sinkins S (2012) Wolbachia strain wMel induces cytoplasmic incompatibility and blocks dengue transmission in *Aedes albopictus*. Proc Natl Acad Sci 109:255–260. doi:10.1073/pnas.1112021108

Bracco J, Capurro M, Lourenço-de-Oliveira R, Sallum M (2007) Genetic variability of Aedes aegypti in the Americas using a mitochondrial gene: evidence of multiple introductions. Mem Inst Oswaldo Cruz 102:573–580

Bull JJ, Turelli M (2013) Wolbachia versus dengue: evolutionary forecasts. Evol Med Public Health 2013:197–207

Calzolari M, Zé-Zé L, Růžek D et al (2012) Detection of mosquito-only flaviviruses in Europe. J Gen Virol 93:1215–1225. doi:10.1099/vir.0.040485-0

Carrington LB, Lipkowitz JR, Hoffmann AA, Turelli M (2011) A re-examination of Wolbachia-induced cytoplasmic incompatibility in California Drosophila simulans. PLoS ONE 6:e22565. doi:10.1371/journal.pone.0022565

Caspari E, Watson GS (1959) On the evolutionary Importance of cytoplasmic sterility in mosquitoes. Evolution 13:568–570

de Castro MG, de Nogueira FB, Nogueira RMR et al (2013) Genetic variation in the 3' untranslated region of dengue virus serotype 3 strains isolated from mosquitoes and humans in Brazil. Virol J 10:3. doi:10.1186/1743-422X-10-3

Charlat S, Le Chat L, Merçot H (2003) Characterization of non- cytoplasmic incompatibility inducing Wolbachia in two continental African populations of Drosophila simulans. Heredity 90:49–55

Chrostek E, Marialva MSP, Esteves SS et al (2013) Wolbachia variants induce differential protection to viruses in *Drosophila melanogaster*: a phenotypic and phylogenomic analysis. PLoS Genet 9:e1003896. doi:10.1371/journal.pgen.1003896

Chrostek E, Marialva MSP, Yamada R et al (2014) High anti-viral protection without immune upregulation after interspecies Wolbachia transfer. PLoS ONE 9:e99025. doi:10.1371/journal.pone.0099025

da Costa V, Marques-Silva A, Floriano V, Moreli M (2014) Safety, immunogenicity and efficacy of a recombinant tetravalent dengue vaccine: a meta-analysis of randomized trials. Vaccine 32:4885–4892. doi:10.1016/j.vaccine.2014.07.008

Cummings DAT, Schwartz IB, Billings L et al (2005) Dynamic effects of antibody-dependent enhancement on the fitness of viruses. Proc Natl Acad Sci U S A 102:15259–15264. doi: 10.1073/pnas.0507320102

Hotopp JC, Clark ME, Oliveira DCSG et al (2007) Widespread lateral gene transfer from intracellular bacteria to multicellular eukaryotes. Science 317:1753–1756. doi:10.1126/science.1142490

Esteva L, Vargas C (1999) A model for dengue disease with variable human population. J Math Biol 38:220–240

Faria NRDC, Nogueira RMR, de Filippis AMB et al (2013) Twenty years of DENV-2 activity in Brazil: molecular characterization and phylogeny of strains isolated from 1990 to 2010. PLoS Negl Trop Dis 7:e2095. doi:10.1371/journal.pntd.0002095

Faria VG, Sucena E (2014) Novel endosymbioses as a catalyst of fast speciation In: Gontier N (ed) Reticulate evolution: symbiogenesis and lateral gen transfer. Springer, Dordrecht. (Series: Interdisciplinary Evolution Research)

Fenton A, Johnson KN, Brownlie JC, Hurst GDD (2011) Solving the Wolbachia paradox: modeling the tripartite interaction between host, Wolbachia, and a natural enemy. Am Nat 178:333–342. doi:10.1086/661247

Fine P (1978) Calculation of vertical transmission rates of infection, illustrated with data on an aphid-borne virus. Am Nat 112:781–786

Frost SDW, Volz EM (2010) Viral phylodynamics and the search for an effective number of infections. Philos Trans R Soc Lond B Biol Sci 365:1879–1890. doi:10.1098/rstb.2010.0060

Gandon S, Day T (2007) The evolutionary epidemiology of vaccination. J R Soc Interface 4:803–817. doi:10.1098/rsif.2006.0207

Gomes MGM, Lipsitch M, Wargo AR et al (2014) A missing dimension in measures of vaccination impacts. PLoS Pathog 10:e1003849. doi:10.1371/journal.ppat.1003849

Gordo I, Campos PRA (2007) Patterns of genetic variation in populations of infectious agents. BMC Evol Biol 7:116. doi:10.1186/1471-2148-7-116

Gordo I, Gomes MGM, Reis DG, Campos PRA (2009) Genetic diversity in the sir model of pathogen evolution. PLoS ONE 4:e4876. doi:10.1371/journal.pone.0004876

Gunawardena J (2014) Models in biology: accurate descriptions of our pathetic thinking. BMC Biol 12(1):29. doi:10.1186/1741-7007-12-29

Guzman MG, Halstead SB, Artsob H et al (2010) Dengue: a continuing global threat. Nat Rev Microbiol 8:S7–16. doi:10.1038/nrmicro2460

Haldane J (1949) Disease and evolution. Ric Sci 19:1–11. Available: http://www.springerlink.com/index/jv2nj18571312218.pdf. Accessed 29 April 2012

Halstead SB (2007a) Dengue. Lancet 370:1644–1652. doi:10.1016/S0140-6736(07)61687-0

Halstead SB (2007b) Dengue virus-mosquito interactions. Annu Rev Entomol 53:273–291. doi:10.1146/annurev.ento.53.103106.093326

Hancock PA, Godfray HCJ (2012) Modelling the spread of Wolbachia in spatially heterogeneous environments. J R Soc Interface 9:3045–3054. doi:10.1098/rsif.2012.0253

Hancock PA, Sinkins SP, Godfray HCJ (2011) Population dynamic models of the spread of Wolbachia. Am Nat 177:323–333. doi:10.1086/658121

Haygood R, Turelli M (2009) Evolution of incompatibility-inducing microbes in subdivided host populations. Evolution (N Y) 63:432–447. doi:10.1111/j.1558-5646.2008.00550.x

Hedges LM, Brownlie JC, O'Neill SL, Johnson KN (2008) Wolbachia and virus protection in insects. Science (80)322:702. doi:10.1126/science.1162418

Heller R, Chikhi L, Siegismund HR (2013) The confounding effect of population structure on Bayesian skyline plot inferences of demographic history. PLoS ONE 8:e62992. doi:10.1371/journal.pone.0062992

Hilgenboecker K, Hammerstein P, Schlattmann P et al (2008) How many species are infected with Wolbachia?–a statistical analysis of current data. FEMS Microbiol Lett 281:215–220. doi:10.1111/j.1574-6968.2008.01110.x

Hoffmann AA, Turelli M, Harshman LG (1990) Factors affecting the distribution of cytoplasmic incompatibility in Drosophila simulans. Genetics 126:933–948

Hoffmann AA, Turelli M, Simmons GM (1986) Unidirectional incompatibility between populations of Drosophila simulans. Evol 40:692–701. doi:10.2307/2408456

Hosokawa T, Koga R, Kikuchi Y et al (2010) Wolbachia as a bacteriocyte-associated nutritional mutualist. Proc Natl Acad Sci 107:769–774. doi:10.1073/pnas.0911476107

Jansen VAA, Turelli M, Godfray HCJ (2008) Stochastic spread of Wolbachia. Proc R Soc B Biol Sci 275:2769–2776. doi:10.1098/rspb.2008.0914

Johansson M, Hombach J, Cummings DT (2011) Models of the impact of dengue vaccines: a review of current research and potential approaches. Vaccine 29:5860–5868. doi:10.1016/j.vaccine.2011.06.042

Jones EO, White A, Boots M (2007) Interference and the persistence of vertically transmitted parasites. J Theor Biol 246:10–17. doi:10.1016/j.jtbi.2006.12.007

Keeling MJ, Rohani P (2008) Modeling infectious diseases in humans and animals. Princeton University Press

Koch R (1880) Investigations into the etiology of traumatic infective diseases. The New Sydenham Society. Available: http://ocp.hul.harvard.edu/contagion/koch.html, accessed 6 Sept 2014

Lambrechts L, Scott TW, Gubler DJ (2010) Consequences of the expanding global distribution of *Aedes albopictus* for dengue virus transmission. PLoS Negl Trop Dis 4:e646. doi:10.1371/journal.pntd.0000646

Lavine M (1999) What is Bayesian statistics and why everything else is wrong. J Undergrad Math Appl 20:165–174. Available: http://people.math.umass.edu/~lavine/whatisbayes.pdf, accessed September 6, 2014

Lipsitch M, Nowak MA, Ebert D, May RM (1995) The population dynamics of vertically and horizontally transmitted parasites. Proc Biol Sci 260:321–327. doi:10.1098/rspb.1995.0099

Lively CM, Clay K, Wade MJ, Fuqua C (2005) Competitive co-existence of vertically and horizontally transmitted parasites. Evol Ecol Res 7:1183–1190

Lourenço-de-Oliveira R, Vazeille M, de Filippis AM, Failloux A (2004) Aedes aegypti in Brazil: genetically differentiated populations with high susceptibility to dengue and yellow fever viruses. Trans R Soc Trop Med Hyg 98:43–54. doi:10.1016/S0035-9203(03)00006-3

Lu P, Bian G, Pan X, Xi Z (2012) Wolbachia induces density-dependent inhibition to dengue virus in mosquito cells. PLoS Neglected Tro Dis 6:e1754. doi:10.1371/journal.pntd.0001754

Luo S, Koelle K (2013) Navigating the devious course of evolution : the importance of mechanistic models for identifying eco-evolutionary dynamics in nature. Am Nat 181:S58–S75. doi:10.1086/669952

Luz PM, Codeço CT, Massad E, Struchiner CJ (2003) Uncertainties regarding dengue modeling in Rio de Janeiro, Brazil. Mem Inst Oswaldo Cruz 98:871–878. doi:10.1590/S0074-02762003000700002

Maciel-de-Freitas R, Koella JC, Lourenço-de-Oliveira R (2011) Lower survival rate, longevity and fecundity of aedes aegypti (Diptera: Culicidae) females orally challenged with dengue virus serotype 2. Trans R Soc Trop Med Hyg 105:452–458. doi:10.1016/j.trstmh.2011.05.006

Martinez J, Longdon B, Bauer S, Chan Y, Miller WJ, Bourtzis K, Teixeira L, Jiggins FM (2014) Symbionts commonly provide broad spectrum resistance to viruses in insects: a comparative analysis of Wolbachia strains, PloS Pathogens, (in press). doi:10.1371/journal.ppat.1004369

Martins NE, Faria VG, Teixeira L et al (2013) Host adaptation is contingent upon the infection route taken by pathogens. PLoS Pathog 9:e1003601. doi:10.1371/journal.ppat.1003601

Medlock J, Luz PM, Struchiner CJ, Galvani AP (2009) The impact of transgenic mosquitoes on dengue virulence to humans and mosquitoes. Am Nat 174:565–577. doi:10.1086/605403

Merçot H, Poinsot D (1998) And discovered on Mount Kilimanjaro, Nature 391:853

Merçot H, Poinsot D (2009) Infection by Wolbachia: from passengers to residents. CR Biol 332(2–3):284–297. doi:10.1016/j.crvi.2008.09.010

Miagostovich MP, dos Santos FB, Fumian TM et al (2006) Complete genetic characterization of a Brazilian dengue virus type 3 strain isolated from a fatal outcome. Mem Inst Oswaldo Cruz 101:307–313. doi:10.1590/S0074-02762006000300015

Mondini A, de Moraes Bronzoni RV, Nunes SHP et al (2009) Spatio-temporal tracking and phylodynamics of an urban dengue 3 outbreak in São Paulo. Brazil. PLoS Negl Trop Dis 3:e448. doi:10.1371/journal.pntd.0000448

Moran NA (2006) Symbiosis. Curr Biol 16:R866–R871. doi:10.1016/j.cub.2006.09.019

Moreira LA, Iturbe-Ormaetxe I, Jeffery JA et al (2009) A Wolbachia symbiont in Aedes aegypti limits infection with Dengue, Chikungunya, and Plasmodium. Cell 139:1268–1278. doi:10.1016/j.cell.2009.11.042

Myung I (2003) Tutorial on maximum likelihood estimation. J Math Psychol 47:90–100. doi:10.1016/S0022-2496(02)00028-7

Nagao Y, Koelle K (2008) Decreases in dengue transmission may act to increase the incidence of dengue hemorrhagic fever. Proc Natl Acad Sci U S A 105:2238–2243. doi:10.1073/pnas.0709029105

Nguyet MN, Duong THK, Trung VT et al (2013) Host and viral features of human dengue cases shape the population of infected and infectious Aedes aegypti mosquitoes. Proc Natl Acad Sci U S A 110:9072–9077. doi:10.1073/pnas.1303395110

Nikoh N, Tanaka K, Shibata F et al. (2008) Wolbachia genome integrated in an insect chromosome: evolution and fate of laterally transferred endosymbiont genes. Genome Res 18:272–280. doi:10.1101/gr.7144908

Nishiura H (2006) Mathematical and statistical analyses of the spread of Dengue. Dengue Bull 30:51–67

OhAinle M, Balmaseda A, Macalalad AR et al. (2011) Dynamics of dengue disease severity determined by the interplay between viral genetics and serotype-specific immunity. Sci Transl Med 3:114ra128. doi:10.1126/scitranslmed.3003084

Pasteur L (2014) The germ theory and its application to medicine and surgery. Read before the French Academy of Sciences, April 29th, 1878. Published in Comptes Rendus de l'Academie des Sciences, lxxxvi., pp. 1037–43. Translated By H. C. Ernst; available at http://biotech.law.lsu.edu/cphl/history/articles/pasteur.htm#paperII, Accessed August 24th, 2014

Perelson AS (2002) Modelling viral and immune system dynamics. Nat Rev Immunol 2:28–36. doi:10.1038/nri700

Pessoa D, Souto-Maior C, Gjini E et al (2014) Unveiling time in dose-response models to infer host susceptibility to pathogens. PLoS Comput Biol 10:e1003773. doi:10.1371/journal.pcbi.1003773

Price GR (1970) Selection and covariance. Nature 227:520–521

Grenfell BT, Pybus OG, Gog JR et al (2004) Unifying the epidemiological and evolutionary dynamics of pathogens. Science 303:327–332. doi:10.1126/science.1090727

Quéméré E, Amelot X, Piersond J et al (2012) Genetic data suggest a natural prehuman origin of open habitats in northern Madagascar and question the deforestation narrative in this region. Proc Natl Acad Sci 109:13028–13033. doi:10.1073/pnas.1200153109

Rasmussen D, Boni M, Koelle K (2014) Reconciling phylodynamics with epidemiology: the case of dengue virus in southern Vietnam. Mol Biol Evol 31:258–271. doi:10.1093/molbev/mst203

Reznick DN (2013) A critical look at reciprocity in ecology and evolution: introduction to the symposium. Am Nat 181:S1–S8. doi:10.1086/670030

Richardson MF, Weinert LA, Welch JJ et al (2012) Population genomics of the Wolbachia endosymbiont in *Drosophila melanogaster*. PLoS Genet 8:e1003129. doi:10.1371/journal.pgen.1003129

Riegler M, Sidhu M, Miller WJ, O'Neill SL (2005) Evidence for a global Wolbachia replacement in *Drosophila melanogaster*. Curr Biol 15:1428–1433

Rosenberg NA, Nordborg M (2002) Genealogical trees, coalescent theory and the analysis of genetic polymorphisms. Nat Rev Genet 3:380–390. doi:10.1038/nrg795

Ross R (1916) An application of the theory of probabilities to the study of a priori pathometry. Part I. Proc R Soc A Math Phys Eng Sci 92:204–230

Schneider DS (2011) Tracing personalized health curves during infections. PLoS Biol 9:e1001158. doi:10.1371/journal.pbio.1001158

SISTEMA DE INFORMAÇÃO DE AGRAVOS DE NOTIFICAÇÃO (SINAN), http://dtr2004.saude.gov.br/sinanweb/, Accessed 5 Sept 2014

Smith DL, Battle KE, Hay SI, Barker CM, Scott TW et al (2012) Ross, macdonald, and a theory for the dynamics and control of mosquito-transmitted pathogens. PLoS Pathog 8:e1002588

Souto-Maior C, Lopes JS, Gjini E, Struchiner CJ, Teixeira L, Gomes MGM (2015) Heterogeneity in symbiotic effects facilitates Wolbachia establishment in insect populations. Theor Ecol 8:53–65. doi:10.1007/s12080-014-0235-7

Teixeira L, Ferreira A, Ashburner M (2008) The bacterial symbiont Wolbachia induces resistance to RNA viral infections in *Drosophila melanogaster*. PLoS Biol 6:e1000002. doi:10.1371/journal.pbio.1000002

Tortosa P, Charlat S, Labbé P et al (2010) Wolbachia age-sex-specific density in *Aedes albopictus*: a host evolutionary response to cytoplasmic incompatibility? PLoS ONE 5:e9700. doi:10.1371/journal.pone.0009700

Turelli M (1994) Evolution of incompatibility-inducing microbes and their hosts. Evolution (N Y) 48:1500–1513. doi:10.1111/j.1558-5646.2008.00550.x

Turelli M (2010) Cytoplasmic incompatibility in populations with overlapping generations. Evolution 64:232–241. doi:10.1111/j.1558-5646.2009.00822.x

Turelli M, Hoffmann AA (1991) Rapid spread of an inherited incompatibility factor in California Drosophila. Nature 353:440–442. doi:10.1038/353440a0

Turelli M, Hoffmann AA (1995) Cytoplasmic incompatibility in Drosophila simulans: dynamics and parameter estimates from natural populations. Genetics 140:1319–1338

Vasilakis N, Cardosa J, Hanley K a, et al. (2011) Fever from the forest: prospects for the continued emergence of sylvatic dengue virus and its impact on public health. Nat Rev Microbiol 9:532–41. doi:10.1038/nrmicro2595

Vavre F, Charlat S (2012) Making (good) use of Wolbachia: what the models say. Curr Opin Microbiol 15:263–268. doi:10.1016/j.mib.2012.03.005

Volz EM (2012) Complex population dynamics and the coalescent under neutrality. Genetics 190:187–201. doi:10.1534/genetics.111.134627

Volz EM, Koelle K, Bedford T (2013) Viral phylodynamics. PLoS Comput Biol 9:e1002947. doi:10.1371/journal.pcbi.1002947

Wakeley J (2005) The limits of theoretical population genetics. Genetics 7:1–7

Wakeley J, Sargsyan O (2009) Extensions of the coalescent effective population size. Genetics 181:341–345. doi:10.1534/genetics.108.092460

Walker T, Johnson PH, Moreira LA et al (2011) The wMel Wolbachia strain blocks dengue and invades caged Aedes aegypti populations. Nature 476:450–453. doi:10.1038/nature10355

Wearing HJ, Rohani P (2006) Ecological and immunological determinants of dengue epidemics. Proc Natl Acad Sci USA 103(31):11802–11807. doi:10.1073/pnas.0602960103

Weeks AR, Turelli M, Harcombe WR et al (2007) From parasite to mutualist: rapid evolution of Wolbachia in natural populations of Drosophila. PLoS Biol 5:e114. doi:10.1371/journal.pbio.0050114

Werren JH, Baldo L, Clark ME (2008) Wolbachia: master manipulators of invertebrate biology. Nat Rev Microbiol 6:741–751. doi:10.1038/nrmicro1969

WHO Health topics: Dengue, http://www.who.int/topics/dengue/en/. Accessed 16 August 2014

WHO-VMI Dengue Vaccine Modeling Group (2012) Assessing the potential of a candidate dengue vaccine with mathematical modeling. PLoS Negl Trop Dis 6(3):e1450. doi:10.1371/journal.pntd.0001450

Ye YH, Woolfit M, Rancès E et al (2013) Wolbachia-associated bacterial protection in the mosquito Aedes aegypti. PLoS Negl Trop Dis 7:e2362. doi:10.1371/journal.pntd.0002362

Zabalou S, Charlat S, Nirgianaki A et al (2004) Natural Wolbachia infections in the Drosophila yakuba species complex do not induce cytoplasmic incompatibility but fully rescue the wRi modification. Genetics 167:827–834. doi:10.1534/genetics.103.015990

Evolution of the Human Microbiome and Impacts on Human Health, Infectious Disease, and Hominid Evolution

Laura S. Weyrich

Abstract Because the diverse microbial communities on and in the body (microbiome) are tightly coevolved with human cells, it is vital to explore the evolutionary history of the human microbiome. With the advent of new methodologies and sequencing technologies, researchers can now explore different factors that influence the bacterial community structure and colonization of specific species in the human microbiome. Using distant out-groups, such as chimpanzees, and human populations with unique lifestyles, such as Amerindians, the history and formation of the modern human microbiome in Westernized societies can be elucidated, providing vital information into how to these microbial communities were impacted by past events. Large cultural and dietary revolutions, such as the Neolithic Revolution (~7500 years ago) and the Industrial Revolution (~200 years ago), largely impacted these microbial communities. Ancient events, such as interbreeding and admixture with our closest ancient relatives, such as Neanderthals or Denisovans, could also have impacted the microbiome that modern humans carry today. Reconstructing the evolutionary history of the human microbiome has proven to be an intricate tale, with impacts on modern health and disease, as well as the evolutionary fate of modern humans.

Keywords Microbiome · Neolithic revolution · Industrial revolution · Hominid evolution · Health · Disease

The human microbiome is tightly linked with many basic physiological functions, including digestion, immune system development, and hormone production (Consortium 2012). This relationship between a host and its commensal

L.S. Weyrich (✉)
Australian Centre for Ancient DNA (ACAD),
University of Adelaide, Adelaide, Australia
e-mail: laura.weyrich@adelaide.edu.au

N. Gontier (ed.), *Reticulate Evolution*, Interdisciplinary Evolution Research 3,
DOI 10.1007/978-3-319-16345-1_9

microorganisms, including bacteria, fungi, and viruses, is certainly a symbiotic one. For example, microorganisms assist the human body in food digestion by breaking down difficult molecules, such as cellulose, allowing the body to obtain nutrients from otherwise nutrient inaccessible foods (Chassard et al. 2010). In return, bacterial organisms living in the digestive track produce essential vitamins that are absorbed through the colon, such as vitamin K (K_2MK7) that cannot otherwise be easily obtained (Ramotar et al. 1984).

These simple observations imply that the human body and the microbiome have coevolved over millions of years, creating a mutualistic relationship where both the microorganisms and the host are dependent upon one another in the following ways. First, the diverse microbial community in the body is dependent on the host for an ecological environment in which to live and is therefore also susceptible to the environments and factors exposed to the host. Second, the microorganisms within the human body can also adapt to change much more quickly than the human genome, potentially providing quick adaptive advantages to its host. This means that each player in this symbiotic relationship can be under different selection regimes. For example, bacterial communities are susceptible to modification when the human diet is altered, when environments are rapidly changed, or when chemical treatments are applied, such as antibiotics (Cho et al. 2012; David et al. 2013; Lax et al. 2014), whereas the human genome can only adapt through genetic mutations or epigenetic modifications from one generation to the next. It may be easier to think about the body as a national park and the microbes as the plants and animals that inhabit that space. As the park is altered over geological time scales, the species present in the park are susceptible to disease, weather, or invasions of new species, which are all factors that can impact the inhabitants and alter that ecological structure at a comparatively fast rate. Nevertheless, this unique dynamic between a single host and the diverse communities that inhabit it provides a remarkable system to investigate the impacts of coevolution under different scenarios. Because microbes are under different selection pressures than their host, the tightly linked evolutionary relationship between humans and their microbes can be disrupted. The exact ramifications and impacts of this disruption are a hotly debated and researched topic that rapidly changes based upon new findings.

It is critical to understand how the modern human microbiome was established, as understanding the pressures and events that moulded the human microbiome may provide vital information in modern medicine, providing insight into how to shape or change the microbiome during disease or infection. Furthermore, investigating the evolution of the human microbiome may provide insight into our human history, and how microorganisms could have contributed to what makes us human.

This chapter will focus on current knowledge surrounding the evolution of the human microbiome, analysing studies that have investigated the evolutionary history of the human microbiome and how these long-term changes in the microbiome have impacted human health, the evolution of infectious disease, and hominid evolution.

1 Comparison Between Ape and Human Microbiomes to Establish the Evolutionary History of Commensal Microorganisms

Because the human microbiome plays an essential role in human physiology, one could assume that the body and the microbiome had evolved together and that the metagenome (the bacterial genomes of organisms inhabiting the body) and the human genome would have responded similarly over evolutionary time. However, bacteria divide and regenerate at incredible speeds compared to humans. *Escherichia coli*, for example, can divide in 20 min compared to the generational time of humans, which is usually estimated on average at 28 years. We also know that bacteria are capable of transferring genetic information horizontally, or between species, allowing access to millions of other genes and the ability to change rapidly when necessary. These observations suggest that the microbiome and metagenome may not have evolved at the same rate or been susceptible to the same evolutionary pressures as the human genome.

A simple way to examine the evolutionary history of the human microbiome is to examine the microorganisms present in related host species. Identifying differences between the chimpanzee microbiome and the human microbiome can provide insight into how the two bacterial communities have changed for the past 5 million years, or the estimated time of evolutionary host speciation (Soares et al. 2009).

Initial studies examining the gut microbiomes of wild, non-human primates were able to detect species-specific differences between the mantled colobus (*Colobus guereza*), red colobus (*Piliocolobus tephrosceles*), and red-tailed guenon (*Cercopithecus)* (Yildirim et al. 2010). Next, several groups utilized the closest living human relative, the chimpanzee, comparing the chimpanzee gut microbiome to human microbiomes from similar locations and Western, industrialized countries. A research group at Yale analysed gut microbiomes from 32 chimpanzees (*Pan troglodytes*) in the Gombe Stream National Park, Tanzania (Moeller et al. 2012). These researchers observed similar gut 'enterotypes' in chimpanzees and humans (enterotypes are simply a way of classifying microbiomes based on the largest proportions of dominant bacterial phyla). Earlier, human gut microbiomes were revealed to be classified into one of three enterotypes, dominated each by a unique bacterial taxa: *Bacteriodetes, Prevotella*, and *Ruminococcus* (Arumugam et al. 2011). When analysing the chimpanzee microbiome, these Yale researchers found that microbiomes from chimpanzees fell into the same three gut enterotypes that were previously identified in humans. For example, one enterotype was dominated by *Bacteriodetes, Faecalibacterium*, and *Parabacteroides* taxa, whereas another type is only dominated by *Lachnospiraceae*. The variation in the last enterotype contained overrepresented numbers of *Dialister*, *Ruminococcus*, *Subdoligranulum*, and *Collinsella* taxa. These groups simply allow researchers to categorize each type, as they try to make sense out of the patterns they see of hundreds of bacterial species from one individual sample. Similar species identified in the gut and the presence of the same three enterotype groupings highly suggest that the hominid microbiome is evolutionarily conserved, or is in concordance with the differences that have occurred since humans and chimpanzees split from an ancient progenitor primate species. However, when these chimpanzees were followed over long time spans, each chimpanzee gut microbiome surveyed over time was capable of switching between different gut enterotypes, demonstrating that factors outside of host designation were capable of altering these bacterial communities.

We must therefore also consider the environmental selection pressures that drive diversity in the hominid gut. Although ancestral factors corresponding to the hominid gut environment, i.e. pH, light availability, and an anaerobic environment, may limit which bacterial species can be successful, other factors, such as diet or environmental exposures, must be able to alter the proportions and subset of microorganisms inhabiting chimpanzees and humans alike. These additional factors are also likely to drive variation present in the microbiome, capable of blurring the lines between these three distinct enterotypes (Yong 2012). After this initial study, the Yale research group and others compared chimpanzee, bonobo (*Pan paniscus*), gorilla (*Gorilla gorilla*) and human microbiomes from the same (sympatric) and different geographical locations (allopatric) to understand how much environmental exposure may contribute to alterations in the hominid microbiome (Moeller et al. 2013). It is plausible that although gut microbiomes may be evolutionarily distinct in concordance with their hosts, cross contact or exposure to certain diets and environments may play a role in the microbial species that colonize the gut. Although each host species could still be distinguished by their microbiomes, this study found that gorillas and chimpanzees living in sympatry did share more bacterial taxa than species living separately, suggesting that environmental exposure contributes to which microbial species are successful in hominids. This was subsequently confirmed to be true in the oral cavity as well. The oral microbial communities in bonobos and chimpanzees were compared to humans from the same location. Humans and apes could still be distinguished, but humans and chimpanzees living in the same location had more similar microbiomes than their allopatric counterparts (Li et al. 2013). Together, these studies demonstrate that environmental and dietary factors can contribute to alterations of the human microbiome, even if humans and chimpanzees once shared a more similar microbiome.

2 Comparison of Microbiomes of Western, Urbanized and Indigenous Populations

Because it was apparent that the human microbiome could adapt to diet and environment, differences between microbiomes from unique human cultural groups were also investigated. In a landmark study, American researchers identified differences in the microbiomes from Amerindians from the Amazon basin, rural residents from Malawi, and metropolitan American children and adults (Yatsunenko et al. 2012). These differences included not only differences in bacterial species, but also in functional repertoires, indicating that the bacterial communities were acting in different ways inside the host. This finding was corroborated when researchers examined the gut microbiomes of children from Europe and Burkina Faso in West Africa, primarily to investigate the impacts of diet on the human microbiome (Filippo et al. 2010). Children from Burkina Faso had increased bacterial diversity, including unique species not obtained from European children. The diet in these African children is quite distinct from that observed in Western, industrialized countries in the European Union and included unique *Prevotella* and *Xylanibacter taxa*. Further studies identified similar findings in other body sites as well, including the oral cavity (Contreras et al. 2010). Recently, microbial differences were also observed between males and females within a single rural culture that play different roles in their communities and eat different daily diets, i.e. women that gather fibrous tubers had increased levels of *Treponema* taxa in comparison to the male hunters in Hadza African tribes (Schnorr et al. 2014). Although these studies suggest that diet, environment, and even culture can alter the human microbiome, they do not disentangle the different evolutionary histories of these different microbiomes and their functions. These studies also do not examine the role that these microbial differences play in health and disease in these different modern human populations.

3 Utilizing Ancient Samples to Understand Changes in the Microbiome Over Vast Time Scales

Two major sample types, faeces and calcified dental plaque (Fig. 1), can be utilized to examine extinct microbiomes. Both sample types therefore help determine how the human microbiome may have evolved and adapted through time. Researchers initially started looking at microorganisms that were preserved in ancient faeces (coprolite) as a source of ancient gut microorganisms. Faeces from living individuals are generally a representative sample of the microorganisms living within the human gut (Consortium 2012). However, there are only a few instances where coprolites are exquisitely preserved (Tito et al. 2012). There is an example of coprolites from La Cueva de los Chiquitos Muertos that resemble human gut microbiomes. However, most coprolites can contain large amounts of environmental contamination or post-depositional bias. For example, coprolites obtained from inside mummies match bacterial 'fingerprints' from compost piles, while coprolites from wet caves contain microorganisms more commonly found in the soil, rather than the human gut. This can be explained by taphonomy, or the degradation of a biological sample after death of the organism has occurred, which likely occurs much more rapidly in exposed coprolites than bone or teeth (Allentoft et al. 2012). Obviously, fresh faeces contains a high proportion of organic matter, which can be decomposed and infiltrated by environmental microorganisms, therefore swamping and killing the original gut microorganisms that were present when the sample was

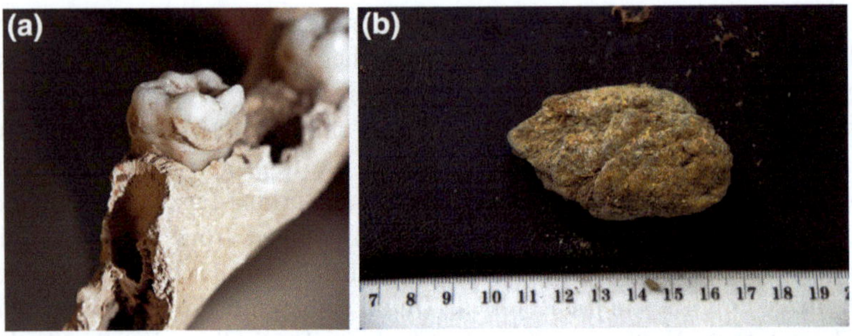

Fig. 1 **a** A preserved bacterial biofilm (dental calculus) on the tooth present in the mandible of a medieval individual. **b** A coprolite identified from a New Zealand cave. These samples contain preserved human-associated microorganisms and can therefore provide information on how ancient human microbiomes have evolved through time

deposited. These examples highlight a few of the technical issues that researchers must overcome when obtaining material to analyse the evolution of past microbiomes and microorganisms (Cooper and Poinar 2000).

In 2012, a team of Chilean researchers were the first to use DNA techniques and identify human-associated microorganisms trapped within calcified dental plaque (calculus) (De La Fuente et al. 2013). Dental calculus is an oral bacterial biofilm that is calcified in place by calcium phosphate, trapping and preserving microorganisms associated with the oral cavity (Ennever et al. 1973). Because the organisms are preserved in place during the life of the individual, they provide a fossil record of human-associated microorganisms, allowing researchers to understand how the human microbiome has changed through time. This finding supported scanning electron microscopy evidence that identified small bacterial cocci and rod shapes within calculus (Dobney and Brothwell 1988). This also supported years of archaeological research on plant fragments and environmental debris (i.e. cotton fibres) identified in calculus (Henry and Piperno 2008; Blatt et al. 2011), demonstrating that this calcified bacterial matrix may also provide information on past environments.

Although Chilean researchers were able to identify several microorganisms that are known to primarily exist in the mouth, such as *Streptococcus*, *Porphyromonas, Actinomyces*, and *Fusobacterium* species (De La Fuente et al. 2013), this initial study on the bacterial DNA preserved in dental calculus was limited to looking at individual species utilizing Sanger DNA sequencing techniques. Shortly after this initial discovery, researchers at the Australian Centre for Ancient DNA used a different technique to sequence and identify all of the bacterial species present in ancient dental calculus specimens over the past 8000 years of human history (Adler et al. 2013). Using next-generation DNA sequencing techniques, such as 454 or Illumina, these researchers used a technique called metabarcoding, where they can sequence a gene (16S ribosomal RNA) that provides a bacterial fingerprinting for each of the thousands of species in one single sample (Caporaso et al. 2012). This new technique provided insight into how the microorganisms within the human body have changed, and identified two major events in human history that significantly impacted these communities: the Neolithic Revolution (the onset of farming) and the Industrial Revolution (the invention of machine) (Adler et al. 2013). During these time periods, the overall bacterial community structure shifted, and several oral pathogens increased in prevalence through time, with the largest level of oral pathogens occurring in modern populations. So, why would two very large cultural revolutions impact the human microbiome?

4 How Did the Neolithic Revolution Impact the Human Microbiome?

The Neolithic Revolution was not only a change in lifestyle practices, but brought on an onslaught of dietary, environmental, and cultural change. Hunting and gathering populations that once ate large amounts of protein (dietary isotope analysis demonstrated that hunting and gathering Neanderthals were as carnivorous as some bears! (Bocherens et al. 2001)) were either converted or replaced by cultures such as the Linear Pottery culture (LBK) (Brandt et al. 2013), which subsisted largely on carbohydrates obtained from planting grains. This occurred about 7500 years before present (yBP) in Europe, even though the transition to agriculture happened at different time points and in different ways around the globe. Maize domestication in the New World, for example, likely occurred about 3000 yBP (Scarre 2013). The Neolithic Revolution was a very dynamic process, taking years and generations to take hold. Although today many researchers view farming and the invention of agriculture as an adaptive advantage, hunter-gatherer groups likely disagreed with this new living strategy, remarkably delaying the adaptation to agricultural techniques in different locations.

Nevertheless, the transition to agriculture was quite successful, and as a result, had widespread consequences around the globe, including changes in diet staples, increases in population size, adaptation to sedentary lifestyles, the development of class and social structures, and exposure to domesticated animals (Scarre 2013). For example, Turnbaugh et al. demonstrated last year that a dietary switch from a protein-based diet to one focused on carbohydrates can alter the microbial communities in the body in as little as 1 week (David et al. 2013). Furthermore, modern hunter-gatherer groups in Burkina Fasso and the Amazon basin also have altered gut microbiomes compared to adults from industrialized nations, which is believed to be primarily due to dietary differences (Filippo et al. 2010; Blaser et al. 2012). As ancient humans switched from hunting and gathering to farming, archaeological evidence suggests a radical dietary change involving decrease in protein intake. Interestingly, an increased incidence of pathologies related to iron-deficiency *anaemia*, likely due to changes in diet, has also been reported (Haviland et al. 2010). Collectively, this suggests that massive changes in health occurred when humans switched to agricultural-based lifestyles.

Even though carbohydrates dominated the diet, increased volumes and reliability of food led to great population booms, and with increased population sizes, infectious diseases also likely concurrently evolved. Larger population sizes allowed pathogens access to increased numbers of susceptible individuals. Recent evidence has suggested that several common modern pathogens, such as *Streptococcus mutans* (oral caries or cavities) (Cornejo et al. 2013; Adler et al. 2013) or *Mycobacterium tuberculosis* (tuberculosis) (Comas et al. 2013), originated during the early years of this transition. Interestingly, modern research has demonstrated that the presence of several pathogens in the human microbiome can result in remarkable shifts in the overall bacterial community structure in the body. For example, colonization of the respiratory pathogen *Bordetella bronchiseptica* (kennel cough) (Weyrich et al. 2013) or the gut pathogen *Salmonella enterica* (food poisoning) (Stecher et al. 2007) can displace common host microorganisms, significantly altering the microbial communities in the body after infection. In addition, colonization by an oral pathogen *Porphyromonas gingivalis* (periodontal disease) can alter communities in the gastrointestinal tract (Arimatsu et al. 2014) when the pathogen is introduced through the bloodstream during oral disease. These pathogens, and likely many that arose during the Neolithic Revolution, likely influenced the microbial communities in the body during this period by in particular manipulating the immune response directed at commensal or host microorganisms. Therefore, it is likely that the introduction of specific pathogens during the Neolithic period into the human population had impacts on which bacterial species were successful in a post-Neolithic microbiome.

This dietary transition during the Neolithic Revolution also meant that people could be settled in one location, without following herds or foraging for food in different locations during different seasons. Villages, cities, and entire civilizations were now able to form with a temporally and geographically stable food supply, which almost meant increased exposure to waste. In human waste, many of the infectious microorganisms transmitted through the faecal–oral route persist, which plays a significant role later in history. Without the agricultural revolution, civilizations would arguably not have been formed, and pathogens, such as *Yersinia pestis* (Black Plague or Black Death), could not have been successful. The Plague or Black Death wrecked several large European cities during the medieval period (1200 to ~500 yBP), eliminating up to 70 % of London during a 1356 epidemic (Bos et al. 2011). Today, modern plumbing and water purification standards protect most individuals in industrialized countries, although waterborne diseases are still the second biggest killer worldwide (WHO 2004). Living in these close communities with increased exposure to human waste also likely resulted in alterations to the microbiome, as species from one human were more easily transferred to another through water contamination, as well as the increased exposure to pathogens and the ability of these gastrointestinal pathogens to rapidly infect large proportions of the human population.

During the Neolithic Revolution, several significant instances of animal domestication also occurred, as animals such as cattle were needed to pull ploughs required to sow seeds (Scarre 2013). It is thought that humans intimately interacted with domesticated animals, sharing houses and sometimes beds with the animals. It is easy then to imagine how the transfer of microorganisms or pathogens present in the bovine microbiome could have been transferred into humans during this period. It was once hypothesized that the establishment of *M. tuberculosis* must have occurred from exposure to cattle, as a closely related organism *Mycobacterium bovis* caused a similar disease in bovids (Behr 2013). However, genome analysis identified that these two microorganisms were the result of convergent evolution, evolving a similar disease in different hosts separately but under the same selective pressures (Comas et al. 2013). We can also blame our exposure to several other gastrointestinal pathogens due to the development of our close relationships with animals during this time. For example, non-pathogenic bacteria in mammals may cause disease in humans, such as food poisoning caused by *E. coli*, which was likely first introduced into populations around this time, although arguably, ancient humans probably had a microbiome more resistant to these gastrointestinal diseases (Willing et al. 2011), unlike modern humans today.

In comparison, evolutionary changes also occurred in the human genome during this time period, albeit at a much slower rate. This exemplifies that the same selective pressures act on the human genome and microbiome similarly, although the outcome can be different and adaptive solutions happen uniquely at an individual and ecological scale. For example, the genetic information encoding lactose tolerance likely arose in early agricultural and cattle domestication centres (Burger et al. 2007). As humans began domesticating bovids and drinking milk, this would have provided increased exposure to vitamins and sugars, such as lactose. Interestingly, several microorganisms that inhabit the mammalian oral cavity, gut, and breast milk, such as *Lactobacillus*, are capable of digesting lactose (Hokama et al. 1996; Cabrera-Rubio et al. 2012). The coevolution that occurred of the host and its microbes is quite interesting, although not yet fully investigated. Another example became apparent when full genomes of archaic and ancient humans were sequenced and compared to modern individuals, revealing several alterations in immunoregulatory genes (Abi-Rached et al. 2011). This suggests that the human genome adapted to various infectious diseases, i.e. mutations in immune genes, likely presented an adaptive advantage to survive certain infectious epidemics that wiped out populations. However, we also know that the immune system is tightly linked to the microbial communities in the body through the intestinal wall (Hooper and Gordon 2001), even though it also remains unclear how these kinds of genomic adaptations and alterations would impact the microbiome, although evidence from modern studies suggests it most certainly should.

Even though the Neolithic Revolution resulted in cultural, environmental, and dietary changes, this event had large impacts on the human microbiome. Cultural and social changes can have large impacts on the human microbiome, simply because human behaviour is changed. Our daily routines, including personal hygiene, interpersonal interactions, and your lunch, can impact the organisms in our body and ultimately our health. Researchers are still trying to figure out how these changes that occurred during the Neolithic Revolution influence modern health, as our diets continue to change, we are exposed to new chemicals, and the climate continues to change. Further in-depth studies of ancient civilizations through time will reveal how specific social and cultural changes impacted our microorganisms.

5 How Did Dietary and Environmental Alterations During the Industrial Revolution Impact the Human Microbiome?

Just as the Neolithic Revolution irreversibly altering mankind, the Industrial Revolution marked additional changes in human behaviour, health, and our microbiomes. The Industrial Revolution was demarcated by the development of mechanical manufacturing, following the invention of the machine tools (Berg and Hudson 1992). Although some of the changes were similar to those identified during the Neolithic Revolution, such as increased population growth and alterations to daily behaviours, it also brought new changes, such as pollution and manufactured food (Berg and Hudson 1992; Hermanussen 2006). For the first time, coal was being burned in large quantities to fuel machinery, and food was being canned for long-term survival.

The invention of the engine was arguably one of the biggest achievements in human history over the past 1000 years. This one invention led to a flood of downstream applications, including locomotion, textile and chemical manufacturing, iron production, and a laundry list of others (Berg and Hudson 1992; Scarre 2013). However, each of these needed a source of energy, which could be achieved through steam power or by burning of coal. Specifically, iron production largely benefited from coal burning, because high temperatures required to make iron, could be achieved without producing sulphur and ash, which was produced during the burning of wood and contaminated the iron (Jones 2009). Large quantities of coal were burned for iron production during the Industrial Revolution, and first-hand descriptions of several British cities during that time period spoke of the black skies and named the area 'Black Country' for the pollution in the air (Jones 2009). Consequential to this, coal burning also releases heavy metals into the atmosphere, including lead, mercury, and nickel. In environmental settings, we know that heavy metal contamination can severely affect microbial communities (Breton et al. 2013), as bacterial microorganisms are similarly sensitive to heavy metals. Several unique biochemical pathways exist in environmental microbes that increase their tolerance to heavy metals (Hu et al. 2005); microorganisms within the human microbiome would have a significant advantage during the Industrial Revolution if they had these pathways.

Another implication of the Industrial Revolution was the processing and preservation of foods. While dried or dehydrated foods had been used to preserve meats prior to this period, the Industrial Revolution allowed manufacturing to develop canning, preserving vegetables and fruits in metal containers for later use (Graham 1981). The impacts on the microbiome are twofold. First, this invention is tightly linked to not only population increases, but also an increased life expectancy and better nutrition (Hermanussen 2006). Famines were the norm prior to industrialization, but canning and subsequent alterations to transportation networks during this period allowed access to food produced in different locations and during different growing seasons. Nutrition, especially in young children, markedly improved during this time period. The percentage of the children born in London who died before the age of five decreased over 40 % in London in the early 1800s (Berg and Hudson 1992). This dietary change alone can have significant impacts on the human microbiome, but overall nutritional health has also been shown to impact the microorganisms in the body (Kau et al. 2011a). Bacteria not only eat what their host eats, but they are also susceptible to the functionality of the immune system, which can be largely dependent upon dietary nutrition (Kau et al. 2011b). Just think about how your mother told you that you were going to get sick if you did not eat your green vegetables; the same principles apply here at both a micro- and macro-scale.

The second potential alteration from food processing during the Industrial Revolution was the removal of microorganisms from food during the pasteurization or sterilization steps of canning. Before a food can be canned, it must be pasteurized or sterilized in such a way that bacteria that would spoil the food after canning are removed. During the Industrial Revolution, pasteurization was commonly achieved by boiling the food. Boiling alters the chemical composition of several key nutrients and kills potentially harmful microorganisms that would be otherwise introduced into the body. This process of food preservation can significantly alter the microbial content, especially of pasteurized foods such as milk (Quigley et al. 2013). Recent modern studies have identified whole suites of microorganisms present on fruits and vegetables, as well as identified significant differences on organic verses non-organic produce (Leff and Fierer 2013). Although it is unclear which microorganisms survive as food passes through your stomach and into your gastrointestinal tract, studies have observed microorganisms present in cured meats survive into the gut and are biochemically active for up to one week (David et al. 2013). This suggests that microbes on the food we eat can be inoculated into the gut and certainly are exposed to the microbiome in the mouth. Altering the flora of our food could therefore impact which species are successful in the body over long time periods.

There were also a few additional alterations to daily human life that occurred around the Industrial Revolution, although arguably may not be tied to industrialization. Several European countries, namely England, Spain, and France, embarked on the colonial period, spanning the sixteenth to nineteenth centuries. As globalization occurred through increased means of travel, i.e. steam powered railways and ships, access to distant lands diversified the diet and brought modern staples, such as sugar, back to wealthier nations (Davis 1962). Sugarcane, sugar beets, and the refinement of sugar are possibly the best examples of this. Even though sugar cane was cultivated nearly 10,000 years ago in New Guinea, access to industrialized nations only occurred in the past 300–500 years, during the colonial period (Deer 1949). As shipping and trade increased from the Pacific Islands and India to Europe, sugar was more easily transported back to Europe. In the early 1800s during the Industrial Revolution, steam engines began to mechanize the process of refining sugar, producing larger quantities to supply an increased demand. Because sugar is an ideal source of carbohydrates for bacteria, as it is for humans, it should be no surprise that the addition of sugar into the Western, industrialized diet would have impacts on the microbiome. *S. mutans*, and other caries-associated bacteria, can use the carbohydrates present to make lactic acid, which is ultimately what destroys tooth enamel and causes cavities (Ajdić et al. 2002). Therefore, microorganisms such as *S. mutans* (dental caries or cavities) have adaptive advantages in the presence of increased sugar, significantly impacting the microbial community structure in the mouth (Wade 2013). The incidence and use of sugar is believed to explain the gross changes in oral health that are observed in human populations through time, i.e. from 7500 yBP to present day (Aufderheide and Conrado 1998; Sajantila 2013), which are tightly associated with changes in the oral microbiome (Adler et al. 2013). Similar influences of sugar on gut communities have also been observed in modern populations (Payne et al. 2012), especially in relationship to the rise of obesity in Western countries (Cameron et al. 2003; Ley et al. 2005), indicating that the introduction of sugar into the human diet likely had drastic impacts on the human microbiome.

6 Impacts of Microbial Alterations on Hominid Evolution

In this chapter, I have described how dietary, cultural, behavioural, and environmental alterations can impact the human microbiome and have given specific examples of how these have likely contributed to the formation of the modern human microbiome over the past 8000 years. However, this only scratches the surface when considering additional alterations that the human microbiome has undergone during the entire course of human evolution since our separation from chimpanzees (Soares et al. 2009). This is compounded by the fact that the evolutionary history of anatomically modern humans is a complex story, full of large migrations across the globe, adaptation events to climate change, and interbreeding with related species.

The taxonomic resolution of the human species is one of the most contentious and hotly debated topics in modern evolutionary studies. It is generally accepted that the homininae family can be broken down into at least the following genera: *Pan* (chimpanzees and bonobos), *Gorilla* (gorillas), and *Homo* (humans). However, controversy arises in what regards the delineation of additional genera (Wood and Richmond 2000) and the number of species that belong to such a genus. Gorillas split from other hominids about 8 million years ago, and chimpanzees and bonobos split from the lineage that would evolve into modern humans nearly 5 million years ago. It is after this split 5 million years ago that the hominid tree becomes much less resolved. During the last few million years, several homininae genera or species have existed, including (along with their estimated extinction date) *Paranthropus* (2.7 MyBP), *Australopithecus* (2 MyBP), *Homo habilis* (1.4 MyBP), *Homo heidelbergensis* (350 KyBP), *Homo erectus* (200 KyBP), *Homo floresiensis* (12 KyBP, although unpublished reports now estimate this date to be much older, i.e. 45 KyBP), *Homo neanderthalensis* (40 KyBP; Neandertals), and Denisova (~40 KyBP), a newly discovered species only identified by genomic techniques applied to teeth and a finger bone (Meyer et al. 2012). While most of these species were identified by mere morphological analyses of retrieved fossils, scholars were able to abstract ancient DNA from *H. heidelbergensis*, Neanderthals, and Denisovans, providing a molecular prospective that shed more light on the recent evolutionary history of the genus *Homo* (Green et al. 2010; Meyer et al. 2012, 2014). Perhaps most surprisingly, ancient DNA sequencing has revealed interbreeding and admixture between Neandertals, Denisovans, and anatomically modern humans, likely occurring 60–40 KyBP, suggesting that we successfully mated with our close relatives before they went extinct around 40,000 years ago (Sankararaman et al. 2012; Meyer et al. 2012; Cooper and Stringer 2013). Why these species went extinct is another highly debated topic. Possible explanations are the ability to adapt in the face of climate change, expansions of other hominid lineages, and diseases.

If we assume that modern humans inherited a microbiome from a hominid progenitor species and that this microbiome then adapted according to cultural, environmental, and dietary alterations, we must investigate the factors that could have resulted in significant changes of the human microbiome over the past 2 million years. So can we identify some of the cultural, behavioural, and environmental factors that would have impacted the microbiome of early hominid species (i.e. before 8000 yBP) and would these changes then have impacted the lives and evolution of modern hominids?

First, we must think about the movement of different hominid groups, specifically focusing on the most recent *Homo* species, including Neanderthals, Denisovans, and anatomically modern humans. Again, this is a highly debated topic, but it is generally accepted that earlier *Homo* species, such as *H. erectus*, left Africa about one million years ago (Jin and Su 2000). However, there are two theories about the timing and events that result in the transformation and arrival of anatomically modern humans outside of Africa: the multiregional model and the out-of-Africa model (Hollox et al. 2013). The multiregional model proposes that the evolutionary changes from *H. erectus* to modern humans occurred in different global locations at different times after *H. erectus* left Africa. For example, the progenitor species of modern humans would have left Africa a million years ago, colonized different locations in Eurasia, and evolved into modern humans in these different parts of the globe (Wolpoff et al. 2000). The out-of-Africa model on the other hand proposes that anatomically modern humans evolved in Africa and then more recently left Africa around 60 KyBP to disperse and colonize the different parts of the world, then interacting with other hominid species that left Africa earlier (Tattersall 2009). Recent genome sequences from two extinct *Homo* species, Neanderthals (Green et al. 2010) and Denisovans (Meyer et al. 2012), indicate that modern humans interbred with these species around 60–40 KyBP in

Eurasia. This finding, as well as other molecular evidence, has presented more current evidence to support the out-of-Africa model, making it now more generally accepted (Science Daily, 2014). For ease, we will assume in this chapter that modern humans evolved during climate shifts in Africa and that modern humans left Africa into Eurasia around 60 KyBP, then interacting with other ancient hominid species that had already colonized Europe, Asia, and the Pacific hundreds of thousands of years ago.

Considering this history, we can then estimate how alterations to the hominid microbiome have occurred over the past 2 million years. As humans headed north out of Africa and traversed across Eurasia, they would have encountered different climates, vegetation, and environments. As discussed earlier, humans would have eaten different food items as they were available in various locations, suggesting that the microbiome would have been altered from changes in diet alone. Interestingly, several studies analysing the dietary contributions of similar diets across different species of zoo animals indicate that diet alone can drive similarities and convergence between microbiomes (Muegge et al. 2011). So we would expect that humans living in similar environments would also adapt to having similar microbiomes, even if they were in different geographical locations. Additionally, exposure to different environments can impact the human microbiome. Modern studies have examined microorganisms present in home dwellings when humans move into a new house and determined that humans and their environment immediately start integrating (Lax et al. 2014). This suggests that humans would acquire microorganisms from different environments that they encountered, and as the environments changed from desserts to forests and coastlines, the microorganisms being exchanged with the human microbiome would be different.

The interbreeding identified between different *Homo* species would have also likely impacted the human microbiome, and conversely, the human microbiome may have even impacted the mating success between these species. First, we can assume that the microbiomes of different bands or even species of hominids would have been distinguishable, as each of these host species is also genetically distinguishable. This suggests that their intermixing and interbreeding between individuals would have altered their microbiomes. As an example, we know that cohabitating modern humans share more microbes with their partners than other people, as well as share bacterial species with pets, such as dogs, that live in the same household (Song et al. 2013). Although the findings were less significant in the oral and gut microbiomes, studies such as this suggest that mixing with other individuals and cohabitation can impact the human microbiome. In addition, the human microbiome is established in children from their mothers and primary caregivers (Caufield et al. 1982; Dominguez-Bello et al. 2010), so that interbreeding and child rearing between Neanderthals and humans would have also likely impacted which microorganisms ended up in their hybrid offspring. These intermixing and interbreeding events are yet another example of how behaviour can impact the microorganisms in the body, but because this is a mutualistic relationship between host and bacteria, it is also possible that bacteria influenced the success of interbreeding between these two host species. Recently, researchers at

Vanderbilt University demonstrated that microbial differences can determine the outcome of successful hybridization between different wasp species (Brucker and Bordenstein 2013). Although highly speculative and untested in larger mammals, it is then feasible that differences in microbiomes from Neanderthals and humans could have contributed to mating success.

There are additional ways that specific microbiomes and alterations to different microbiomes could have influenced hominid evolution. Modern research has shown that the bacterial content in the gut can affect nearly every aspect of human health. One recent example suggests that gut microorganisms can influence multiple aspects of mental health, including anxiety, depression, and diseases, such as schizophrenia and autism (Flight 2014). In the case of autism, an overall decrease of diversity and a decrease of *Prevotella* microorganisms in the gut are linked with the onset of this disease, directly linking mental illness to the presence of specific microbial diversity in the gut (Adams et al. 2011). Similarly, the link between microorganisms and schizophrenia became apparent when patients suffering from this disease were helped when they were treated with antibiotics (Jhamnani et al. 2013). Although this does not mean that Neanderthals had autism or schizophrenia, it does mean that the presence and alterations of microorganisms can influence mental health on a daily basis. For example, the presence of a probiotic microorganism called *Lactobacillus rhamnosus* can reduce anxiety- and depression-related behaviour (Bravo et al. 2011), indicating that single microorganisms introduced into the body can influence your daily behaviour. It is then easy to see why slight differences in the microbiome could have influenced a Neanderthal's mood or attitude in daily life, which can influence an individual's fitness or ability to adapt. This may then have significant impacts on a species' evolutionary success.

Several studies are currently underway to sequence ancient Neanderthal microbiomes and understand the similarities and differences between humans and Neanderthals. These sequencing efforts will allows researchers to understand how the microbiome of our closest relative adapted to climate change and to their highly carnivorous diets (Bocherens et al. 2001), shedding light on ecological scenarios during the time of their lives and extinctions. These projects will also identify whether or not some bacterial pathogens were shared between these closely related and physically intertwined species. Analysis of food particles trapped within the calculus has already providing information on the foraging component of a Neanderthal diet (Henry et al. 2010), and DNA analysis of these particles will also provide information on the edible plants and the palaeoecological environment surrounding Neanderthals nearly 50,000 years ago.

7 Ramification of Microbial Extinctions on Human Health

Now that we understand that the microbiome is a fluid, responsive ecological system that can be modified by diet, environment, culture, and behaviour, researchers are trying to understand how modifications to the human microbiome contribute to health. As discussed earlier, microbiomes from Western, industrialized nations are distinct from hunter-gatherers in both Africa and the Amazon Basin (Filippo et al. 2010; Yatsunenko et al. 2012), and we also know that the modern human microbiome has changed significantly in response to changes in human diet, environment, and culture over time. Does this mean that the changes in the human microbiome make individuals more susceptible to certain diseases? Or alternatively, are we healthier because of these microbial adaptation events?

If we assume that the microbiome has been evolving and adapting to each of these environments independently, then one would assume that a host's microbiome should have adapted to protect the host from disease. Take the classic example of tourists from the USA visiting Central America. Visitors are advised against drinking the tap water that local people drink on a daily basis. Researcher from the University of British Columbia are investigating whether or not local Central Americans have a gut microbiome that provides protection against certain gastrointestinal diseases, suggesting that this protective portion of the microbiome is absent in US tourists (Willing et al. 2011). It is likely that the loss of certain microorganisms or changes to bacterial community structure in the body may be associated with the loss of protection against certain diseases.

On the other hand, there are several specific examples of the microbiome providing protection against disease, although many of these experiments were done in mice to avoid obvious bioethical complications in humans. Mice were protected against *Bordetella pertussis* (whooping cough or pertussis) infection if given a single dose of a commensal nasal cavity microorganism prior to infection (Weyrich et al. 2013). Similarly, streptococcal and staphylococcal infections can also be minimized or avoided entirely if probiotics and commensal microorganisms are given to mice before they were infected (Cangemi de Gutierrez et al. 2001; Gonzalez et al. 2011). Young mice can also be protected from other non-infectious diseases, such as respiratory allergies, when first given commensal microorganisms found in dust (Fujimura et al. 2014), suggesting that obtaining bugs from our environments may be important to prevent many common Western diseases, such as obesity, allergies, and Type II diabetes. Further research will need to investigate each disease and scenario to understand the contributions of microbial interactions with other microbes and the host, ensuring that the road to understanding all of the health impacts of the microbiome alterations will undoubtedly be a long one.

A common thread in most microbiomes from Western, industrialized individuals is the loss of specific microorganisms in comparison with the hunter-gatherer cultures from other nations. Because we know that the microbiome plays a significant role in the progression and protection against certain diseases, we must ask about the roles of these missing microorganisms. Perhaps, this is best demonstrated by discussing the hygiene hypothesis (Okada et al. 2010). Because of the modern Westernized cultural practices that include home cleaning, sterilization, and hand sanitizers, it is believed that we have limited our exposure to many environmental microorganisms that in turn prime the immune system and help keep us protected against infection, allergies, and autoimmune diseases. The more we disinfect, with the ideology that we are killing bad bacteria, the more we also kill good bacteria, limiting the pool of potentially helpful microorganisms that can enter and colonize the human body. In addition to the hygiene hypothesis, the frequent and often overuse of antibiotics in Westernized cultures also fuels the elimination of good, protective bacterial species from the human body (Cho et al. 2012). Future research is needed to not only identify which microorganism are missing out of the modern human microbiome, but also understand how to reinstate these bacterial species and identify the specific roles they play in health and disease (Blaser 2014).

8 Conclusion

The human body and the microorganisms that inhabit it have been coevolving for millions of years. Through the use of new methodologies and sequencing techniques, we can now study these diverse microbial communities on and within the human body, and begin to understand the role they play in health, disease, and evolution. Tracing the evolutionary history of the human microbiome will have significant impacts in how we treat modern diseases associated with these microbes, as well as how we protect and reinstate microorganisms that we may have lost in the recent past. Recent events, such as the onset of farming during the Neolithic Revolution or the invention of the machine at the beginning of the Industrial Revolution, can have large impacts on the microorganisms that live within the body, leading us to question which future events will significantly impact the microorganisms that live within us. Similarly, events in our ancient past could also have influenced these bacteria, leading to significant downstream repercussions for the survival and adaptation of mankind. Understanding how our history, and ultimately our actions, has impacted our bacterial communities will provide a valuable guide on how to avoid negatively impacting our microbiomes in the future.

Acknowledgments Many thanks for the photographs of calculus and a coprolite that were provided by Dr. Julien Soubrier and Alexander Boast at the University of Adelaide.

Glossary

Allopatric A descriptive term of processes, such as evolution, that happen to two different populations of organisms separated into different geographical locations

Amerindians Indigenous populations that are currently living in North or South America, which are typically living without contact with Westernized civilizations

Anatomically modern humans Hominids that are morphologically similar to modern *Homo sapiens sapiens*, i.e. modern humans, and likely originated from Africa around 200,000 years ago

Archaic humans Hominids that share a lineage with *Homo sapiens* after the evolutionary split from chimpanzees, including Neandertals and Denisovans

Bacterial 'fingerprints' or fingerprinting This is a technique that sequences a single gene conserved across all bacterial species, typically the gene encoding the 16S ribosomal RNA subunit, which provides a picture of all the bacterial species present and their abundances. These sequences can be highly specific to an individual, acting the same way a fingerprint does in forensics

Biofilm This is a diverse bacterial community that forms works together as a single organism and can form on surfaces of hot springs, pipes, or even human teeth

Convergent evolution An evolutionary process that results in similar outcomes, i.e. function, even though the species that are evolving are in different locations or are unrelated

Enterotypes A simple way of classifying gut microbiomes based on the largest proportions of dominant bacterial phyla present in a sample

Hygiene hypothesis The theory that a lack of exposure to microbes (or an increased exposure to sterile environments) increases one's susceptibility to disease by failing to stimulate the immune system during development

Indigenous A modern human population that is native to a specific place over a long time span, usually prior to European colonization during the colonial period

Industrial Revolution A cultural change that occurred the eighteenth and nineteenth centuries, following the invention of the machine, and lead to numerous social and economic changes

Isotope analysis The identification of chemical elements within compounds that are present in material to determine differences in composition, variation, or change, i.e. using carbon isotopes to track the diet of humans

Linear Pottery culture (LBK) Linearbandkermik (LBK) is a culture existing about ~7500 years before present, which were Europe's first farmers and are demarcated by a distinct banding pattern on the pottery

Microbiome Any microbial community that exists in one space, i.e. the human body

Metagenome The defining term of all of the genomes that exist in the human body, which includes the human genome and the bacterial or microbial genomes present

Metabarcoding A technique that sequences one 'barcoding' or identifying gene conserved across a wide array of species as a means of surveying biodiversity

Next-generation DNA sequencing techniques DNA sequencing by any diverse means to sequencing full genomes or millions of different sequencings in one experiment

Neolithic Revolution A cultural and revolutionary process in which humans adopt farming techniques, leading to a whole host of social and economic changes

Oral caries (cavities) An oral disease caused primarily by *S. mutans*, which deteriorates the enamel on the tooth surface and exposes nervous tissue

Post-depositional bias Alterations to microbial communities after the sample is deposited or collected that occur due to environmental or decompositional factors

Periodontal disease An oral disease of the gingiva or gums, in which diverse bacterial communities stimulate inflammation and destroy gingival tissue

Sanger DNA sequencing techniques Chain termination DNA sequencing invented by Peter Sanger in 1977, which was a common method to determine the sequencing of deoxynucleotides in a single strand of only one DNA molecule

Scanning electron microscopy A form of visualization that identifies the shapes of objects via a focused beam of electrons, illuminating the size and shape of many microorganisms

Sedentary lifestyles Modern human habits that revolve around activities that do not involve physical movement (walking or running), resulting in the majority of a person's time spent sitting or not moving

Sympatric A descriptive term of processes, such as evolution, that happen to organisms located in the same location

Taphonomy A process of degradation that ancient samples undergo, where DNA is enzymatically broken down and degraded, limiting the genetic information that can be obtained from a material. The process is highly dynamic and is dependent upon a vast array of variables, including temperature, water content, and soil chemistry

References

Abi-Rached L, Jobin MJ, Kulkarni S et al (2011) The shaping of modern human immune systems by multiregional admixture with archaic humans. Science 334:89–94. doi:10.1126/science.1209202

Adams JB, Johansen LJ, Powell LD et al (2011) Gastrointestinal flora and gastrointestinal status in children with autism—comparisons to typical children and correlation with autism severity. 22

Adler CJ, Dobney K, Weyrich LS et al (2013) Sequencing ancient calcified dental plaque shows changes in oral microbiota with dietary shifts of the Neolithic and Industrial revolutions. Nat Genet 45:450–455. doi:10.1038/ng.2536

Ajdić D, McShan WM, McLaughlin RE et al (2002) Genome sequence of *Streptococcus mutans* UA159, a cariogenic dental pathogen 99:14434–14439

Allentoft ME, Collins M, Harker D et al (2012) The half-life of DNA in bone: measuring decay kinetics in 158 dated fossils. Proc R Soc B Biol Sci 279:4724–4733. doi:10.1098/rspb.2012.1745

Arimatsu K, Yamada H, Miyazawa H et al (2014) Oral pathobiont induces systemic inflammation and metabolic changes associated with alteration of gut microbiota. Sci Rep. doi:10.1038/srep04828

Arumugam M, Raes J, Pelletier E et al (2011) Enterotypes of the human gut microbiome. Nature 473:174–180. doi:10.1038/nature09944

Aufderheide AC, Conrado R (1998) The cambridge encyclopedia of human paleopathology. Cambridge University Press, Cambridge

Behr MA (2013) Evolution of *Mycobacterium tuberculosis*. Adv Exp Med Biol 783:81–91. doi:10.1007/978-1-4614-6111-1_4

Berg M, Hudson P (1992) Rehabilitating the industrial revolution. Econ Hist Rev 45:24. doi:10.2307/2598327

Blaser MJ (2014) Missing microbes: how the overuse of antibiotics is fueling our modern plagues, 1st edn. Henry Holt and Co., New York

Blaser MJ, Dominguez-Bello MG, Contreras M et al (2012) Distinct cutaneous bacterial assemblages in a sampling of South American Amerindians and US residents

Blatt SH, Redmond BG, Cassman V, Sciulli PW (2011) Dirty teeth and ancient trade: evidence of cotton fibres in human dental calculus from late Woodland, Ohio. Int J Osteoarchaeol 21:669–678. doi:10.1002/oa.1173

Bocherens H, Billiou D, Mariotti A et al (2001) New isotopic evidence for dietary habits of Neandertals from Belgium. http://adelaideaus.library.ingentaconnect.com/content/ap/hu/2001/00000040/00000006/art00452. Accessed 12 Dec 2013

Bos KI, Schuenemann VJ, Golding GB et al (2011) A draft genome of Yersinia pestis from victims of the black death 478:506–510

Brandt G, Haak W, Adler CJ et al (2013) Ancient DNA reveals key stages in the formation of central European mitochondrial genetic diversity. Science 342:257–261. doi:10.1126/science.1241844

Bravo JA, Forsythe P, Chew MV et al (2011) Ingestion of Lactobacillus strain regulates emotional behavior and central GABA receptor expression in a mouse via the vagus nerve. Proc Natl Acad Sci 108:16050–16055. doi:10.1073/pnas.1102999108

Breton J, Massart S, Vandamme P et al (2013) Ecotoxicology inside the gut: impact of heavy metals on the mouse microbiome. BMC Pharmacol Toxicol 14:62. doi:10.1186/2050-6511-14-62

Brucker RM, Bordenstein SR (2013) The hologenomic basis of speciation: gut bacteria cause hybrid lethality in the genus *nasonia*. Science 341:667–669. doi:10.1126/science.1240659

Burger J, Kirchner M, Bramanti B et al (2007) Absence of the lactase-persistence-associated allele in early Neolithic Europeans. Proc Natl Acad Sci USA 104:3736–3741

Cabrera-Rubio R, Collado MC, Laitinen K et al (2012) The human milk microbiome changes over lactation and is shaped by maternal weight and mode of delivery. Am J Clin Nutr 96:544–551. doi:10.3945/ajcn.112.037382

Cameron AJ, Zimmet PZ, Dunstan DW et al (2003) Overweight and obesity in Australia: the 1999–2000 Australian diabetes, obesity and lifestyle study (AusDiab). Med J Aust 178:427–432

Cangemi de Gutierrez R, Santos V, Nader-Macías ME (2001) Protective effect of intranasally inoculated Lactobacillus fermentum against *Streptococcus pneumoniae* challenge on the mouse respiratory tract. FEMS Immunol Med Microbiol 31:187–195. doi:10.1111/j.1574-695X.2001.tb00519.x

Caporaso JG, Lauber CL, Walters WA et al (2012) Ultra-high-throughput microbial community analysis on the Illumina HiSeq and MiSeq platforms. ISME J 6:1621–1624. doi:10.1038/ismej.2012.8

Caufield PW, Wannemuehler YM, Hansen JB (1982) Familial clustering of the *Streptococcus mutans* cryptic plasmid strain in a dental clinic population. Infect Immun 38:785–787

Chassard C, Delmas E, Robert C, Bernalier-Donadille A (2010) The cellulose-degrading microbial community of the human gut varies according to the presence or absence of methanogens. FEMS Microbiol Ecol 74:205–213. doi:10.1111/j.1574-6941.2010.00941.x

Cho I, Yamanishi S, Cox L et al (2012) Antibiotics in early life alter the murine colonic microbiome and adiposity. Nature 488:621–626. doi:10.1038/nature11400

Comas I, Coscolla M, Luo T et al (2013) Out-of-Africa migration and Neolithic coexpansion of *Mycobacterium tuberculosis* with modern humans. Nat Genet 45:1176–1182. doi:10.1038/ng.2744

Consortium THMP (2012) Structure, function and diversity of the healthy human microbiome. Nature 486:207–214

Contreras M, Costello EK, Hidalgo G et al (2010) The bacterial microbiota in the oral mucosa of rural Amerindians. Microbiol Read Engl 156:3282–3287. doi:10.1099/mic.0.043174-0

Cooper A, Poinar HN (2000) Ancient DNA: do it right or not at all. Science 289:1139–1139. doi:10.1126/science.289.5482.1139b

Cooper A, Stringer CB (2013) Did the Denisovans cross wallace's line? Science 342:321–323. doi:10.1126/science.1244869

Cornejo OE, Lefébure T, Pavinski Bitar PD et al (2013) Evolutionary and population genomics of the cavity causing bacteria streptococcus mutans. Mol Biol Evol. doi:10.1093/molbev/mss278

David LA, Maurice CF, Carmody RN et al (2013) Diet rapidly and reproducibly alters the human gut microbiome. Nature. doi:10.1038/nature12820

Davis R (1962) English Foreign trade, 1700–17741. Econ Hist Rev 15:285–303. doi:10.1111/j.1468-0289.1962.tb02239.x

Deer N (1949) The history of sugar, vol 1. Chapman and Hall Ltd., London

De La Fuente C, Flores S, Moraga M (2013) DNA from human ancient bacteria: a novel source of genetic evidence from archaeological dental calculus. Archaeometry 55:767–778. doi:10.1111/j.1475-4754.2012.00707.x

Dobney K, Brothwell D (1988) A scanning electron microscope study of archaeological dental calculus. In: Scanning electron microscopy in archaeology. Oxford, pp 372–385

Dominguez-Bello MG, Costello EK, Contreras M et al (2010) Delivery mode shapes the acquisition and structure of the initial microbiota across multiple body habitats in newborns. Proc Natl Acad Sci 201002601. doi:10.1073/pnas.1002601107

Ennever J, Vogel JJ, Benson LA (1973) Lipid and calculus matrix calcification in vitro. J Dent Res 52:1056–1059

Filippo CD, Cavalieri D, Paola MD et al (2010) Impact of diet in shaping gut microbiota revealed by a comparative study in children from Europe and rural Africa. Proc Natl Acad Sci 107:14691–14696. doi:10.1073/pnas.1005963107

Flight MH (2014) Neurodevelopmental disorders: the gut-microbiome-brain connection. Nat Rev Neurosci 15:65–65. doi:10.1038/nrn3669

Fujimura KE, Demoor T, Rauch M et al (2014) House dust exposure mediates gut microbiome Lactobacillus enrichment and airway immune defense against allergens and virus infection. Proc Natl Acad Sci 111:805–810. doi:10.1073/pnas.1310750111

Gonzalez DJ, Haste NM, Hollands A et al (2011) Microbial competition between *Bacillus subtilis* and *Staphylococcus aureus* monitored by imaging mass spectrometry. Microbiology 157:2485–2492. doi:10.1099/mic.0.048736-0

Graham JC (1981) The French connection in the early history of canning. J R Soc Med 74:374–381. doi:10.1177/014107688107400511

Green RE, Krause J, Briggs AW et al (2010) A draft sequence of the Neandertal genome. Science 328:710–722. doi:10.1126/science.1188021

Haviland W, Walrath D, Prins H, McBride B (2010) Evolution and prehistory: the human challenge. Cengage Learning

Henry AG, Brooks AS, Piperno DR (2010) Microfossils in calculus demonstrate consumption of plants and cooked foods in Neanderthal diets (Shanidar III, Iraq. Proc Natl Acad Sci, Spy I and II, Belgium). doi:10.1073/pnas.1016868108

Henry AG, Piperno DR (2008) Using plant microfossils from dental calculus to recover human diet: a case study from Tell al-Raqā'i, Syria. J Archaeol Sci 35:1943–1950. doi:10.1016/j.jas.2007.12.005

Hermanussen M (2006) In: Fogel RW The escape from hunger and premature death, 1700–2100. Europe, America, and the third world. (Cambridge University Press, Cambridge, 2004), p 191. (£16.99, ISBN 0-521-00488-8, Paperback; £40.00, ISBN 0-521-80878-2, Hardback. J Biosoc Sci 38:571–572)

Hokama T, Hamamoto I, Takenaka S et al (1996) Throat microflora in breastfed and formula-fed infants. J Trop Pediatr 42:324–326

Hollox E, Hurles M, Kivisild T, Tyler-Smith C (2013) Human evolutionary genetics, 2nd edn. Garland Science, New York

Hooper LV, Gordon JI (2001) Commensal host-bacterial relationships in the gut. Science 292:1115–1118

Hu P, Brodie EL, Suzuki Y et al (2005) Whole-genome transcriptional analysis of heavy metal stresses in *Caulobacter crescentus*. J Bacteriol 187:8437–8449. doi:10.1128/JB.187.24.8437-8449.2005

Jhamnani K, Shivakumar V, Kalmady S et al (2013) Successful use of add-on minocycline for treatment of persistent negative symptoms in schizophrenia. J Neuropsychiatry Clin Neurosci 25:E06–07. doi:10.1176/appi.neuropsych.11120376

Jin L, Su B (2000) Natives or immigrants: modern human origin in east Asia. Nat Rev Genet 1:126–133. doi:10.1038/35038565

Jones PM (2009) Industrial enlightenment: science, technology and culture in Birmingham and the West Midlands, 1760–1820. Manchester University Press, Manchester

Kau AL, Ahern PP, Griffin NW et al (2011a) Human nutrition, the gut microbiome, and immune system: envisioning the future. Nature 474:327–336. doi:10.1038/nature10213

Kau AL, Ahern PP, Griffin NW et al (2011b) Human nutrition, the gut microbiome and the immune system. Nature 474:327–336

Lax S, Smith DP, Hampton-Marcell J et al (2014) Longitudinal analysis of microbial interaction between humans and the indoor environment. Science 345:1048–1052. doi:10.1126/science.1254529

Leff JW, Fierer N (2013) Bacterial communities associated with the surfaces of fresh fruits and vegetables. PLoS ONE 8:e59310. doi:10.1371/journal.pone.0059310

Ley RE, Bäckhed F, Turnbaugh P et al (2005) Obesity alters gut microbial ecology. Proc Natl Acad Sci USA 102:11070–11075. doi:10.1073/pnas.0504978102

Li J, Nasidze I, Quinque D et al (2013) The saliva microbiome of pan and homo. BMC Microbiol 13:204. doi:10.1186/1471-2180-13-204

Meyer M, Fu Q, Aximu-Petri A et al (2014) A mitochondrial genome sequence of a hominin from Sima de los Huesos. Nature 505:403–406. doi:10.1038/nature12788

Meyer M, Kircher M, Gansauge M-T et al (2012) A high-coverage genome sequence from an archaic denisovan individual. Science 338:222–226. doi:10.1126/science.1224344

Moeller AH, Degnan PH, Pusey AE et al (2012) Chimpanzees and humans harbour compositionally similar gut enterotypes. Nat Commun 3:1179. doi:10.1038/ncomms2159

Moeller AH, Peeters M, Ndjango J-B et al (2013) Sympatric chimpanzees and gorillas harbor convergent gut microbial communities. Genome Res 23:1715–1720. doi:10.1101/gr.154773.113

Muegge BD, Kuczynski J, Knights D et al (2011) Diet drives convergence in gut microbiome functions across mammalian phylogeny and within humans. Science 332:970–974. doi:10.1126/science.1198719

New Research Confirms "Out Of Africa" Theory of human evolution. In: ScienceDaily. http://www.sciencedaily.com/releases/2007/05/070509161829.htm. Accessed 10 Sep 2014

Okada H, Kuhn C, Feillet H, Bach J-F (2010) The "hygiene hypothesis" for autoimmune and allergic diseases: an update. Clin Exp Immunol 160:1–9. doi:10.1111/j.1365-2249.2010.04139.x

Payne AN, Chassard C, Lacroix C (2012) Gut microbial adaptation to dietary consumption of fructose, artificial sweeteners and sugar alcohols: implications for host–microbe interactions contributing to obesity. Obes Rev 13:799–809. doi:10.1111/j.1467-789X.2012.01009.x

Quigley L, McCarthy R, O'Sullivan O et al (2013) The microbial content of raw and pasteurized cow milk as determined by molecular approaches. J Dairy Sci 96:4928–4937. doi:10.3168/jds.2013-6688

Ramotar K, Conly JM, Chubb H, Louie TJ (1984) Production of menaquinones by intestinal anaerobes. J Infect Dis 150:213–218

Sajantila A (2013) Major historical dietary changes are reflected in the dental microbiome of ancient skeletons. Investig Genet 4:1–2. doi:10.1186/2041-2223-4-10

Sankararaman S, Patterson N, Li H et al (2012) The date of interbreeding between Neandertals and modern humans. PLoS Genet 8:e1002947. doi:10.1371/journal.pgen.1002947

Scarre C (2013) The human past: world prehistory and the development of human societies, 3rd edn edn. Thames & Hudson, New York

Schnorr SL, Candela M, Rampelli S et al (2014) Gut microbiome of the Hadza hunter-gatherers. Nat Commun. doi:10.1038/ncomms4654

Soares P, Ermini L, Thomson N et al (2009) Correcting for purifying selection: an improved human mitochondrial molecular clock. Am J Hum Genet 84:740–759. doi:10.1016/j.ajhg.2009.05.001

Song SJ, Lauber C, Costello EK et al (2013) Cohabiting family members share microbiota with one another and with their dogs. eLife 2:e00458. doi:10.7554/eLife.00458

Stecher B, Robbiani R, Walker AW et al (2007) *Salmonella enterica* Serovar Typhimurium exploits inflammation to compete with the intestinal microbiota. PLoS Biol 5:e244. doi:10.1371/journal.pbio.0050244

Tattersall I (2009) Human origins: out of Africa. Proc Natl Acad Sci 106:16018–16021. doi: 10.1073/pnas.0903207106

Tito RY, Knights D, Metcalf J et al (2012) Insights from characterizing extinct human gut microbiomes. PLoS ONE 7:e51146. doi:10.1371/journal.pone.0051146

Wade WG (2013) The oral microbiome in health and disease. Pharmacol Res 69:137–143. doi:10.1016/j.phrs.2012.11.006

Weyrich LS, Feaga HA, Park J et al (2013) Resident microbiota affect *Bordetella pertussis* infectious dose and host specificity. J Infect Dis jit597. doi:10.1093/infdis/jit597

Willing BP, Vacharaksa A, Croxen M et al (2011) Altering host resistance to infections through microbial transplantation. PLoS ONE 6:e26988. doi:10.1371/journal.pone.0026988

Wolpoff MH, Hawks J, Caspari R (2000) Multiregional, not multiple origins. Am J Phys Anthropol 112:129–136. doi:10.1002/(SICI)1096-8644(200005)112:1<129:AID-AJPA11>3.0.CO;2-K
Wood B, Richmond BG (2000) Human evolution: taxonomy and paleobiology. J Anat 197:19–60. doi:10.1046/j.1469-7580.2000.19710019.x
Yatsunenko T, Rey FE, Manary MJ et al (2012) Human gut microbiome viewed across age and geography. Nature 486:222–227. doi:10.1038/nature11053
Yildirim S, Yeoman CJ, Sipos M et al (2010) Characterization of the fecal microbiome from non-human wild primates reveals species specific microbial communities. PLoS ONE 5:e13963. doi:10.1371/journal.pone.0013963
Yong E (2012) Gut microbial "enterotypes" become less clear-cut. Nature. doi:10.1038/nature.2012.10276
WHO (2004) | The world health report 2004—changing history. http://www.who.int/whr/2004/en/. Accessed 12 Jan 2012

Divergence-with-Gene-Flow—What Humans and Other Mammals Got up to

Michael L. Arnold, Amanda N. Brothers, Jennafer A.P. Hamlin, Sunni J. Taylor and Noland H. Martin

Abstract In this review, we posit the hypothesis that divergence-with-gene-flow or, using the terminology that has historical precedence, evolution-with-hybridization is not the exception as argued by the neo-Darwinian architects, but rather the rule. In particular, we will discuss briefly how the definition of species and the process of speciation that emerged from the Modern Synthesis limited greatly how evolutionary diversification was perceived to occur. This, in turn, resulted in only certain hypotheses and research directions being deemed legitimate for evolutionary biologists. Yet, we will also argue that in general, the assumptions that resulted in the definitions and concepts surrounding speciation were made much stronger by a dearth, rather than a wealth, of the data needed to test hypotheses. Specifically, these alternative hypotheses were divergence in allopatry versus divergence with at least some genetic exchange. We will point to the observation made by Anderson—the architect of studies of introgressive hybridization—that to test for contemporaneous or ancient gene flow between diverging/divergent lineages requires discrete markers. Thus, not until the advent of methods for analyzing the genetic constitution of individual organisms was it possible to test rigorously the alternate modes of divergence. To illustrate our hypotheses and conclusions, we will focus on mammalian lineages, including our own species. This focus reflects our desire to emphasize that not only prokaryotes and plants, but animals as well, reflect the process of reticulate evolution.

M.L. Arnold (✉) · A.N. Brothers · J.A.P. Hamlin
Department of Genetics, Fred C. Davison Life Sciences Complex,
University of Georgia, Athens, GA 30602, USA
e-mail: arnold@plantbio.uga.edu

S.J. Taylor · N.H. Martin
Department of Biology, Texas State University, San Marcos, TX 78666, USA

N. Gontier (ed.), *Reticulate Evolution*, Interdisciplinary Evolution Research 3,
DOI 10.1007/978-3-319-16345-1_10

1 Brief (and Thus Incomplete) Review of Speciation and Hybridization *ala* the Modern Synthesis

The neo-Darwinian (or Modern) Synthesis was a watershed for evolutionary biology, providing a rich and varied series of concepts (Dobzhansky 1937; Mayr 1942; Stebbins 1950) that formed the basis for succeeding analyses of topics as diverse as anthropology and population genetics. One of the overarching aims espoused by the architects of this synthesis was to discern and define how new lineages formed (Dobzhansky 1940; Mayr 1942). Specifically, they wanted to understand whether these novel forms known as 'species' arose due to selection, genetic drift, gene flow, or some combination of these processes. Yet, before it was possible to begin to understand the process of species formation—or speciation—a consensus was necessary in defining species. The definition arrived at determined the type of questions asked, the type of data gathered, and even the interpretation of the data gathered by subsequent generations of evolutionary biologists.

The emergence of the biological species concept, reflecting the point of view that only with reproductive isolation could evolution progress (Dobzhansky 1937; Mayr 1942), as the preeminent species definition immediately placed a premium on the development of reproductive isolating barriers for evolutionary diversification. For example, Hennig (1966) wrote 'Is not the species concept that the species includes all individuals that together are capable of producing completely fertile offspring, and must we not then consider groups whose individuals can produce new species by hybridization as partial groups of one species?' This quote by Hennig reflects well the neo-Darwinian species concept—that species must be reproductively isolated from other such groups to be *real*.

With the emphasis placed on protecting the coadapted gene combinations that made species recognizable from other species from being lost by recombination, it was natural that the process of speciation would be seen as necessitating the geographic isolation of the diverging lineages (i.e., allopatric divergence), at least until such time as the diverging lineages were protected by the cocoon of their respective reproductive isolating barriers. For a number of decades, evolutionary biologists applied the neo-Darwinian conceptual framework to their own studies, often constraining themselves to testing when, where, and how their organism of choice had been geographically isolated from related lineages during their divergence. Unfortunately, these constraints often neglected to take into account the dynamic nature of the geographical distribution of diverging lineages, or the observation that many recognized species were currently exchanging genes with congeners in areas of sympatry or parapatry (Arnold 1997, 2006). Thus, we evolutionary biologists did not often stop to ask how likely it was that diverging lineages actually remained allopatric during, for example, oscillating glacial maxima and minima or why we were continually encountering examples of organisms termed different at the specific level that were exchanging genes at a frequency that theory suggested should quickly lead to their evolutionary demise (Wright 1931; Slatkin 1985).

In retrospect, the conceptual framework proceeding from the neo-Darwinian synthesis was actually seen as powerfully explanatory mostly due to a lack of data to test the various hypotheses. In the context of speciation, the data necessary to test the alternative hypotheses of allopatric divergence versus reticulate evolution are discrete, genetic markers rather than quantitative (e.g., morphological) characters. During the early days of the Modern Synthesis, Anderson and Hubricht (1938) in an article defining the process of introgressive hybridization (i.e., introgression) emphasized the lack of such data. Specifically, these authors pointed out that in natural hybrid populations, 'After a few back-crosses most of the individuals cannot be distinguished by morphological means from the pure species…'

Without discrete markers, the assumptions surrounding how species formed—particularly the need (or lack thereof) for allopatric distributions of diverging lineages—went untested. Thus, the ability to produce data such as allozymes and restriction fragment length polymorphisms of even a few genes revolutionized the understanding of both contemporaneous introgression in hybrid zones and more ancient introgression and hybrid speciation. Furthermore, the wholesale testing for allopatric divergence versus divergence-with-gene-flow across many clades has been catalyzed by the capacity to assay hundreds if not thousands of loci simultaneously using recent developments in sequencing and genotyping (e.g., next-generation sequencing; Metzker 2010; Stapley et al. 2010).

2 Have Mammary Glands and Will Diverge While Exchanging Genes—Rationale for Mammals as the Exemplar of Reticulate Evolution

The question might be asked, given the wealth of examples of divergence-with-gene-flow among plant and microbial lineages (see Arnold 2006, 2008 for examples), why we have chosen to focus our attention on class Mammalia. There are two major reasons for this taxonomic choice. The first is historical, once again reflecting the predominant viewpoint that proceeded from the neo-Darwinian synthesis (and which in some quarters has continued through the present day; Coyne and Orr 2004). To exemplify this viewpoint, we can consider a conclusion from Mayr concerning the supposed rarity of hybridization among animal lineages. Specifically, he assumed that if hybridization between individuals belonging to different species occurred 'The majority of such hybrids are totally sterile… Even those hybrids that produce normal gametes in one or both sexes are nevertheless unsuccessful in most cases and do not participate in reproduction. Finally, when they do backcross to the parental species, they normally produce genotypes of inferior viability that are eliminated by natural selection' (Mayr 1963). This obviously emphasizes the unlikely role of divergence-with-gene-flow as a major component of speciation. Furthermore, Mayr's quote reflects two of the major philosophical arguments against divergence-with-gene-flow as an important part

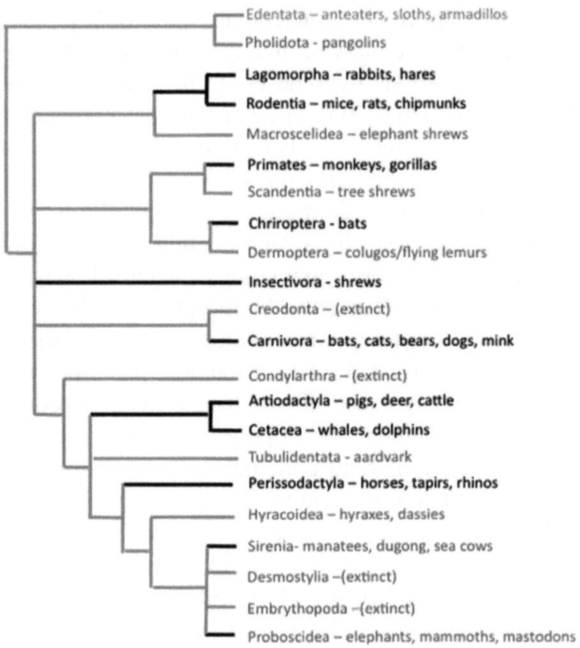

Fig. 1 The phylogenetic distribution of natural hybridization, introgressive hybridization, and hybrid speciation among mammals. The lineages denoted with *dark lines* are those for which reticulate evolution has been inferred. The references from which this figure is derived can be found in the present chapter as well as Arnold (2006, 2008)

of speciation—hybridization is an extremely rare phenomenon, and when rare hybrids are formed, they are less fit than offspring from within species crosses.

For the majority of instances of hybridization, both of the above assumptions have been shown to be inaccurate. Thus, hybrid fitness can vary from less than to greater than that of conspecific progeny (see Arnold and Hodges 1995; Arnold 1997 for reviews) and rarity of hybrid formation, just as with rarity of point mutations that cause increases in fitness, does not indicate unimportance in evolutionary diversification (Arnold and Hodges 1995). As mentioned above, an extension of the neo-Darwinian concept of the maladaptive nature of hybridization and introgression was the assumption—based upon theoretical considerations (Wright 1931; Slatkin 1985)—that even negligible gene flow was enough to homogenize gene frequencies between populations. The assumption of the efficacy of gene flow to disrupt gene combinations and thus create homogeneous gene frequencies from previously divergent populations fell well within the conceptual framework that saw hybridization and introgression as a 'violation' of 'species integrity' (Arnold 1997).

The second rationale for selecting mammals is that this clade contains an ever-increasing set of examples (Fig. 1) of the various outcomes of reticulate evolution—hybrid speciation, adaptive introgression, and loss of biodiversity through genetic assimilation and reinforcement.

3 Really Ancient Genetic Exchange and Mammalian Origins

Not only recent, but also very ancient reticulate evolution has been inferred for members of the Mammalian clade. Evidence for the role of genetic exchange—hybridization and horizontal gene transfer—during the evolutionary diversification of mammals is reflected well by an analysis of nearly three million base pairs of DNA sequence from 31 lineages (Hallström and Janke 2010) and studies of two mammalian genes co-opted from retroviral lineages, *CGIN1* and *syncytin-Car1* (Marco and Marín 2009; Cornelis et al. 2012).

In the large-scale phylogenomic analysis, several portions of the placental clade remained unresolved, even though the data were of excellent quality and quantity (Hallström and Janke 2010). These included the most ancient divergence event involving the relative placement of Xenarthra, Boreoplacentalia, and Afrotheria. Likewise, the position of the bat clade and that of tree shrews also remained unresolved. Hallström and Janke (2010) suggested that these unresolved relationships were not due to poor taxa or sequence sampling, but rather more likely due to incomplete lineage sorting or reticulate evolutionary processes such as introgression.

As with the above study, the analysis of genes hypothesized to protect mammals against retroviral infections, in particular those related to *GIN1* (*Gypsy integrase 1*), uncovered evidence for ancient recombination events, in this case between retroviruses and mammalian genomes. Specifically, Marco and Marín (2009) defined a new gene from the *GIN1* family that they named *CGIN1* or *Cousin of GIN1* that possessed sequence similarity for portions of both retrovirus and mammalian genes. Notwithstanding the name of this new gene, reminiscent of an animated doll from a horror movie, the findings suggested strongly that its origin involved the fusion of a mammalian gene with sequences from a retroviral lineage and that this fusion occurred in the ancestor of the marsupial and eutherian clades (Marco and Marín 2009).

Like *CGIN1*, the mammalian gene family known as *Syncytins* have been co-opted from retroviral sequences, with a number of lineage-specific genes identified for mammalian groups as divergent as primates, lagomorphs, and murids. Though the members of this particular group of genes all contribute to placentation, their evolutionary origins involved independent 'capture' events from retroviruses into the various mammalian lineages (Cornelis et al. 2012). To date, the most ancient of these capture events detected is that of the carnivore-specific *syncytin-Car1*; this gene was apparently integrated into the mammalian genome before the radiation of Carnivora some 60–85 mya (Cornelis et al. 2012). The numerous retroviral capture events—both for *GIN1* and *syncytin* genes—indicate the degree to which highly divergent genomes may contribute to evolutionary diversification through genetic exchange (Arnold 2006, 2008).

4 Divergence-with-Gene-Flow Among Mammals—Climbing Our Way Through the Mammalian 'Tree' of Life

That evolutionary diversification is best represented by a Web (Arnold 2006) is the premise of this chapter. Yet, it is very useful to utilize the tree metaphor used by Darwin (1859) as a structure for discussing the lineages of mammals that demonstrate reticulate evolution (Fig. 1). This may seem counterintuitive, but the fact of the matter is that most of us think of a treelike pattern without gene flow when we consider species relationships. Furthermore, notwithstanding the reticulate nature detected in many phylogenies, it is still correct that divergent evolution results in portions of phylogenies reflecting branch-like patterning. We have thus used the information at the Web site, http://tolweb.org/Mammalia/ as the basis for determining taxonomic categories used in the following discussion.

We have also mainly limited our discussions of taxonomic examples to studies that have been reported since 2008. This decision means that the details of some cases of reticulate evolution in mammalian lineages will have been omitted (Fig. 1). Though this may seem extremely subjective, we have tried to avoid repetition with our earlier works, particularly by the senior author (see Arnold 1997, 2006, 2008 for additional cases not discussed in the present chapter). Thus, some classic groups will not be included because we have recently written at length about these. An example of this is the Canidae. With sincere apologies to Robert Wayne and his colleagues, we will only include references to these detailed and elegant studies in tabular form (but see Arnold et al. 2012 for a recent discussion of some of these data). Even with limiting our discussions to recent studies, and ones of which we have not written about at length, there is a wealth of examples. Indeed, we will limit the following case studies to only a few, representative exemplars that reflect the various outcomes of hybridization. Finally, we also realize that our methodology for identifying pertinent studies—through searches of a set of ca. 30 'major' journals (e.g., *Molecular Ecology*, *Science*, *Nature*, *Evolution*)—will lead to the inadvertent omission of relevant studies.

4.1 Marsupial Divergence-with-Gene-Flow: Interspecific Mating in the Australian Bush

Nothing is more evocative of the Australian interior, or Australia per se, than the silhouette and bounding gate of members of the kangaroo and wallaby clade. In particular, the largest members of this assemblage, species of red and gray kangaroos, are scattered across the Australian landscape from the east to west coasts (Short et al. 1983). Gray kangaroos are classified into western and eastern forms, *Macropus fuliginosus* and eastern *M. giganteus*, respectively, with a zone of sympatry between the two species located in portions of the states of Victoria,

New South Wales, and Queensland (Caughley et al. 1984; Neaves et al. 2010). Notwithstanding such a large area of overlap (ca. 0.68 km^2; Caughley et al. 1984; Neaves et al. 2010) and the production of hybrids in captivity (Kirsch and Pool 1972), there was little evidence of natural hybridization between these macropod species (Coulson and Coulson 2001).

In contrast to previous analyses, Neaves et al. (2010) did indeed find evidence consistent with introgressive hybridization between *M. fuliginosus* and *M. giganteus* in the area of overlap in eastern Australia. Based upon variation at both nuclear and mitochondrial loci, 7.6 % of the >200 kangaroos sampled in the sympatric region were found to be hybrids. As is the case with the majority of cases of introgression among both plants and animals, there was an asymmetry in the genetic exchange, in this case with 14 of the 17 hybrid individuals identified as backcrosses toward *M. giganteus* and the remaining three hybrids assigned to the class of backcrosses to *M. fuliginosus* (Neaves et al. 2010). None of the hybrid genotypes were consistent with an F_1 generation; thus, these findings reflect divergence-with-gene-flow between these marsupial species. Furthermore, Neaves et al. (2010) argued for the importance of genetic exchange in potentially allowing the transfer of genomic components affecting adaptive traits—such as those associated with aridity. Testing for such adaptive trait transfer has been a major emphasis in many systems and was pointed out as possibly the most important outcome of introgressive hybridization by Anderson (1949).

4.2 Rodents and Divergence-with-Gene-Flow: Mice in the House…and Hybrid Zone

One of the truly classic mammalian systems typifying divergence-with-introgression involves the house mice taxa *Mus musculus musculus* and *Mus musculus domesticus* (sometimes referred to as *M. musculus* and *M. domesticus*). The vast majority of studies involving hybridization between these lineages have addressed the genomic architecture of reproductive isolation. For example, Janousek et al. (2012) examined ca. 1400 loci distributed throughout the genome in individual mice from two natural hybrid zones. By comparing the patterns of variation, they were able to identify specific loci that contributed differentially to reproductive isolation, some of which were associated with such factors as hybrid male sterility.

Though an emphasis has been placed on identifying the genomic regions that provide various levels of reproductive isolation between the house mouse lineages, a number of authors have discussed the significance of introgression. Thus, patterns of introgression have alternatively been utilized to infer such disparate factors as colonization histories and adaptive trait transfer (Payseur et al. 2004; Gompert and Buerkle 2009; Jones et al. 2010). For example, Jones et al. (2010) analyzed a variety of DNA (Y chromosome, autosomal, and mitochondrial DNA) sequences and a morphological trait to construct the history of the human-mediated dispersal of *M. m. musculus* and *M. m. domesticus* into Norway. Involved in this dispersal was the establishment of ancient and recent hybrid zones when their human carriers brought the two subspecies into contact. The colonization scenario favored by Jones et al. (2010) involved *M. m. domesticus* being brought to Norway during the Viking era and encountering the previously dispersed *M. m. musculus*. Regardless of when these forms arrived, their dispersal into what is now Norway was accompanied by extensive hybridization and introgression.

Numerous zones of hybridization and introgression, such as those studied by Jones et al. (2010) and Janousek et al. (2012), have been described across the overlapping ranges of the two *Mus* subspecies. As with all cases of hybridization (Barton and Hewitt 1985; Arnold 1997), each of the *Mus* hybrid zones reflects selection against some hybrid genotypes. Thus, some regions of the genomes of these two subspecies are resistant to introgression (Janousek et al. 2012). However, also as found for all cases of hybridization that proceed past the initial F_1 generation (Arnold 2006), some regions of the genomes of the *Mus* subspecies are available for recombination with the alternate taxon.

Though some of the introgression between the house mouse taxa is consistent with 'neutral diffusion,' some instances of gene transfer reflect apparent adaptive trait transfers. Gompert and Buerkle (2009), in an analysis of variation across X chromosome loci in a *M. m. musculus/M. m. domesticus* hybrid zone (data reported by Payseur et al. 2004), inferred directional selection favoring introgression mainly from *M. m. musculus* into *M. m. domesticus*. Such directional introgression supports the hypothesis that alleles from one lineage are adaptive in the alternate lineage as well. Indeed, Staubach et al. (2012) inferred just such adaptive trait introgression from an analysis of high-density single nucleotide polymorphism (i.e., SNP) typing arrays. These authors concluded that '...natural genomes are subject to complex adaptive processes, including the introgression of haplotypes from other differentiated populations or species at a larger scale than previously assumed for animals' (Staubach et al. 2012).

4.3 Rodents and Divergence-with-Gene-Flow: Traveling Rats

As with mice, members of the genus *Rattus* represent commensal lineages that have accompanied humans around the globe. Indeed, Lack et al. (2012) have concluded that so-called black rats (i.e., those taxa belonging to the *Rattus rattus* species group) 'are arguably the most successful mammalian invaders on the planet' with established populations on all continents except Antarctica. Of particular interest for the current discussion is the finding that the six currently recognized species have been brought into contact on various continents and thus have the potential to exemplify divergence-with-gene-flow. Consistent with the process of reticulate evolution, Lack et al. (2012) detected genetic variability at a combination of nuclear and mitochondrial loci indicative of introgressive hybridization. However, in contrast to the model of continued genetic and phenotypic divergence in the face of gene flow, Lack et al. (2012) also inferred genetic extinction of *Rattus tanezumi* resulting from asymmetric introgression from *R. rattus* lineages into the genome/populations of *R. tanezumi* located in North America. This latter species is also apparently hybridizing with a second *Rattus* species in Asia, though the degree of introgression may not be as great as that seen in the North American samples (Lack et al. 2012). These findings thus suggest that divergence-with-gene-flow, as well as the loss of biodiversity through introgression, is occurring simultaneously in this mammalian invader.

4.4 Rodents and Divergence-with-Gene-Flow: Chipmunks Arose with Ancient and Recent Introgression

Work by Sullivan and his colleagues on the phylogenetics and population genetics of members of the North American genus of chipmunks, *Tamias* have identified taxonomically and temporally widespread divergence-with-gene-flow (Good et al. 2008; Hird and Sullivan 2009; Reid et al. 2012). For example, Hird and Sullivan (2009) carried out analyses of DNA sequence (i.e., mtDNA) and morphological variation within a hybrid zone between two subspecies of *Tamias ruficaudus*. Their results supported the hypothesis that introgressive hybridization was an important stage of divergence in this species (Hird and Sullivan 2009), likely producing the type of evolutionary novelty expected when divergent genomes recombine (Arnold 1997, 2006). This latter conclusion was drawn from the observation that populations within the zone of overlap were genotypically differentiated from parental populations (Hird and Sullivan 2009).

Not only recent, but also ancient introgression has impacted the North American chipmunk clade. This inference comes from additional studies by Sullivan et al. in which they sampled broadly the genotypic and phenotypic patterns of diversity within the *Tamias* species complex. In the first of these studies, Good et al. (2008) tested for genetic exchange between *T. ruficaudus* and *T. amoenus*. By comparing data from nuclear genes and microsatellites, with that of previously published mtDNA sequences, Good et al. (2008) detected the signature of ancient introgressive hybridization. Specifically, coalescent analyses dated mtDNA introgression between these two species at 1–3 million years ago (mya; Good et al. 2008).

Similar to Good et al.'s (2008) findings, Reid et al. (2012) detected widespread introgression among various species pairs. Most of these episodes of divergence-with-gene-flow were inferred to be deep within the phylogenetic assemblages, indicative of ancient genetic exchange, while a few apparently involved recent/contemporaneous introgression (Reid et al. 2012). All of the instances of ancient introgression where relationships of the introgressing lineages were resolved involved non-sister taxa; four of the six cases involved asymmetric introgression from *Tamias minimus* into another lineage. Both sister and non-sister lineages contributed to the recent exchanges. Finally, it was suggested that the numerous examples of recent introgression within the southern Rocky Mountains clade of *Tamias* might be due to ecological partitioning of the various taxa resulting in an abundance of contact zones scattered throughout this geographic region (Reid et al. 2012).

4.5 Lagamorphs and Divergence-with-Gene-Flow: Hopping and Hybridizing Through the Landscape of Europe and Asia

4.5.1 European and Asian Hares

Alves, Melo-Ferreira, and their colleagues have developed the species complex of European hares (genus *Lepus*) into a classic example of reticulate evolution in animals. In particular, they have described the complex pattern of divergence-with-gene-flow resulting in introgression among a number of species, some of which have left their genetic legacy in areas from which they are now extinct. For example, Melo-Ferreira et al. (2009, 2012) used population genetic and phylogenetic approaches, respectively, to test for the source and directionality of genetic exchange. Their findings indicated that the arctic/arboreal species, *Lepus timidus*—now extinct from the Iberian Peninsula—had left its genetic signature behind through introgression with several extant, temperate species (i.e., *L. granatensis*, *L. europaeus*, *L. castroviejoi*, *L. corsicanus*). Interestingly, as with the examples from the chipmunks, the introgression events reflected in the mtDNA of the Iberian hare spe cies resulted from both ancient and more recent genetic exchange. Melo-Ferreira (2012) reflected this conclusion by stating: 'Despite the many uncertainties on divergence time estimates and the difficulty of finding paleontological calibration points within *Lepus*, it seems clear that mtDNA introgression occurred at 2 different epochs, first presumably into the ancestor of *L. castroviejoi* and *L. corsicanus* and then more recently into the former, in the Iberian Peninsula.'

As with the Iberian species, Asian *Lepus* species also reflect the important role of ancient and contemporaneous hybridization and introgression. Significantly, and as predicted by Alves et al. (2008), the genetic legacy of *L. timidus* was also detected in Asian species as well (Liu et al. 2011). For example, the mtDNA haplotypes present in *Lepus mandshuricus* and *L. capensis* consisted of those from *L. timidus* and a second species, *L. sinensis*. Recent hybridization among a number of Asian *Lepus* species was also indicated by the presence of a number of individuals that were heterozygous for species–specific alleles at nuclear loci (Liu et al. 2011). Similarly, a population genetic/phylogenetic analysis of the Asian species, *L. capensis* and *L. yarkandensis*, revealed bidirectional introgression of Y chromosome and mtDNA between these ecologically and morphologically distinct species (Wu et al. 2011). Thus, as predicted by the landmark studies of Alves, Melo-Ferreira et al., *Lepus* species in general are a paradigm of mammalian complexes that have, and are, evolving via divergence-with-introgression.

4.5.2 European Rabbits

Like their sister taxa (i.e., members of the family Leporidae) the hares, rabbits also demonstrate patterns of ancient and recent reticulate evolution. In particular, Carneiro et al. (2009, 2010, 2011) detected the genetic signature of divergence-with-gene-flow in both the wild and domesticated European rabbit, *Oryctolagus cuniculus*.

Carneiro et al. (2009, 2010) analyzed sequence variation across natural populations of the two parapatrically distributed subspecies, *O. c. algirus* and *O. c. cuniculus*. Both of these analyses revealed a wide range of recombination frequencies at the nuclear loci studied; loci near centromeres and on the X chromosome demonstrated high levels of divergence (i.e., low levels of introgression) between the two subspecies, while those autosomal loci distal to centromeres generally reflected high levels of introgression (Carneiro et al. 2009, 2010). The widely varying estimates of introgression and haplotype sharing between *O. c. algirus* and *O. c. cuniculus* were consistent with (1) a non-allopatric model of divergence and (2) the presence of genomic regions that contribute differentially to reproductive isolation (Carneiro et al. 2009, 2010).

Using nine autosomal and seven X chromosome loci, Carneiro et al. (2011) were able to determine the region of origin (France) and the direct ancestor (French populations of *O. c. cuniculus*) of the domestic rabbit. Though there was no evidence that *O. c. algirus* populations were used in the domestication process some 1200 years ago, this subspecies did have an impact on the genomic makeup of the domesticated form. Natural introgressive hybridization between the two wild subspecies, predating the domestication event, thus resulted in *O. cuniculus* possessing a mosaic genome made up of elements from both subspecies (Carneiro et al. 2011).

4.6 Insectivores and Divergence-with-Gene-Flow: Introgress Fast and Die Young…Reticulate Evolution Among the Shrews

The shrew genus, *Sorex*, has been used as a model to determine the number and type of chromosomal rearrangements present and their effects on gene flow within species and introgression between species. For example, White et al. (2010) used chromosomal rearrangements and a network methodology to reconstruct the evolutionary relationships among the 72 chromosomal races of the common shrew, *Sorex araneus*. From this analysis, they were able to infer both the minimum number of intermediate chromosomal rearrangements that linked the various races and the mode of raciation—i.e., whether or not the races were generated within hybrid zones between races or in isolation from other races ('zonal raciation'). White et al. (2010) concluded that both zonal and hybrid race formation (divergence-with-introgression) had led to the extensive array of *S. araneus* chromosomal forms. Hybrid formation of chromosomal races within *S. araneus*, as inferred by White et al. (2010), is somewhat unexpected, given that such introgression should result in highly sterile heterozygous hybrids. However, this expectation has not been borne out by analyses of contemporaneous hybrid zones within *S. araneus*. Both Horn et al. (2012) and Polyakov et al. (2011) failed to detect reproductive isolation between races that, at meiosis I, formed aberrant configurations made up of up to nine chromosomes (but see Bulatova et al. 2011). Likewise, studies of contact zones between subspecies of the lesser white-toothed shrew, *Crocidura suaveolens*, also detected advanced-generation hybrids, indicating a lack of strong reproductive isolation between lineages defined by relatively high levels of genetic divergence (Dubey et al. 2008). These results indicate the ongoing impact of divergence-with-gene-flow between the *Sorex* and *Crocidura* intraspecific lineages.

In contrast to intraspecific hybridization, interspecific contact zones do indeed demonstrate elevated levels of reproductive isolation (e.g., Yannic et al. 2008a), but there is also evidence of both ancient and recent introgression among species of *Sorex*. For example, discordance in the phylogenetic placement of *Sorex granarius* relative to *S. araneus* and *S. coronatus* was, alternately, taken as evidence of ancient introgressive hybridization resulting in the transfer of the Y chromosome from *S. coronatus* into *S. granarius* (Yannic et al. 2008b) and/or the introgression of mtDNA from *S. araneus* into *S. granarius* (Yannic et al. 2010). Contemporaneous introgression was also noted in the species complex studied by Yannic et al. (2008b) (e.g., between *S. antinorii* and *S. araneus*). A major role for reticulate evolution was thus inferred for this group as reflected by the following quote: 'The evolutionary history of the southwestern European populations of the *S. araneus* group can only be understood considering secondary contacts between taxa after their divergence, implying genetic exchanges by means of hybridization and/or introgression' (Yannic et al. 2008b).

4.7 Carnivores and Divergence-with-Gene-Flow: Meat Eaters—Conservation Concerns and Adaptive Genetic Transfers

4.7.1 American Mink

The American mink, alternatively called *Neovison vison* or *Mustela vison* (we will refer to this taxon as *N. vison*), has been utilized as a source of fur from both wild-caught and captive-reared animals (Kidd et al. 2009; Bifolchi et al. 2010). The utilization of so-called farmed animals in fur production in North America and Europe has resulted in feral populations formed from escaped individuals. In North America, the escapee populations have introgressed with natural populations of *N. vison*, leading to concerns over the conservation of the native populations. For example, in two populations in Ontario (Canada) consisting of escaped, wild, and hybrid mink, Kidd et al. (2009) detected only 36 % wild individuals, with the remainder being hybrid (10 and 46 %) or escaped genotypes. Thus, hybridization between the escaped and wild populations could lead to the genetic extirpation of the native gene pool, or alternatively, the production of novel lineages that form the basis of further evolutionary innovation (Arnold 1997, 2006; Seehausen 2004).

In Europe, where escaped populations of the non-native *N. vison* can be viewed as an invasive species, a different scenario has been detected. Specifically, Bifolchi et al. (2010) reported that the genetic makeup of feral populations of American mink in Brittany reflected hybridization and introgression among genetically divergent lineages. This pattern of admixture was likely a reflection of both inbreeding in the farmed populations leading to divergent subpopulations that then escaped and also the history of introductions of captive animals from North America that derived from three separate *N. vison* subspecies (Bifolchi et al. 2010). Thus, as with the North American native x wild hybrids, the admixed populations in France reflect novel genotypes due to introgressive hybridization between divergent mammalian lineages.

4.7.2 European Mink and Polecats

Introgression among naturally occurring species of European mustelids has also been detected. In particular, Cabria et al. (2011) investigated the genetic makeup of populations of the endangered European mink (*Mustela lutreola*) and the more numerous European polecats (*Mustela putorius*). By utilizing nuclear (i.e., microsatellite), mtDNA and Y chromosome loci, it was possible to test for both the frequency and direction of introgression between these species. As with the findings of an earlier study in which only nuclear loci were used (Lodé et al. 2005), the analysis by Cabria et al. (2011) detected low levels of introgression (ca. 1 %).

The maternal and paternal genetic markers incorporated into the latter study also indicated that the introgression was asymmetric; 'only pure polecat males mate with pure European mink females. Furthermore, backcrossing and genetic introgression was detected only from female first-generation (F_1) hybrids of European mink to polecats' (Cabria et al. 2011). The observation that only female F_1s were contributing to later generation hybrids was taken as evidence of the role of Haldane's rule (i.e., the heterogametic sex is sterile in the F_1 generation—in this case the male hybrids; Cabria et al. 2011).

4.7.3 Polar and Brown Bears

The polar bear, *Ursus maritimus*, is one of the poster species of the arctic ecosystem conservation effort. As such, the reports of ancient, recent, and ongoing hybridization events between polar and brown/grizzly bears (*Ursus arctos*) have caused some scientists to argue that habitat loss and climate change will lead to more genetic exchange and thus the loss of biodiversity (Kelly et al. 2010). Though loss of intraspecific adaptive gene combinations is one possible outcome of introgressive hybridization, especially when rare and endangered forms hybridize with a more numerous species, genetic enrichment of the rare form by the more numerous lineages (Arnold 1992) is another likely outcome that may ultimately be advantageous. Indeed, so-called genetic rescue, whereby rare forms are intentionally crossed with related taxa to provide a means by which fitness can be increased and adaptive evolution catalyzed in genetically impoverished forms (e.g., see discussion of the Florida panther, below), is a model for what might take place in cases such as polar bear x brown bear introgressive hybridization. Often those who argue for protecting the genetic 'purity' of species do so because (even as rationale scientists) they have a philosophical stance that leads them to argue against the 'pollution' of a phenotype by introgression, even if such 'contamination' elevates the fitness of the introgressed form (Arnold 1997). In discussions of lineages such as *U. maritimus,* it should be remembered that the outcome of the genetic exchange may actually be that the lineages are able to adapt to a changing environment through the wholesale introduction of mutations, especially in the face of rapid habitat changes.

Introgressive hybridization, such as that seen between polar bears and brown bears (e.g., Miller et al. 2012), may be important for facilitating the long-term survival of the polar bear lineage. Recent changes in climatic conditions have led to rapid habitat loss and increased temperatures across the arctic. While polar bears are a genetically differentiated group, there is evidence that hybridization among brown bears and polar bears in previous glacial cycles was important for sufficient genetic variation, which may have contributed to their survival (Edwards et al. 2011; Hailer et al. 2012). The ongoing loss of suitable habitat, in addition to other human-induced stressors, will impact the survival of polar bears (Hailer et al. 2012). Monitoring of brown bears, polar bears, and their hybrids should be continued as genetic exchange among these species may be informative for other instances of introgressive hybridization involving threatened species. For example, Edwards et al. (2011) in reference to the evidence for ancient mtDNA exchange between polar and brown bears concluded, 'This suggests that interspecific hybridization not only may be more common than previously considered but may be a mechanism by which species deal with marginal habitats during periods of environmental deterioration.'

4.7.4 Florida Panthers

As mentioned in the previous section, the Florida panther, *Puma concolor coryi*, is an example of conservation efforts that have included human-mediated reticulate evolution. Specifically, individuals from divergent lineages of this species (from Texas) were introduced into the range of the Florida panther in 1995 (Johnson et al. 2010). As expected from the hypothesis that introgression can result in elevated fitness in hybrids relative to inbred parents, F_1 and backcross genotypes demonstrated increased survivorship as kittens, subadults, and adults (Johnson et al. 2010). Significantly, the elevated fitness of hybrid individuals was correlated with an increase in population size of 14 % per year between 1996 and 2003 (Johnson et al. 2010). These results suggest that hybridization and introgression, involving individuals from the divergent *P. concolor* lineages, are helping to pull the Florida panther back from the brink of extinction.

4.7.5 South American Wild Cats

The Neotropical felid genus, *Leopardus*, consists of seven species thought to have evolved from a common ancestor that crossed the Isthmus of Panama into South America during the Pliocene (ca. 3 mya; Trigo et al. 2008). Based upon mtDNA sequences, a phylogenetically well-supported clade within this genus includes *Leopardus tigrinus*, *L. geoffroyi*, *L. guigna*, *L. colocolo,* and *L. jacobita* (Johnson et al. 1999). However, the earlier analysis of mtDNA coupled with Y chromosome loci identified some discordant genotypes suggesting hybridization between *L. tigrinus* and *L. colocolo* (Johnson et al. 1999).

Trigo et al. (2008), using a combination of mtDNA and nuclear loci, tested the hypothesis of divergence-with-gene-flow among the species of *Leopardus*. The data gathered from this analysis contained a pattern indicative of past and ongoing introgression involving *L. tigrinus*, *L. geoffroyi,* and *L. colocolo*. In particular, a hybrid zone in southern Brazil was detected between *L. tigrinus* and *L. geoffroyi*, with introgression from the latter into the former species extending away from the zone of sympatry (Trigo et al. 2008). Furthermore, the *L. tigrinus* populations also contained mtDNA variation apparently captured through introgressive hybridization with *L. colocolo*. Thus, the diversification of *Leopardus* into numerous evolutionary lineages has involved the admixture of divergent felid lineages leading to complex, mosaic genomes (Trigo et al. 2008).

4.8 Artiodactyla and Divergence-with-Gene-Flow: Meat on the Hoof Develops Through Divergence-with-Introgression

4.8.1 Wild Boar and Domestic Pigs

As humans expanded their populations across the globe, they both took previously domesticated companions with them and began new domestication projects. Most of these projects were associated with producing consumable protein. One of the most widely used animal protein sources has been various lineages of the wild and domesticated pig genus *Sus*. In particular, lineages of *Sus scrofa* from Europe, the Near East, and the Far East formed the basis of the current genetic diversity in wild and domesticated populations (Ramírez et al. 2009). An analysis of Y chromosome, mtDNA, and autosomal loci revealed genomic patterns suggesting (1) two centers of pig domestication (Europe and the Far East), (2) human transportation of these domesticated varieties to Africa and South America with (3) subsequent introgression between the domesticated lineages and the nearby wild boar populations (Ramírez et al. 2009). Thus, as the domesticated and wild lineages of *S. scrofa* were adapting to human-altered and natural habitats, respectively, they were simultaneously exchanging portions of their genomes. This scenario is thought to at least partially explain both the lack of divergence between wild boar and domesticated pig populations and the high level of genetic variability in domesticated *S. scrofa* (Ramírez et al. 2009).

Also consistent with the hypothesis of at least sporadic introgression between domesticated and wild *Sus* lineages (Ramírez et al. 2009), as well as between different wild lineages that have been introduced for hunting, are a set of population genetic/phylogenetic analyses of wild and domesticated samples from various European regions. Thus, although Alves et al. (2010) and Scandura et al. (2008) detected very limited amounts of mtDNA introgression between domestic pigs and wild boar Iberian lineages and different wild boar lineages brought into contact by human-mediated translocation, respectively, Scandura et al. (2011) detected a major impact from such translocations on the island of Sardinia. In particular, this latter analysis produced data indicating that the subspecies of Sardinian wild pigs (i.e., *Sus scrofa meridionalis*) possessed introgressed alleles from Italian peninsula, central European and domesticated lineages (Scandura et al. 2011). Taken together, the data from the genus *Sus* indicate the large role humans can have in catalyzing periods of divergence-with-gene-flow, particularly when the organisms are edible.

4.8.2 Wildebeest

As Basil Fawlty points out to a difficult guest in an episode of the British sitcom, *Fawlty Towers*, there are no wildebeest sweeping majestically across the landscape of Torquay, England. However, these ungainly African bovids are a great example of reticulate evolutionary processes. In particular, during both ancient and historical periods, the two species known as the 'blue' and 'black' wildebeest—*Connochaetes taurinus* and *C. gnou*, respectively—have overlapped and, especially during recent human-mediated perturbations, introgressed (Fabricius et al. 1988).

Ackermann et al. (2010) have recently described morphological anomalies in hybrid offspring relative to parental species individuals found in a hybrid wildebeest population collected at Spioenkop Dam Nature Reserve, KwaZulu-Natal, South Africa. Ackermann et al. (2010) were particularly interested in the types of morphological aberrations evidenced by extant hybrid populations in order to predict the types of divergent phenotypes that might be found in mammalian, hybrid fossils. Specifically, they detected a number of cranial features in the hybrid wildebeest not found in the parental species. In addition, a low frequency of dental anomalies was also detected in the hybrid animals.

The study of the *C. taurinus* and *C. gnou* admixed population reflected what could be defined as transgressive/novel phenotypic variability routinely apparent in introgressed individuals. Such variability can form the basis for additional evolution, as these novel phenotypes sometimes reflect novel adaptations (e.g., see Rieseberg et al. 2003). Though Ackermann et al. (2010) achieved their main goal of testing whether aberrant morphological characteristics are typical of mammalian hybrids in general, their study also indicated the role played by introgressive hybridization in generating potentially important novelties as noted in a broad array of plant, animal, and microorganisms (see Arnold 1997, 2006 for examples).

4.8.3 Chamois

The chamois, genus *Rupicapra*, has been utilized as a *Homo sapiens* protein source for hundreds of thousands of years (Rivals and Deniaux 2005). Indeed, chamois meat continues to be consumed by contemporary human populations as well (e.g., Hofbauer et al. 2006). Significantly, the interest in *Rupicapra* as a source of food has likely resulted in some of the detected patterns of both intra- and interspecific reticulate evolution.

Crestanello et al. (2009) argued against the inference of Rodríquez et al. (2009) that admixtures of highly divergent mtDNA haplotypes reflected ancient introgressive hybridization during glacial cycles. Instead, the former authors concluded that 'a more likely explanation for the patchy presence of *pyrenaica* haplotypes in 3 Alpine populations (WA1, EA6, and EA7) is an undocumented restocking or reintroduction from the Pyrenees to the Western Alps (WA1) within the last 150 years followed by a documented introduction of descendents of these animals from the Western to the Eastern Alps (EA6, EA7) in the early 1970s…' However, Crestanello et al. (2009) also recognized the likely affect of population expansions and contractions due to climatic events, rather than human translocations, to the admixed genetic structure of various chamois populations.

In a subsequent analysis that included sampling not only mtDNA, but also nuclear variation across all subspecies of *Rupicapra*, Rodriquez et al. (2010) again detected genomic signatures indicative of both ancient and more recent introgression. Thus, the chamois lineages have apparently diverged in ancient and recent episodes of contact and genetic exchange as predicted by a non-allopatric model of divergence.

4.8.4 Caribou

Caribou, the domesticated form known as reindeer in North America, has been used as a beast of burden, a food item, and a clothing source for ancient and modern human populations (Pryde 1971). The *Rangifer tarandus* clade consists of numerous morphological forms placed into a series of taxa, many of which overlap and introgress. For example, McDevitt et al. (2009) found discordances between variation at mtDNA and nuclear loci indicating the presence of a hybrid swarm between the barren-ground caribou (*Rangifer tarandus groenlandicus*) and the woodland caribou (*Rangifer tarandus caribou)*. This result was surprising given that the admixed population occurs in the Rocky Mountains, hundreds of kilometers south of the range of *R. t. groenlandicus* (McDevitt et al. 2009). These results suggested that though the woodland and barren ground forms likely diverged in separate refugia during glacial maxima, they have subsequently overlapped and introgressed. Specifically, McDevitt et al. (2009) suggested 'An ice-free corridor at the end of the last glaciation likely allowed, for the first time, for barren-ground caribou to migrate from the North and overlap with woodland caribou expanding from the South.'

In a subsequent analysis, Weckworth et al. (2012) confirmed the admixture reflecting the hybrid swarm in Alberta. Furthermore, these latter data revealed previously unrecognized and geographically widespread introgression in Alberta boreal caribou herds. As with the earlier study, the possible catalyst inferred for the formation of admixed individuals, populations, and ecotypes was the recession of glaciers after the last glacial maxima (Weckworth et al. 2012).

4.8.5 Mule Deer and Black-Tailed Deer

Pease et al. (2009) utilized a landscape genetics approach to determine the spatial and ecological distribution of deer belonging to the species *Odocoileus hemionus* (i.e., mule and black-tailed deer). The patterns of variation at the mtDNA and nuclear loci resulted in the definition of five divergent ecotypes within the state of California. This observation was a surprising one given the ability of these deer to move large distances. In fact, the degree of divergence in both the genetic and ecological characters associated with the various lineages indicated that the diversification seen in this relatively small geographical area occurred recently and was likely caused by habitat selection (Pease et al. 2009).

One of the divisions among the five lineages detected by Pease et al. (2009) involved a clade consisting of deer from northwestern California and one made up of genotypes from the eastern, central, and southern California deer. The northwestern group was associated with the black-tailed subspecies, *O. hemionus columbianus*. However, the genetic analysis of deer from the northwestern region also detected a relatively high frequency (i.e., 23.5 %) of genotypes inferred to be of hybrid origin; the putative hybrid individuals all came from areas of sympatry between the mule deer (*O. h. hemionus*) and black-tailed deer.

Latch et al. (2011), in an analysis of genetic and ecological characteristics across a zone of overlap between mule deer and black-tailed deer occurring along the east and west slopes of the Cascade Mountains (in the states of Washington and Oregon), likewise detected admixture. Indeed, this analysis resulted in the conclusion that the area of overlap contained a hybrid swarm between these genetically, morphologically, and ecologically distinct lineages (Latch et al. 2011). Given the substantial difference in size of individuals from these two subspecies (a characteristic associated with mating success) and the fact that the zone of overlap encompasses a strongly differentiated ecotone, the observation of extensive, bidirectional introgression was surprising (Latch et al. 2011). Furthermore, the absence of linkage disequilibrium in the center of the zone and the random association of genotypes within the area of overlap was further evidence of the existence of a hybrid swarm and thus ongoing divergence-with-introgression between these two *O. hemionus* subspecies.

4.9 Cetaceans and Divergence-with-Gene-Flow: Interspecies Sex at Sea

There are few (at least as found through our searches) examples yet reported of reticulate evolution in marine mammals. Some of these have been discussed previously (e.g., sea lions—see Lancaster et al. 2007). Our hypothesis is that, as with many animal groups, more cases may not have been detected because they are assumed not to occur.

Of course, if extended to non-mammalian animals, there is a wealth of examples of divergence-with-gene-flow in organisms as diverse as bivalves and sharks (e.g., Stuckas et al. 2009; Morgan et al. 2012; see Arnold and Fogarty 2009 for additional examples). However, in addition to 'older' studies such as those of the sea lions, there are examples of marine mammals that reflect reticulate evolutionary processes.

4.9.1 Dolphins

One group of marine mammals possessing both genomic and morphological indications of divergence-with-introgression is the clade of Delphinidae dolphins, in particular members of the genus *Stenella*. Both DNA sequences and morphology suggest paraphyly among the members of this genus (LeDuc et al. 1999; Kingston et al. 2009). Explanations given for this paraphyletic arrangement include incorrect taxonomic placement of the members in this genus, lack of phylogenetic resolution due to a rapid radiation (thus resulting in unresolved branching patterns), and/or the origin of some taxa through hybridization (LeDuc et al. 1999; Kingston et al. 2009).

In regard to the origin of lineages through hybridization, LeDuc et al. (1999) suggested that the Clymene dolphin, *Stenella clymene*, might reflect such an evolutionary history. This inference was based upon the discordance in phylogenetic placement of this species based on different morphological characters as well as morphological versus DNA sequences (i.e., mtDNA). Comparisons of the morphological and molecular data thus indicated that the Clymene dolphin was more closely allied to *Stenella longirostris* or *S. coeruleoalba*, respectively (LeDuc et al. 1999).

Though Kingston et al. (2009) argued against LeDuc et al.'s (1999) inference of a hybrid origin for *S. clymene*, they did so because phylogenies based upon data from nuclear loci did not place this species intermediate to the putative parents, *S. longirostris* and *S. coeruleoalba*. Instead, unlike the mtDNA data that indicated a sister-species relationship between the Clymene dolphin and *S. coeruleoalba*, their nuclear loci data placed *S. clymene* in a clade with *S. longirostris*. In contrast, this type of phylogenetic discordance between the two molecular data sets (as well as that seen between morphological and molecular characters) is the pattern expected if the Clymene dolphin lineage has a reticulate origin (see Arnold 2006 for a discussion). Furthermore, Kingston et al. (2009) detection of a number of hybrid individuals between *S. frontalis* and *S. attenuata* indicates the potential role of divergence-with-gene-flow such as inferred by LeDuc et al. (1999) for the Clymene dolphin species.

4.9.2 Whales

As mentioned at the beginning of this section on cetaceans, there are relatively few examples of marine mammal hybridization in nature. Once again, we would suggest that this is likely due to a lack of sampling/reporting and that as more molecular data are gathered, more examples of reticulate evolution will be noted. Regardless of whether this hypothesis is supported with additional data, various species of whales, like dolphins, do demonstrate interspecific introgression. For example, Berube and Aguilar (1998) reported a hybrid between the blue whale, *Balaenoptera musculus*, and the fin whale, *Balaenoptera physalus*. This was not the first report of hybridization between these two species leading Berube and Aguilar (1998) to conclude 'Examination of data for the five fin-blue whale hybrids in the literature, plus other anecdotal reports, indicates that hybridization between these two species occurs in various geographic regions and is relatively frequent…'

A more recent report by Glover et al. (2010) noted hybrid formation between the two species of minke whale, the Antarctic *Balaenoptera bonaerensis*, which had been thought to occur only in the Southern Hemisphere, and the common minke whale, *Balaenoptera acutorostrata*, which is found in both hemispheres. In a genetic analysis of minke whales harvested between 1996 and 2008, two individuals were detected that were aberrant in regard to their genetic profile and where they were harvested. Specifically, a *B. bonaerensis* individual and a hybrid individual with a mtDNA haplotype from *B. bonaerensis* and a nuclear DNA profile from *B. acutorostrata* were harvested from the Arctic Northeast Atlantic (Glover et al. 2010). Though these two species had been thought to not overlap due to non-overlapping seasonal migrations to different poles, rare, possibly episodic, genetic exchange can and does occur.

4.10 Perissodactyla and Divergence-with-Gene-Flow: Horses Looking for Interspecific Love on the Range

4.10.1 Equus Ferus/Equus Caballus

Likely because we humans are fascinated with ourselves and those organisms with which we frequently interact, studies describing the domestication of our food sources, clothing sources, companions, etc., have been a main focus of many research groups (see Arnold 2008 for examples). Yet, much is left to learn about the how, when, and where of many of the domestication events. Like many such research programs, the history of studies concerning the evolution of *Equus caballus* (i.e., the domestic horse) has been replete with competing hypotheses, especially concerning whether there was a single point of origin for the domestication process or alternatively many different foci of domestication from wild lineages.

To test explicitly the various scenarios, Warmuth et al. (2012) utilized nuclear, autosomal loci as the data for modeling the demographic parameters involved in horse domestication. Furthermore, these same data allowed inferences concerning the population genetic structure of the extinct, wild progenitor of *E. caballus*, *Equus ferus*. Several conclusions were reached from this analysis, including the inference that horse domestication occurred in the western Eurasian steppe (Warmuth et al. 2012). Most applicable to the topic of this review, however, is that this analysis detected signatures of introgression between populations of *E. ferus* and those of the domesticated horse as the latter expanded from the defined region of domestication. In addition, Warmuth et al. (2012) concluded that the high levels of mtDNA diversity detected previously in *E. caballus* suggested that the introgression derived from the capture and introduction of *E. ferus* mares into domestic populations. They reasoned 'Because stallions are inherently more difficult to handle than mares, the easiest way to maintain or grow herd sizes would have been to restock existing herds with wild females' (Warmuth et al. 2012).

4.10.2 Equus Przewalskii/Equus Ferus/Equus Caballus

As indicated by the previous section, interactions between the domestic horse and *E. ferus* resulted in introgression between these two species. Likewise, the genomes of *E. caballus* and Przewalski's horse, *Equus przewalskii*, are mosaics of shared alleles. McCue et al. (2012) detected a pattern of SNP variation supporting this conclusion. Thus, both breeding programs designed to save *E. przewalskii* by crossing this species with the domestic horse, as well as likely natural overlaps between the two species resulted in the observation that 'surviving Przewalski's horses today are truly *E. przewalskii* and *E. caballus* hybrids' (McCue et al. 2012).

The geographical pattern of the SNP variation assayed by McCue et al. (2012) also allowed an explicit test of where *E. przewalskii* and *E. caballus* may have overlapped and introgressed without direct human intervention. In this regard, the finding of an extremely close association of genotypes of *E. przewalskii* and those of Mongolian horse populations suggests this general region as a source of past introgression. Indeed, these two species are known to have overlapped in the region of China, Russia, and Mongolia (see McCue et al. 2012 for a discussion of this topic as well as additional references).

4.11 Chiroptera and Divergence-with-Gene-Flow: Hanging Around with Hybridizing Bats

4.11.1 Artibeus

Like shrews, the bat clade has been a focus of researchers interested in deciphering evolution catalyzed by chromosomal rearrangements. In particular, Baker, Bickham, and their colleagues have carried out numerous, wonderfully detailed analyses aimed at understanding how evolutionary diversification may have been influenced by the structural reorganization of the genome (e.g., Hoffmann et al. 2003; Baird et al. 2009). A recent analysis by these workers has also pointed to the role that reticulate evolution has played in the origin of new species.

Larsen et al. (2010) collected molecular (nuclear and mtDNA) and morphological data from both mainland South America and Caribbean species of the Neotropical bat genus *Artibeus*. All seven recognized Caribbean species were analyzed. From the analysis of these island taxa, a zone of hybridization and introgression was detected among *Artibeus jamaicensis*, *A. planirostris,* and *A. schwartzi*; the hybrid zone extended north to south across the island chain from St. Lucia to Grenada (Larsen et al. 2010). Interestingly, not only did the data support admixture among these three *Artibeus* species, but also a hybrid origin for *A. schwartzi*. The findings supporting this latter hypothesis were that this species had (1) a nuclear genome made up of alleles from *A. jamaicensis* and *A. planirostris*, (2) mtDNA haplotypes from another, as yet unidentified, lineage, and (3) an apparently transgressive morphology that did not overlap with that of either *A. jamaicensis* or *A. planirostris* (Larsen et al. 2010). Thus, ongoing and ancient divergence-with-gene-flow characterizes at least the Caribbean Island lineages of this Neotropical bat genus.

4.11.2 Rhinolophus/Nyctimene

As with the inference of hybrid speciation and introgressive hybridization by Larsen et al. (2010) in Neotropical bats, Old World chiropteran clades also demonstrate genomic signatures of reticulate evolution. Two such assemblages are the horseshoe bat and tube-nosed fruit bat species complexes belonging to the genera *Rhinolophus* and *Nyctimene*, respectively. For both of these clades, divergence accompanied by introgressive hybridization was inferred from discordant phylogenetic patterns derived from different data sets. As reviewed elsewhere (Arnold 1997, 2006), phylogenetic discordance of this type has been used as one of the main indicators of potential divergence-with-gene-flow.

In the case of the horseshoe bats, samples of individuals were collected across the area in which taxa overlap. Mitochondrial DNA sequences from these samples resolved phylogenetic trees concordant with expected taxonomic placements of individuals of the sibling species, *Rhinolophus yunanensis* and *R. pearsoni*, but not of the two subspecies of *R. pearsoni* (i.e., *R. p. pearsoni* and *R. p. chinensis*; Mao et al. 2010). Contrasting with the mtDNA results, nuclear gene sequences—as expected from taxonomic considerations—resulted in phylogenetic trees placing individuals of the two subspecies into separate clades. However, sequences from these same genes resolved a clade containing *R. yunanensis* and *R. p. pearsoni*, separate from *R. p. chinensis* (Mao et al. 2010). Taken together, the sequence data and taxonomic placement of the *Rhinolophus* species and subspecies suggested ancient male-mediated gene flow between *R. pearsoni* and *R. yunanensis* and past or contemporaneous asymmetric introgression of mtDNA from *R. p. chinensis* into *R. p. pearsoni* (Mao et al. 2010).

Samples of tube-nosed fruit bats from the eastern Indonesian archipelago islands known as the Moluccas were screened for both mtDNA and nuclear markers. Though three species had been recorded from these islands based upon morphological characters, the genetic analyses indicated the presence of only *Nyctimene cephalotes* and *N. albiventer* (Newbound et al. 2008). Yet, the two genetic data sets also were in conflict regarding which of the six islands sampled were occupied by these two species. The nuclear markers placed *N. albiventer* on Wokam Island, with *N. cephalotes* occurring on the other five islands. In contrast, mtDNA sequences clustered into two very divergent clades, one including the bats from Wokam and Yamdena Islands and the other the individuals from the remaining four islands (Newbound et al. 2008).

The discordant patterns obtained from the nuclear and mtDNA data for the tube-nosed fruit bats as pointed out by Newbound et al. (2008) could be explained by either retention of ancestral polymorphisms (i.e., incomplete lineage sorting) or, alternatively, by introgressive hybridization between *N. cephalotes* and *N. albiventer*. Given the large nuclear sequence divergence, the geographic distribution of genetic variation, and the estimated divergence times between the two species, Newbound et al. (2008) concluded that the most likely explanation for the observed discordance between the genetic data sets was a single mtDNA introgression event from *N. albiventer* into *N. cephalotes*.

4.12 Primates and Divergence-with-Gene-Flow: Lemurs, Monkeys, Apes and What the Parrot Saw

The primate clade in general is a rich source of examples of reticulate evolution (Arnold and Meyer 2006; Arnold 2008). This may seem surprising to those who hold to the neo-Darwinian paradigm of allopatric speciation and the biological species concept, coupled with the intuition that primates should have well-developed behaviors that keep them from making such 'mistakes' as breeding with members of another, divergent lineage. In fact, like canids, ducks, and numerous other animal clades, members of Primates seem to have a particular propensity for intertaxa matings (Jolly 2001; Arnold 2006, 2008; Arnold and Meyer 2006).

In the following sections, we will illustrate the role of hybridization, introgression, hybrid speciation, etc., in groups from lemurs to Old World apes. Yet, as with most of the other mammalian groups we discuss, we will only scratch the surface of species complexes that could be included. As one example of this, only Old World clades will be discussed even though there are data indicating the occurrence of reticulate evolution in a variety of New World clades as well (e.g., Bicca-Marques et al. 2008).

4.12.1 Lemurs

The diversity of lemuriforms on the island continent of Madagascar is truly astounding. For example, Mittermeier et al. (2008) listed five families, 15 genera, and 99 species and subspecies of lemurs. Amazingly, their list included 39 species described since 2000. Significantly, these workers also recognized a series of species and subspecies known to have formed hybrid zones (Mittermeier et al. 2008).

Because many of the lemur lineages have been shown to be associated with specific habitats, with ecotones defining the margins of their distributions, it has been concluded that selection due to environmental parameters has played a significant role in the radiation of this clade (Rakotondranary et al. 2011). As with any divergence due to habitats (Endler 1973, 1977), radiations catalyzed by environmental gradients would be likely to reflect gene flow during divergence across the ecotonal habitats. As predicted by Endler (1973, 1977) and Moore (1977), some cases of hybridization between lemur species occupying different ecological settings may have resulted in stabilized hybrid lineages adapted to the ecotonal environment (e.g., Delmore et al. 2011).

In addition to putative stabilized hybrid lineages in ecotonal regions, genetic exchange outside of the areas of overlap and into the parental taxa has been documented for a number of the instances of lemur hybridization. For example, both Gligor et al. (2009) and Pastorini et al. (2009) reported introgressive hybridization between lemur species. In the first of these studies, individuals from outside and within an ecotonal zone of overlap between mouse lemur species (*Microcebus griseorufus* and *M. murinus*) were characterized for morphological characters as well as mtDNA and nuclear genetic markers. These data indicated that most individuals within the area of sympatry were admixtures of the *M. griseorufus* and *M. murinus* genomes/morphological traits (Gligor et al. 2009). Furthermore, introgression outside of the hybrid zone was indicated by a sharp transition of mtDNA variation at the ecotone edge, but a broader cline across the ecotonal boundaries for the nuclear markers. Taken together, these results were consistent with a lack of mtDNA introgression and the occurrence of asymmetric, male-mediated nuclear introgression from *M. griseorufus* into *M. murinus* (Gligor et al. 2009).

In contrast to the asymmetric introgression detected for the mouse lemurs, mtDNA and nuclear genetic variation across a hybrid zone between the mongoose lemur (*Eulemur mongoz*) and brown lemur (*Eulemur fulvus*) led to the inference of bidirectional genetic exchange. Pastorini et al. (2009) found that both nuclear and mtDNA markers diagnostic for each of the two species also occurred at varying frequencies in individuals of the alternate species in areas near, or within, regions of sympatry. Though alleles from *E. mongoz* were detected in *E. fulvus* as were nuclear and mtDNA markers from *E. fulvus* detected in *E. mongoz*, there was evidence that introgression may have occurred at a greater frequency from the mongoose lemur into the brown lemur. For example, all eight of the *E. fulvus* individuals from a region of sympatry possessed either nuclear alleles or mtDNA haplotypes characteristic of the mongoose lemur (Pastorini et al. 2009). The pattern and frequency of the mtDNA and nuclear introgression between the brown and mongoose lemurs suggested to Pastorini et al. (2009) that past and contemporaneous reticulate evolution had impacted these two species. More importantly, they also concluded 'Introgressive hybridization may hasten speciation and allow rapid ecological adaptation of taxa, hence be one of the driving forces for the adaptive radiation of lemurs in Madagascar' (Pastorini et al. 2009).

4.12.2 Rungwecebus Kipunji

The kipunji, originally recognized in 2005 as a new species, *Lophocebus kipunji*, was subsequently—based upon a unique set of morphological characteristics and ecological associations, along with nuclear and mitochondrial gene sequences—elevated to generic status, i.e., *Rungwecebus kipunji* (Davenport et al. 2006).

Though the elevation of the kipunji to generic status was originally met with some skepticism by primatologists, subsequent morphological analyses have supported this taxonomic revision. For example, phenetic and phylogenetic analyses of morphological traits have provided evidence indicating that the phenotype possessed by *R. kipunji* is unique, yet also reflects this taxon's close evolutionary affinity to *Papio*, *Lophocebus*, and *Theropithecus* (Singleton 2009; Gilbert et al. 2011). The inferences drawn from these morphological analyses, and the original molecular data (Davenport et al. 2006), suggesting the uniqueness and phylogenetic placement of *R. kipunji* relative to other Old World primate genera also were in accord with the findings of a study that included the collection of DNA sequences from a range of autosomal, X chromosome, Y chromosome, and mtDNA loci (Olson et al. 2008). Significantly, this latter analysis resolved a sister-lineage relationship between *Rungwecebus* and the baboon genus, *Papio,* rather than with the genus *Lophocebus* (the genus in which the kipunji was originally placed; Olson et al. 2008).

Given the findings of Zinner et al. (2009) and Burrell et al. (2009), the indication of the sister-genus relationship between baboons and *R. kipunji* is telling. In particular, both of these analyses, which sampled widely from the geographical distribution of *Papio*, found that rather than being a sister lineage, *Rungwecebus* clustered within the baboon clade. This result indicated that, like other Old World monkeys, such as macaques and indeed many members of the genus *Papio* (e.g., Stevison and Kohn 2009; Osada et al. 2010; Jolly et al. 2011), the evolutionary history of the kipunji included episodes of divergence-with-gene-exchange.

Though their findings were concordant with one another, leading to the inference of reticulate processes during the evolution of the *R. kipunji*, Burrell et al. (2009) and Zinner et al. (2009) drew somewhat different conclusions. The former authors inferred a hybrid speciation event in the origin of the kipunji, occurring ca. 650,000 ya and involving crosses between female *Papio* and male *Lophocebus* individuals (Burrell et al. 2009). In contrast, because they found that the phylogenetically most closely related *Papio* lineages to *Rungwecebus* were also the closest geographically, Zinner et al. (2009) inferred asymmetric introgression of mtDNA from baboons into the kipunji. Regardless of whether the event is ancient or more recent, this newly recognized genus of African monkey has an admixed genome consisting of elements from different genera.

4.12.3 Colobines

As highlighted by Messier and Stewart (1997) '…colobine Old World monkeys are unique among the primates in having a complex foregut in which bacteria ferment leafy plant materials, followed by a true stomach that expresses high levels of the bacteriolytic enzyme, lysozyme.' This clade of primates, throughout their geographic distribution in Africa and Asia, also contains a number of taxa that reflect ancient and more recent genetic exchange.

Discordances in phylogenetic trees derived from different molecular data sets (e.g., Ting et al. 2008), but which only reflected a portion of the extant taxa, led Roos et al. (2011) to undertake an extensive phylogenetic analysis involving all colobine genera. Furthermore, these workers utilized an array of molecular markers, including insertion sites of transposable elements as well as sequence data from mitochondrial, autosomal, X chromosome, and Y chromosome loci. From the extensive taxon and genomic samples, it was possible to test rigorously for the effect of various processes in the evolution of the colobine lineages—including that of divergence-with-gene-flow. In this regard, though Roos et al. (2011) recognized the potential contributions of such processes as insufficient data and incomplete lineage sorting, they argued for a major role of introgressive hybridization in causing the observed phylogenetic discordances. For the African colobine clade, the phylogenetic results suggested a female-based introgression event with genes moving from *Piliocolobus*/*Procolobus* into *Colobus* (Roos et al. 2011). In contrast, within the Asian colobine radiation, a hypothesis of male-mediated introgression from *Semnopithecus* into *Trachypithecus*—followed by backcrossing resulting in nuclear swamping—was favored (Roos et al. 2011). Regardless of the specific events, the colobine evolutionary pathway apparently involved reticulation.

In light of the findings by Roos et al. (2011), it is significant that a series of earlier studies by Karanth and his colleagues (Karanth 2008, 2010; Karanth et al. 2008) concluded that the Asian colobine clade showed evidence of ancient hybrid speciation leading to the origin of the golden leaf monkey (*Trachypithecus geei*) and capped leaf monkey (*Trachypithecus pileatus*). Karanth et al. (2008) summarized these results in the following manner: 'The phylogenetic position of the capped and golden leaf monkeys remains unresolved. It is clear from both nDNA and mtDNA data that these two species are closely related. However, the mtDNA…tree strongly suggests that they belong to the India clade (*Semnopithecus*), whereas the nuclear encoded lysozyme gene suggests that they may belong to the SE Asian clade (*Trachypithecus*). Interestingly, these two species are distributed in an area that is sandwiched between the distributions of *Semnopithecus* and *Trachypithecus*.' It would thus seem likely that not only has ancient and contemporaneous genetic exchange led to admixed genomes, but that such admixture has also led to new colobine lineages.

4.12.4 Gorilla

The genus *Gorilla* contains the sister taxa to both chimpanzees and humans. As with *Pan* and most other non-human primates, all of the lineages within this clade are endangered mainly due to the actions of its sister-genus, *Homo*. Divided into eastern and western species, containing two subspecies each (i.e., *Gorilla beringei beringei*/*Gorilla beringei graueri* and *Gorilla gorilla gorilla*/*Gorilla gorilla diehli*, respectively (see Ackermann and Bishop 2010 for references)), these primates are rapidly being lost due to habitat destruction and market hunting (http://www.iucnredlist.org). Also, as with many other primates, these close evolutionary allies of our own species tell of past and recent intertaxa genetic exchange.

Recently, the whole genome of a *G. g. gorilla* individual, along with a partial genome sequence for a second western gorilla as well as a *G. b. graueri* individual, was reported (Scally et al. 2012). Comparisons of these genome sequences revealed a deep divergence between the western and eastern species occurring ca. 1.75 mya. Notwithstanding the significant genetic distinctiveness of *G. beringei* and *G. gorilla*, the species evidently diverged non-allopatrically. Indeed, Scally et al. (2012) estimated an exchange of 0.2 individuals per generation in each direction over the past 500,000 years, for a total rate of migration (i.e., reflecting introgression) of 5000 animals.

As discussed in a previous section, Ackermann and her colleagues have led the way in deciphering the morphological anomalies expected in extant hybrid mammals and thus in fossils of hybrid origin (e.g., Ackermann 2010). With regard to gorilla lineages, Ackermann and Bishop (2010) utilized both phenotypic and genotypic data to test for gene flow and introgression within and between the various taxa, respectively. Significantly, these authors detected genetic exchange at all taxonomic levels from within subspecies to between *G. beringei* and *G. gorilla* (Ackermann and Bishop 2010). Furthermore, there was directionality to gene flow among the populations of the western species and between the western and eastern species; genes moved from west to east between populations of *G. gorilla* and from *G. gorilla* into *G. beringei*. Likewise, there was asymmetric introgression from the eastern lowland gorilla (*G. b. graueri*) into the mountain gorilla (*G. b. beringei*; Ackermann and Bishop 2010). These findings, along with those from the morphological analyses, indicate dispersal and admixture among divergent lineages of *Gorilla* resulting in taxonomic confusion, particularly for the relatively highly admixed eastern lowland subspecies (Ackermann and Bishop 2010).

4.12.5 Pan—Bonobos and Common Chimpanzees

A number of authors have tested the hypothesis of no divergence-with-gene-flow between the common chimpanzees and bonobo (i.e., *Pan troglodytes* and *P. paniscus*, respectively)—the species considered closest to the human lineage (Gagneux 2004). The results of some of these tests have produced conflicting conclusions, with more studies concluding no introgression between these lineages as they diverged from a common ancestor (see Arnold 2008 for earlier analyses).

Wegmann and Excoffier (2010) utilized 265 microsatellite loci along with sequences from 26 unlinked, intergenic regions from common chimpanzees and bonobos as the data for an approximate Bayesian computation analysis. This analysis was designed to infer the evolutionary history of the genus *Pan*. As a part of this inference exercise were tests for divergence-with-introgression between the various chimpanzee subspecies and between the chimpanzee and bonobo lineages. With regard to gene flow between the diverging lineages that would, respectively, give rise to the common chimp and bonobo, Wegmann and Excoffier (2010) inferred that '…this divergence appears to have been very progressive with the maintenance of relatively high levels of gene flow between the ancestral chimpanzee population and the bonobos.'

An example of a recent analysis that inferred a lack of genetic exchange between the bonobo and common chimpanzee (or their progenitor lineages) was that of Prüfer et al. (2012). These authors reported the sequencing and assembly of the *P. paniscus* genome and its comparison to the genomes of *P. troglodytes* and *H. sapiens*. Though the major emphasis of their analysis was to understand shared and distinctive genomic regions between these three taxa that may help elucidate phenotypic evolution, their analyses also allowed tests for both incomplete lineage sorting and introgressive hybridization. Unlike the earlier study of Wegmann and Excoffier (2010), Prüfer et al. (2012) found a lack of allele sharing and thus concluded that there was '…no indication of preferential gene flow between bonobos and any of the chimpanzee groups tested. Such a complete separation contrasts with reports of hybridization between many other primates. It is, however, consistent with the suggestion that the formation of the Congo River 1.5–2.5 million years ago created a barrier to gene flow that allowed bonobos and chimpanzees to evolve different phenotypes over a relatively short time.' In this context, it should be noted that Patterson et al. (2006) inference of a 'complex speciation' event involving introgression between proto-chimp and proto-human lineages (possibly around 6 mya) catalyzed a series of publications arguing against reticulation between these sister lineages as well (e.g., see Wakeley 2008; Presgraves and Yi 2009; Hvilsom et al. 2012; Yamamichi et al. 2012).

4.12.6 Pan—Common Chimpanzee Subspecies

As mentioned in the previous section, Wegmann and Excoffier (2010), in addition to testing for gene flow between the diverging bonobo and common chimpanzee lineages, tested for introgression among the various chimpanzee subspecies as well. Using the same microsatellite and sequence data, these authors were able to infer multiple instances of genetic exchange during the evolutionary history of *P. troglodytes* (Wegmann and Excoffier 2010). In particular, ancient migration (i.e., introgression) events were detected between the eastern (*Pan troglodytes schweinfurthii*) and central (*Pan troglodytes troglodytes*) and central and western (*Pan troglodytes verus*) subspecies (see also Caswell et al. 2008). Likewise, recent migration between the central and eastern subspecies was inferred. Similarly, a fourth subspecies, *Pan troglodytes ellioti*, whose geographic distribution extends from southern Nigeria to western Cameroon, is thought to form a contemporary hybrid zone with *P. t. troglodytes* (Gonder et al. 2011).

Interestingly, the ancient introgression between *P. t. troglodytes* and *P. t. verus* was strongly asymmetric with genes moving from the latter into the former subspecies (Wegmann and Excoffier 2010), a result agreeing with inferences by Hey (2010). However, Hey's (2010) inference was that the unidirectional introgressive hybridization had actually occurred from the western subspecies into the lineage that gave rise to both the eastern and central subspecies. It seems possible that this 'single' event might explain both the western–central and western–eastern migration events detected by Wegmann and Excoffier (2010).

4.12.7 Homo

Much has been written about the likelihood of introgressive hybridization between *H. sapiens* and other, archaic, species of *Homo*. Indeed, it is arguable that writings on this topic have been responsible for the decimation of numerous tropical forests harvested for paper pulp. For example, Jolly (2009), in a review of one of the senior author's books (Arnold 2008), took exception to the author's arguments for introgression between humans and their archaic sister taxa. He indicated this in the following way: 'By insisting, against the weight of evidence, that *H. sapiens* "must have" interbred with other human species, Arnold misses the opportunity to discuss the interesting paradox in these findings and the many questions arising from it. Hybridization among nonhuman primates and other mammals suggests that *H. sapiens* was, more likely than not, interfertile with any other member of the genus *Homo*, certainly including *Homo neanderthalensis*. Yet no recognizable genetic evidence for reticulation seems to exist' (Jolly 2009).

Jolly might be excused for such a strong stance given that the most recent studies—based upon whole-genome sequencing of archaic lineages (Green et al. 2010; Reich et al. 2010; Meyer et al. 2012)—were not available. However, a wealth of data *was* available previous to 2008, leading many workers other than Arnold to the inference of divergence-with-gene-flow affecting humans and related taxa (e.g., Ackermann et al. 2006; Hayakawa et al. 2006; Evans et al. 2006; Templeton 2007). Even more so at this time, to use Jolly's (2009) phraseology, we would have to argue 'against the weight of evidence' to infer a lack of divergence-with-introgression in the evolution of various *Homo* species.

It is important to note, however, that some recent analyses have agreed with an inference of no admixture between anatomically modern and archaic lineages from this genus. For example, Ghirotto et al. (2011), from an analysis of only mtDNA sequences from modern humans and Neanderthals, argued for a lack of introgression. Thus, there was no clustering of archaic and modern samples as might be expected given divergence-with-gene-flow as *H. sapiens* spread from its African point of origin (Ghirotto et al. 2011). Likewise, Blum and Jakobsson (2011) using 20 autosomal and 20 X-linked loci sequenced from extant human populations inferred a lack of genetic exchange between archaic and modern lineages. Specifically, these authors argued that the presence of 'ancient' alleles in present-day humans could be best explained by the demographic parameters of migrating *H. sapiens* rather than by introgression between this species and its sister taxa such as *H. neanderthalensis* (Blum and Jakobsson 2011; see also Eriksson and Manica 2012 for a similar inference; but see Wall et al. 2009; Alves et al. 2012; Yang et al. 2012 for a contrasting inference). Finally, Schwartz and Tattersall (2010) argued from fossil data that the range of variation in modern forms was more likely due to developmental variation rather than interbreeding between archaic and modern species (but see Ackermann 2010).

Notwithstanding the above arguments for replacement-without-introgression of archaic *Homo* lineages by the wave of advancing *H. sapiens*, to date there have been numerous data sets produced that falsify this hypothesis. Many of the earlier studies (i.e., before 2008) are discussed in Arnold (2008). Yet, some of the most powerful data sets—including the comparison of genes and whole genomes from both extant and extinct species of *Homo*—appeared in the subsequent five years (see Wood and Baker 2011 for a review). Thus, although modeling of the patterns of genetic variation found between genomes of archaic and anatomically modern lineages suggest restricted introgressive hybridization, possibly due to behavioral differences and/or low hybrid fitness (Currat and Excoffier 2011), such genetic exchange has been detected.

Recent analyses by Michael Hammer and his colleagues (in which they surveyed genomic variability in extant human populations) have supported the hypothesis of introgression from archaic species into *H. sapiens*. In particular, Hammer et al. (2011) and Mendez et al. (2012) detected genomic signatures of ancient alleles apparently introgressed from extinct species of *Homo* into the genomes of modern-day humans distributed in Africa and Melanesia, respectively. In the first of these analyses, the pattern of admixture was similar to that seen previously for Eurasian populations, but in this instance, the ancient alleles were exchanged between archaic and modern lineages living exclusively in Africa (Hammer et al. 2011). Likewise, Lachance et al. (2012) and Schlebusch et al. (2012) also detected patterns of genomic variation indicative of introgression among various *Homo* lineages. As with Hammer and his colleagues,

Lachance et al. (2012)—in an analysis of African hunter-gather populations—inferred archaic species x *H. sapiens* hybridization resulting in introgressed genomes belonging to Pygmy, Hadza, and Sandawe lineages. The analysis reported by Schlebusch et al. (2012) also provided evidence for introgression among African populations and indeed suggested a role for adaptive trait transfer (i.e., pigmentation genes). Specifically, there appears to have been introgression from southern African Bantu-speakers into the Nama resulting in the transfer and subsequent selection of alleles that provided greater UV protection for members of the recipient lineage (Schlebusch et al. 2012). All the cases of introgressive hybridization in African populations would have been facilitated by the cohabitation of various lineages of this clade (both archaic and anatomically modern) across long periods of time (Arnold 2008).

The detection by Mendez et al. (2012) of high similarity to an archaic haplotype (i.e., 'Denisovan' haplotype; Reich et al. 2010) and linkage disequilibrium in Melanesian populations of an ca. 90 kb stretch of DNA associated with the innate immune gene *OAS1* falsified the hypothesis of no introgression from an archaic lineage into Melanesian *H. sapiens*. These results substantiated those of Rasmussen et al. (2011) who detected a higher level of similarity between an Aboriginal Australian's genome sequences and those of Denisovans relative to Neanderthal also suggesting introgression from the former archaic lineage into a 'Melanesian' lineage of *H. sapiens*. Significantly, a portion of the divergence-with-gene-flow events has apparently resulted in adaptive trait introgression. In this regard, Abi-Rached et al. (2011) inferred that *HLA-B*73* of the major histocompatibility complex represented an introgressed allele from the Denisovan lineage. Furthermore, more than half the *HLA* alleles of modern Eurasians appear to have been donated by archaic sister lineages, with the high frequency in extant *H. sapiens* populations reflecting the significance of adaptive introgression of archaic alleles in the evolution of the human immune system (Abi-Rached et al. 2011).

Most of the studies discussed in this section (at least those since 2010) have compared their sequence/genotype data sets to the draft genome sequences taken from the DNA isolated from fossils of *H. neanderthalensis* and its sister taxon known as the Denisovans (Green et al. 2010; Reich et al. 2010, respectively). The work by Pääbo and his colleagues has thus revolutionized the way in which many of the long-standing hypotheses concerning human evolution—including adaptive evolution in the human lineage versus sister taxa such as chimpanzees and gorillas—can be tested. In the context of the subject of this review, these whole-genome data sets have once and for all given a means by which the hypothesis of no introgression during the evolution of *Homo* lineages can be tested rigorously. Indeed, the reports of the genome sequences of Neanderthals and Denisovans included a description of the pattern of genetic variation in these archaic genomes and those of extant humans that indicated on the one hand introgression from Neanderthals into Eurasian *H. sapiens* (Green et al. 2010) and on the other, introgression from Denisovans into Melanesian *H. sapiens* lineages (Reich et al. 2010; Meyer et al. 2012). In short, these data falsify the hypothesis of simple replacement of archaic forms by our species and instead favor a scenario of mutual attraction and genetic exchange leading to a human genome that is a mosaic of recent and ancient DNA sequences (Pääbo 2003).

5 Divergence-with-Gene-Flow—What Have We Learned and Where Might We Go from Here?

In retrospect, it would seem intuitively obvious that the dynamic, somewhat cyclical, and rapid (in evolutionary timescales) nature of environmental fluctuation would also lead to dynamic, somewhat cyclical, and rapid geographic range changes in diverging lineages. This should, in turn, have led to repeated opportunities for genetic exchange as the ranges of related—and not so related—organisms overlapped spatially and temporally. The conclusion then reached from this inference is that speciation that can be referred to as 'allopatric' is much less likely than what could be called 'parapatric' or 'sympatric'. Once again, it seems surprising that the canalization of evolutionary thought resulting from the Modern Synthesis would have formed such an impervious barrier to the investigation of how speciation may have occurred, at least with regard to the possibility of genetic exchange accompanying divergence. That this mind-set has carried through to the present time is evidenced by treatises such as that by Coyne and Orr (2004). The degree to which the support of both allopatric speciation and the biological species concept has rested on verbal arguments rather than data are also puzzling, leading one of us to argue that such inferences rest as much (or more) on philosophical rather than scientific grounds (Arnold 1997, 2006, 2008). If nothing else, the above examples of mammalian reticulate evolution, spanning large time periods and the most diverse clades, point to the fact that it was incorrect to conclude that hybridization played an unimportant role in animal evolution (Mayr 1963).

Thankfully, the data sets, along with a younger generation of scientists amenable to challenging the status quo hypotheses, are now available. The mass of genomic information being produced is staggering, but more is needed for whole clades of organisms such as those produced for the mammalian lineages discussed above. As such data become available, it will become more imperative that analyses that allow tests of reticulate versus bifurcating divergence, and that can manage enormous amounts of sequence information, continue to be developed. These types of data sets and analyses will indeed catalogue the frequency of genetic exchange events during the divergence of lineages. Of more importance, they will also allow tests of hypotheses of 'hybrid speciation' and 'adaptive introgression' thereby adding to our understanding of the affect of genetic exchange on both biodiversity and adaptive evolution. Indeed, such analyses have already been undertaken as reflected, for example, in the analyses of our own genus, the mice that we battle as pests, and the hares that provide us with food. These studies, incorporating as they have, the methodologies of coalescence, population genetics and genomics and ecological genetics have provided the most rigorous set of tests to date of the hypothesis of divergence-with-gene-flow.

As important as the data and data analysis methodologies will be in moving the study of speciation into areas not encouraged previously, philosophical changes will also be essential. Thus, students will need to continue to be exposed to hypotheses such as those found in Anderson (1949), Anderson and Stebbins (1954), Arnold (1992, 1997), Rieseberg et al. (2003), Seehausen (2004), and Arnold et al. (2012) to name but a few. Such exposure will prevent a continued canalization of how studies of speciation and evolutionary biology are pursued. This will then hopefully push our field of study into arenas that we do not yet perceive. This can only be a good thing given what has already been achieved in the past two decades in advancing our understanding of the pervasiveness of divergence-with-gene-flow.

Acknowledgments We wish to thank the senior editors, Nathalie Gontier and Olga Pombo, for the invitation to contribute this review and to participate in the International School on Evolution at the Centre for Philosophy of Science of the University of Lisbon during March 2013. MLA was supported by NSF grant DEB-0949479/0949424 (collaborative grant between MLA and NHM) and by a grant from the Chinese Academy of Sciences, Kunming Institute of Zoology (Kunming, PRC). ANB was supported by funds from the Office of the Vice President for Research at the University of Georgia.

Glossary

Allopatric divergence Independent evolution of populations that are completely geographically separated (Mayr 1942)

Allopolyploid hybrid speciation Hybrid speciation in which the stabilized hybrid lineage has an increase in chromosome number by one or more entire haploid set compared to the parental taxa (Stebbins 1947)

Allozymes Any of the variants of an enzyme that are determined by alleles at a single genetic locus (http://www.merriam-webster.com/medical/allozyme)

Biological species concept Species whose members include ‘a group of actually or potentially interbreeding populations’ (Dobzhansky 1937; Mayr 1942)

Coalescent analysis Genealogical methodology for determining the most recent common ancestral state of a gene, thus allowing the dating of evolutionary events such as introgression (Good et al. 2008)

Divergence-with-gene-flow Evolution of diverging populations with some amount of continued genetic exchange between them (Feder et al. 2012)

Ecotone A transitional area of vegetation between two different plant communities (http://www.britannica.com/EBchecked/topic/178617/ecotone)

Ecotype A population of a species that survives as a distinct group through environmental selection (http://www.merriam-webster.com/dictionary/ecotype)

Homoploid hybrid speciation Hybrid speciation in which the stabilized hybrid lineage has the same (or very similar) number of chromosomes as the parental taxa (Rieseberg 1997)

Hybrid swarm A group of hybrid individuals between divergent taxa often includes multiple generations of hybrid offspring and thus a great deal of phenotypic and genetic diversity (Grant 1981)

Hybrid zone Two populations of individuals that are distinguishable on the basis of one or more heritable characters overlap spatially and temporally and cross to form viable and at least partially fertile offspring (Arnold 1997)

Introgressive hybridization (introgression) Transfer of genetic material from one lineage to another through repeated backcrossing (Anderson and Hubricht 1938)

Neo-Darwinian (modern) synthesis The reconciling of Darwinian evolutionary theory with genetics using data from genetics, systematics, and paleontology (Futuyma 2009)

Next-generation sequencing Highly automated DNA sequencing technologies that generate large amounts of sequence data (Metzker 2010)

Parapatry Populations of taxa whose ranges are adjacent and overlap only along a narrow region (Futuyma 2009)

Restriction fragment length polymorphisms DNA fragments that vary in length due to differences in the location of restriction endonuclease cut sites (Avise 1994)

Reticulate evolution Evolution that includes genetic exchange between divergent lineages (Arnold 2006)

Retrovirus Single-stranded RNA viruses that produce reverse transcriptase by means of which DNA is produced using their RNA as a template and incorporated into the genome of infected cells (http://www.merriam-webster.com/dictionary/retrovirus)

Sympatry Populations of taxa occur in the same geographic region (Futuyma 2009)

References

Abi-Rached L, Jobin MJ, Kulkarni S, McWhinnie A, Dalva K, Gragert L et al (2011) The shaping of modern human immune systems by multiregional admixture with archaic humans. Science 334:89–94

Ackermann RR (2010) Phenotypic traits of primate hybrids: recognizing admixture in the fossil record. Evol Anthro 19:258–270

Ackermann RR, Bishop JM (2010) Morphological and molecular evidence reveals recent hybridization between gorilla taxa. Evolution 64:271–290

Ackermann RR, Rogers J, Cheverud JM (2006) Identifying the morphological signatures of hybridization in primate and human evolution. J Hum Evol 51:632–645

Ackermann RR, Brink JS, Vrahimis S, de Klerk B (2010) Hybrid wildebeest (Artiodactyla: Bovidae) provide further evidence for shared signatures of admixture in mammalian crania. S Afr J Sci 106, Art. #423

Alves PC, Melo-Ferreira J, Freitas H, Boursot P (2008) The ubiquitous mountain hare mitochondria: multiple introgressive hybridization in hares, genus *Lepus*. Philos T R Soc B 363:2831–2839

Alves PC, Pinheiro I, Godinho R, Vicente J, Gortázar C, Scandura M (2010) Genetic diversity of wild boar populations and domestic pig breeds (*Sus scrofa*) in South-western Europe. Biol J Linn Soc 101:797–822

Alves I, Hanulová AS, Foll M, Excoffier L (2012) Genomic data reveal a complex making of humans. PLoS Genet 8(7):e1002837

Anderson E (1949) Introgressive hybridization. Wiley, New York

Anderson E, Hubricht L (1938) Hybridization in *Tradescantia*. III. The evidence for introgressive hybridization. Am J Bot 25:396–402

Anderson E, Stebbins GL Jr (1954) Hybridization as an evolutionary stimulus. Evolution 8:378–388

Arnold ML (1992) Natural hybridization as an evolutionary process. Annu Rev Ecol Syst 23:237–261

Arnold ML (1997) Natural hybridization and evolution. Oxford University Press, New York

Arnold ML (2006) Evolution through genetic exchange. Oxford University Press, Oxford

Arnold ML (2008) Reticulate evolution and humans—origins and ecology. Oxford University Press, Oxford

Arnold ML, Fogarty ND (2009) Reticulate evolution and marine organisms: the last frontier? Inter J Mol Sci 10:3836–3860

Arnold ML, Hodges SA (1995) Are natural hybrids fit or unfit relative to their parents? Trends Ecol Evol 10:67–71

Arnold ML, Meyer A (2006) Natural hybridization in primates: one evolutionary mechanism. Zoology 109:261–276

Arnold ML, Hamlin JP, Brothers AN, Ballerini ES (2012) Natural hybridization as a catalyst of rapid evolutionary change. In: Singh RS, Xu J, Kulathinal RJ (eds) Rapidly evolving genes and genetic systems. Oxford University Press, Oxford

Avise JC (1994) Molecular markers, natural history and evolution. Chapman & Hall, New York

Baird AB, Hillis DM, Patton JC, Bickham JW (2009) Speciation by monobrachial centric fusions: a test of the model using nuclear DNA sequences from the bat genus *Rhogeessa*. Mol Phyl Evol 50:256–267

Barton NH, Hewitt GM (1985) Analysis of hybrid zones. Annu Rev Ecol Syst 16:113–148

Bérubé M, Aguilar A (1998) A new hybrid between a blue whale, *Balaenoptera musculus*, and a fin whale, *B. physalus*: frequency and implications of hybridization. Mar Mam Sci 14:82–98

Bicca-Marques JC, Prates HM, Cunha de Aguiar FR, Jones CB (2008) Survey of *Alouatta caraya*, the black-and-gold howler monkey, and *Alouatta guariba clamitans*, the brown howler monkey, in a contact zone, State of Rio Grande do Sul, Brazil: evidence for hybridization. Primates 49:246–252

Bifolchi A, Picard D, Lemaire C, Cormier JP, Pagan A (2010) Evidence of admixture between differentiated genetic pools at a regional scale in an invasive carnivore. Conserv Genet 11:1–9

Blum MGB, Jakobsson M (2011) Deep divergences of human gene trees and models of human origins. Mol Biol Evol 28:889–898

Bulatova N, Jones RM, White TA, Shchipanov NA, Pavlova SV, Searle JB (2011) Natural hybridization between extremely divergent chromosomal races of the common shrew (*Sorex araneus*, Soricidae, Soricomorpha): hybrid zone in European Russia. J Evol Biol 24:573–586

Burrell AS, Jolly CJ, Tosi AJ, Disotell TR (2009) Mitochondrial evidence for the hybrid origin of the kipunji, *Rungwecebus kipunji* (Primates: Papionini). Mol Phyl Evol 51:340–348

Cabria MT, Michaux JR, Gomez-Moliner BJ, Skumatov D, Maran T, Fournier P et al (2011) Bayesian analysis of hybridization and introgression between the endangered European mink (*Mustela lutreola*) and the polecat (*Mustela putorius*). Mol Ecol 20:1176–1190

Carneiro M, Ferrand N, Nachman MW (2009) Recombination and speciation: loci near centromeres are more differentiated than loci near telomeres between subspecies of the European rabbit (*Oryctolagus cuniculus*). Genetics 181:593–606

Carneiro M, Blanco-Aguiar JA, Villafuerte R, Ferrand N, Nachman MW (2010) Speciation in the European rabbit (*Oryctolagus cuniculus*): islands of differentiation on the X chromosome and autosomes. Evolution 64:3443–3460

Carneiro M, Alfonso A, Geraldes A, Garreau H, Bolet G, Boucher S et al (2011) The genetic structure of domestic rabbits. Mol Biol Evol 28:1801–1816

Caswell JL, Mallick S, Richter DJ, Neubauer J, Schirmer C, Gnerre S et al (2008) Analysis of chimpanzee history based on genome sequence alignments. PLoS Genet 4:e1000057

Caughley G, Brown B, Dostine P, Grice D (1984) The grey kangaroo overlap zone. Aust J Wildl Res 11:1–10

Cornelis G, Heidmann O, Bernard-Stoecklin S, Reynaud K, Véron G, Mulot B et al (2012) Ancestral capture of *syncytin-Car1*, a fusogenic endogenous retroviral envelope gene involved in placentation and conserved in Carnivora. Proc Natl Acad Sci USA 109:E432–E441

Coulson G, Coulson R (2001) A natural grey kangaroo hybrid? Aust Zool 31:599–602

Cox MP, Mendez FL, Karafet TM, Pilkington MM, Kingan SB, Destro-Bisol G et al (2008) Testing for archaic hominin admixture on the X chromosome: model likelihoods for the modern human *RRM2P4* region from summaries of genealogical topology under the structured coalescent. Genetics 178:427–437

Coyne JA, Orr HA (2004) Speciation. Sinauer Associates, Inc., Sunderland

Crestanello B, Pecchioli E, Vernesi C, Mona S, Martínková N, Janiga M et al (2009) The genetic impact of translocations and habitat fragmentation in chamois (*Rupicapra*) spp. J Hered 100:691–708

Currat M, Excoffier L (2011) Strong reproductive isolation between humans and Neanderthals inferred from observed patterns of introgression. Proc Natl Acad Sci USA 108:15129–15134

Darwin C (1859) On the origin of species by means of natural selection or the preservation of favoured races in the struggle for life. John Murray, London

Davenport TRB, Stanley WT, Sargis EJ, De Luca DW, Mpunga NE, Machaga SJ et al (2006) A new genus of African monkey, *Rungwecebus*: morphology, ecology, and molecular phylogenetics. Science 312:1378–1381

Delmore KE, Louis EE Jr, Johnson SE (2011) Morphological characterization of a brown lemur hybrid zone (*Eulemur rufifrons* x *E. cinereiceps*). Am J Phys Anthro 145:55–66

Dobzhansky Th (1937) Genetics and the origin of species. Columbia University Press, New York

Dobzhansky Th (1940) Speciation as a stage in evolutionary divergence. Am Nat 74:312–321

Dubey S, Diker E, Kurtonur C, Vogel P (2008) Secondary contact zones and hybridizations: the case of the lesser white-toothed shrew (*Crocidura suaveolens* group, Soricidae). Biol J Linn Soc 95:557–565

Edwards CJ, Suchard MA, Lemey P, Welch JJ, Barnes I, Fulton TL et al (2011) Ancient hybridization and an Irish origin for the modern polar bear matriline. Curr Biol 21:1251–1258

Endler JA (1973) Gene flow and population differentiation. Science 179:243–250

Endler JA (1977) Geographic variation, speciation, and clines. Princeton University Press, Princeton

Eriksson A, Manica A (2012) Effect of ancient population structure on the degree of polymorphism shared between modern human populations and ancient hominins. Proc Natl Acad Sci USA 109:13956–13960

Evans PD, Mekel-Bobrov N, Vallender EJ, Hudson RR, Lahn BT (2006) Evidence that the adaptive allele of the brain size gene *microcephalin* introgressed into *Homo sapiens* from an archaic *Homo* lineage. Proc Natl Acad Sci USA 103:8178–8183

Fabricius C, Lowry D, van den Berg P (1988) Fecund black wildebeest x blue wildebeest hybrids. S Afr J Wildl Res 18:35–37

Feder JL, Egan SP, Nosil P (2012) The genomics of speciation-with-gene-flow. Trends Genet 28:342–350

Futuyma DJ (2009) Evolution, 2nd edn. Sinauer Associates, Inc., Sunderland

Gagneux P (2004) A Pan-oramic view: insights into hominoid evolution through the chimpanzee genome. Tr Ecol Evol 19:571–576

Ghirotto S, Tassi F, Benazzo A, Barbujani G (2011) No evidence of Neandertal admixture in the mitochondrial genomes of early European modern humans and contemporary Europeans. Am J Phys Anthro 146:242–252

Gilbert CC, Stanley WT, Olson LE, Davenport TRB, Sargis EJ (2011) Morphological systematics of the kipunji (*Rungwecebus kipunji*) and the ontogenetic development of phylogenetically informative characters in the Papionini. J Hum Evol 60:731–745

Gligor M, Ganzhorn JU, Rakotondravony D, Ramilijaona OR, Razafimahatratra E, Zischler H et al (2009) Hybridization between mouse lemurs in an ecological transition zone in southern Madagascar. Mol Ecol 18:520–533

Glover KA, Kanda N, Haug T, Pastene LA, Øien N, Goto M et al (2010) Migration of Antarctic minke whales to the Arctic. PLoS ONE 5:e15197

Gompert Z, Buerkle CA (2009) A powerful regression-based method for admixture mapping of isolation across the genome of hybrids. Mol Ecol 18:1207–1224

Gonder MK, Locatelli S, Ghobrial L, Mitchell MW, Kujawski JT, Lankester FJ et al (2011) Evidence from Cameroon reveals differences in the genetic structure and histories of chimpanzee populations. Proc Natl Acad Sci USA 108:4766–4771
Good JM, Hird S, Reid N, Demboski JR, Steppan SJ, Martin-Nims TR et al (2008) Ancient hybridization and mitochondrial capture between two species of chipmunks. Mol Ecol 17:1313–1327
Grant V (1981) Plant speciation. Columbia University Press, New York
Green RE, Krause J, Briggs AW, Maricic T, Stenzel U, Kircher M et al (2010) A draft sequence of the Neandertal genome. Science 328:710–722
Hailer F, Kutschera VE, Hallström BM, Klassert D, Fain SR, Leonard JA et al (2012) Nuclear genomic sequences reveal that polar bears are an old and distinct bear lineage. Science 336:344–347
Hallström BM, Janke A (2010) Mammalian evolution may not be strictly bifurcating. Mol Biol Evol 27:2804–2816
Hammer MF, Woerner AE, Mendez FL, Watkins JC, Wall JD (2011) Genetic evidence for archaic admixture in Africa. Proc Natl Acad Sci USA 108:15123–15128
Hayakawa T, Aki I, Varki A, Satta Y, Takahata N (2006) Fixation of the human-specific CMP-*N*-acetylneuraminic acid hydroxylase pseudogene and implications of haplotype diversity for human evolution. Genetics 172:139–146
Hennig W (1966) Phylogenetic systematics. University of Illinois Press, Urbana
Hey J (2010) The divergence of chimpanzee species and subspecies as revealed in multipopulation isolation-with-migration analyses. Mol Biol Evol 27:921–933
Hird S, Sullivan J (2009) Assessment of gene flow across a hybrid zone in red-tailed chipmunks (*Tamias ruficaudus*). Mol Ecol 18:3097–3109
Hofbauer P, Bauer F, Paulsen P (2006) Meat of chamois—a note on quality traits of the m. longissimus of chamois (*Rupicapra rupicapra* L.) in Austrian sub-alpine regions. Fleischwirtschaft 86:100–102
Hoffmann FG, Owen JG, Baker RJ (2003) mtDNA perspective of chromosomal diversification and hybridization in Peters' tent-making bat (*Uroderma bilobatum*: Phyllostomidae). Mol Ecol 12:2981–2993
Horn A, Basset P, Yannic G, Banaszek A, Borodin PM, Bulatova NS et al (2012) Chromosomal rearrangements do not seem to affect the gene flow in hybrid zones between karyotypic races of the common shrew (*Sorex araneus*). Evolution 66:882–889
Hvilsom C, Qian Y, Bataillon T, Li Y, Mailund T, Sallé B et al (2012) Extensive X-linked adaptive evolution in central chimpanzees. Proc Natl Acad Sci USA 109:2054–2059
Janousek V, Wang L, Luzynski K, Dufková P, Vyskocilová MM, Nachman MW et al (2012) Genome-wide architecture of reproductive isolation in a naturally occurring hybrid zone between *Mus musculus musculus* and *M. m. domesticus*. Mol Ecol 21:3032–3047
Johnson WE, Pecon-Slattery J, Eizirik E, Kim J, Raymond MM, Bonacic C et al (1999) Disparate phylogeography patterns of molecular genetic variation in four closely related South American small cat species. Mol Ecol 8(Suppl):S79–S94
Johnson WE, Onorato DP, Roelke ME, Land ED, Cunningham M, Belden RC et al (2010) Genetic restoration of the Florida panther. Science 329:1641–1645
Jolly CJ (2001) A proper study for mankind: analogies from the papionin monkeys and their implications for human evolution. Year Phys Anthrop 44:177–204
Jolly CJ (2009) Mixed signals: reticulation in human and primate evolution. Evol Anthro 18:275–281
Jolly CJ, Burrell AS, Phillips-Conroy JE, Bergey C, Rogers J (2011) Kinda baboons (*Papio kindae*) and grayfoot chacma baboons (*P. ursinus griseipes*) hybridize in the Kafue River Valley, Zambia. Am J Primatol 73:291–303
Jones EP, Van Der Kooij J, Solheim R, Searle JB (2010) Norwegian house mice (*Mus musculus musculus/domesticus*): distributions, routes of colonization and patterns of hybridization. Mol Ecol 19:5252–5264

Karanth KP (2008) Primate numts and reticulate evolution of capped and golden leaf monkeys (Primates: Colobinae). J Biosci 33:761–770
Karanth KP (2010) Molecular systematics and conservation of the langurs and leaf monkeys of South Asia. J Genet 89:393–399
Karanth KP, Singh L, Collura RV, Stewart C-B (2008) Molecular phylogeny and biogeography of langurs and leaf monkeys of South Asia (Primates: Colobinae). Mol Phyl Evol 46:683–694
Kelly B, Whiteley A, Tallmon D (2010) The Arctic melting pot. Nature 468:891
Kidd AG, Bowman J, Lesbarreres Schulte-Hostedde AI (2009) Hybridization between escaped domestic and wild American mink (*Neovison vison*). Mol Ecol 18:1175–1186
Kingston SE, Adams LD, Rosel PE (2009) Testing mitochondrial sequences and anonymous nuclear markers for phylogeny reconstruction in a rapidly radiating group: molecular systematics of the Delphininae (Cetacea: Odontoceti: Delphinidae). BMC Evol Biol 9:245
Kirsch JAW, Pool WE (1972) Taxonomy and distribution of the grey kangaroos, *Macropus giganteus* Shaw and *Macropus fuliginosus* (Desmarest), and their subspecies (Marsupialia: Macropodidae). Aust J Zool 20:315–339
Lachance J, Vernot B, Elbers CC, Ferwerda B, Froment A, Bodo J-M et al (2012) Evolutionary history and adaptation from high-coverage whole-genome sequences of diverse African hunter-gatherers. Cell 150:457–469
Lack JB, Greene DU, Conroy CJ, Hamilton MJ, Braun JK, Mares MA et al (2012) Invasion facilitates hybridization with introgression in the *Rattus rattus* species complex. Mol Ecol 21:3545–3561
Lancaster ML, Goldsworthy SD, Sunnucks P (2007) Multiple mating strategies explain unexpected genetic mixing of New Zealand fur seals with two congenerics in a recently recolonized population. Mol Ecol 16:5267–5276
Larsen PA, Marchán-Rivadeneira MR, Baker RJ (2010) Natural hybridization generates mammalian lineage with species characteristics. Proc Natl Acad Sci USA 107:11447–11452
Latch EK, Kierepka EM, Heffelfinger JR, Rhodes OE Jr (2011) Hybrid swarm between divergent lineages of mule deer (*Odocoileus hemionus*). Mol Ecol 20:5265–5279
LeDuc RG, Perrin WF, Dizon AE (1999) Phylogenetic relationships among the Delphinid cetaceans based on full Cytochrome *B* sequences. Mar Mammal Sci 15:619–648
Liu J, Yu L, Arnold ML, Wu C-H, Wu S-F, Lu X et al (2011) Reticulate evolution: frequent introgressive hybridization among Chinese hares (genus Lepus) revealed by analyses of multiple mitochondrial and nuclear DNA loci. BMC Evol Biol 11:223
Lodé T, Guiral G, Peltier D (2005) European mink–polecat hybridization events: hazards from natural process? J Hered 96:89–96
Mao X, Zhang J, Zhang S, Rossiter SJ (2010) Historical male-mediated introgression in horseshoe bats revealed by multilocus DNA sequence data. Mol Ecol 19:1352–1366
Marco A, Marín I (2009) *CGIN1*: a retroviral contribution to mammalian genomes. Mol Biol Evol 26:2167–2170
Mayr E (1942) Systematics and the origin of species. Columbia University Press, New York
Mayr E (1963) Animal species and evolution. Belknap Press, Cambridge
McCue ME, Bannasch DL, Petersen JL, Gurr J, Bailey E, Binns MM et al (2012) A high density SNP array for the domestic horse and extant Perissodactyla: utility for association mapping, genetic diversity, and phylogeny studies. PLoS Genet 8:e1002451
McDevitt AD, Mariani S, Hebblewhite M, Decesare NJ, Morgantini L, Seip D et al (2009) Survival in the Rockies of an endangered hybrid swarm from diverged caribou (*Rangifer tarandus*) lineages. Mol Ecol 18:665–679
Melo-Ferreira J, Alves PC, Freitas H, Ferrand N, Boursot P (2009) The genomic legacy from the extinct *Lepus timidus* to the three hare species of Iberia: contrast between mtDNA, sex chromosomes and autosomes. Mol Ecol 18:2643–2658
Melo-Ferreira J, Boursot P, Carneiro M, Esteves PJ, Farelo L, Alves PC (2012) Recurrent introgression of mitochondrial DNA among hares (*Lepus* spp.) revealed by species-tree inference and coalescent simulations. Syst Biol 61:367–381

Mendez FL, Watkins JC, Hammer MF (2012) Global genetic variation at *OAS1* provides evidence of archaic admixture in Melanesian populations. Mol Biol Evol 29:1513–1520

Messier W, Stewart C-B (1997) Episodic adaptive evolution of primate lysozymes. Nature 385:151–154

Metzker ML (2010) Sequencing technologies—the next generation. Nat Rev Genet 11:31–46

Meyer M, Kircher M, Gansauge M-T, Li H, Racimo F, Mallick S et al (2012) A high-coverage genome sequence from an archaic Denisovan individual. Science 338:222–226

Miller W, Schuster SC, Welch AJ, Ratan A, Bedoya-Reina OC, Zhao F et al (2012) Polar and brown bear genomes reveal ancient admixture and demographic footprints of past climate change. Proc Natl Acad Sci USA 109:E2382–E2390

Mittermeier RA, Ganzhorn JU, Konstant WR, Glander K, Tattersall I, Groves CP et al (2008) Lemur diversity in Madagascar. Int J Primatol 29:1607–1656

Moore WS (1977) An evaluation of narrow hybrid zones in vertebrates. Quart Rev Biol 52:263–277

Morgan JAT, Harry AV, Welch DJ, Street R, White J, Geraghty PT et al (2012) Detection of interspecies hybridisation in Chondrichthyes: hybrids and hybrid offspring between Australian (*Carcharhinus tilstoni*) and common (*C. limbatus*) blacktip shark found in an Australian fishery. Conserv Genet 13:455–463

Neaves LE, Zenger KR, Cooper DW, Eldridge MDB (2010) Molecular detection of hybridization between sympatric kangaroo species in south-eastern Australia. Heredity 104:502–512

Newbound CN, Hisheh S, Suyanto A, How RA, Schmitt LH (2008) Markedly discordant mitochondrial DNA and allozyme phylogenies of tube-nosed fruit bats, Nyctimene, at the Australian-Oriental biogeographical interface. Biol J Linn Soc 93:589–602

Olson LE, Sargis EJ, Stanley WT, Hildebrandt KBP, Davenport TRB (2008) Additional molecular evidence strongly supports the distinction between the recently described African primate *Rungwecebus kipunji* (Cercopithecidae, Papionini) and *Lophocebus*. Mol Phyl Evol 48:789–794

Osada N, Uno Y, Mineta K, Kameoka Y, Takahashi I, Terao K (2010) Ancient genome-wide admixture beyond the current hybrid zone between *Macaca fascicularis* and *M. mulatta*. Mol Ecol 19:2884–2895

Pääbo S (2003) The mosaic that is our genome. Nature 421:409–412

Pastorini J, Zaramody A, Curtis DJ, Nievergelt CM, Mundy NI (2009) Genetic analysis of hybridization and introgression between wild mongoose and brown lemurs. BMC Evol Biol 9:32

Patterson N, Richter DJ, Gnerre S, Lander ES, Reich D (2006) Genetic evidence for complex speciation of humans and chimpanzees. Nature 441:1103–1108

Payseur BA, Krenz JG, Nachman MW (2004) Differential patterns of introgression across the X chromosome in a hybrid zone between two species of house mice. Evolution 58:2064–2078

Pease KM, Freedman AH, Pollinger JP, McCormack JE, Buermann W, Rodzen J et al (2009) Landscape genetics of California mule deer (*Odocoileus hemionus*): the roles of ecological and historical factors in generating differentiation. Mol Ecol 18:1848–1862

Polyakov AV, White TA, Jones RM, Borodin PM, Searle JB (2011) Natural hybridization between extremely divergent chromosomal races of the common shrew (*Sorex araneus*, Soricidae, Soricomorpha): hybrid zone in Siberia. J Evol Biol 24:1393–1402

Presgraves DC, Yi SV (2009) Doubts about complex speciation between humans and chimpanzees. Tr Ecol Evol 24:533–540

Prüfer K, Munch K, Hellmann I, Akagi K, Miller JR, Walenz B et al (2012) The bonobo genome compared with the chimpanzee and human genomes. Nature 486:527–531

Pryde D (1971) Nunaga—ten years of Eskimo life. Walker and Company, New York

Rakotondranary SJ, Hapke A, Ganzhorn JU (2011) Distribution and morphological variation of *Microcebus* spp. along an environmental gradient in southeastern Madagascar. Int J Primatol 32:1037–1057

Ramírez O, Ojeda A, Tomàs A, Gallardo D, Huang LS, Folch JM et al (2009) Integrating Y-chromosome, mitochondrial, and autosomal data to analyze the origin of pig breeds. Mol Biol Evol 26:2061–2072

Rasmussen M, Guo X, Wang Y, Lohmueller KE, Rasmussen S, Albrechtsen A et al (2011) An Aboriginal Australian genome reveals separate human dispersals into Asia. Science 334:94–98

Reich D, Green RE, Kircher M, Krause J, Patterson N, Durand EY et al (2010) Genetic history of an archaic hominin group from Denisova Cave in Siberia. Nature 468:1053–1060

Reid N, Demboski JR, Sullivan J (2012) Phylogeny estimation of the radiation of western North American chipmunks (*Tamias*) in the face of introgression using reproductive protein genes. Syst Biol 61:44–62

Rieseberg LH (1997) Hybrid origins of plant species. Annu Rev Ecol Syst 28:359–389

Rieseberg LH, Raymond O, Rosenthal DM, Zhao L, Livingstone K, Nakazato T et al (2003) Major ecological transitions in wild sunflowers facilitated by hybridization. Science 301:1211–1216

Rivals F, Deniaux B (2005) Investigation of human hunting seasonality through dental microwear analysis of two Caprinae in late Pleistocene localities in southern France. J Archae Sci 32:1603–1612

Rodríguez F, Hammer SE, Pérez T, Suchentrunk F, Lorenzini R, Michallet J et al (2009) Cytochrome b phylogeography of Chamois (*Rupicapra spp.*). Population contractions, expansions and hybridizations governed the diversification of the genus. J Hered 100:47–55

Rodríguez F, Pérez T, Hammer SE, Albornoz J, Domínguez A (2010) Integrating phylogeographic patterns of microsatellite and mtDNA divergence to infer the evolutionary history of chamois (genus Rupicapra). BMC Evol Biol 10:222

Roos C, Zinner D, Kubatko LS, Schwarz C, Yang M, Meyer D et al (2011) Nuclear versus mitochondrial DNA: evidence for hybridization in colobine monkeys. BMC Evol Biol 11:77

Scally A, Dutheil JY, Hillier LW, Jordan GE, Goodhead I, Herrero J et al (2012) Insights into hominid evolution from the gorilla genome sequence. Nature 483:169–175

Scandura M, Iacolina L, Crestanello B, Pecchioli E, Di Benedetto MF, Russo V et al (2008) Ancient vs. recent processes as factors shaping the genetic variation of the European wild boar: are the effects of the last glaciation still detectable? Mol Ecol 17:1745–1762

Scandura M, Iacolina L, Cossu A, Apollonio M (2011) Effects of human perturbation on the genetic make-up of an island population: the case of the Sardinian wild boar. Heredity 106:1012–1020

Schlebusch CM, Skoglund P, Sjödin P, Gattepaille LM, Hernandez D, Jay F et al (2012) Genomic variation in seven Khoe-San groups reveals adaptation and complex African history. Science 338:374–379

Schwartz JH, Tattersall I (2010) Fossil evidence for the origin of *Homo sapiens*. Year Phys Anthro 53:94–121

Seehausen O (2004) Hybridization and adaptive radiation. Trends Ecol Evol 19:198–207

Short J, Caughley G, Grice D, Brown B (1983) The distribution and abundance of kangaroos in relation to environment in Western Australia. Aust J Wildl Res 10:435–451

Singleton M (2009) The phenetic affinities of *Rungwecebus kipunji*. J Hum Evol 56:25–42

Slatkin M (1985) Gene flow in natural populations. Annu Rev Ecol Syst 16:393–430

Stapley J, Reger J, Feulner PGD, Smadja C, Galindo J, Ekblom R et al (2010) Adaptation genomics: the next generation. Trends Ecol Evol 25:705–712

Staubach F, Lorenc A, Messer PW, Tang K, Petrov DA, Tautz D (2012) Genome patterns of selection and introgression of haplotypes in natural populations of the house mouse (*Mus musculus*). PLoS Genet 8:e1002891

Stebbins GL Jr (1947) Types of polyploids: their classification and significance. Adv Genet 1:403–429

Stebbins GL Jr (1950) Variation and evolution in plants. Columbia University Press, New York

Stevison LS, Kohn MH (2009) Divergence population genetic analysis of hybridization between rhesus and cynomolgus macaques. Mol Ecol 18:2457–2475

Stuckas H, Stoof K, Quesada H, Tiedemann R (2009) Evolutionary implications of discordant clines across the Baltic Mytilus hybrid zone (*Mytilus edulis* and *Mytilus trossulus*). Heredity 103:146–156

Templeton AR (2007) Genetics and recent human evolution. Evolution 61:1507–1519
Ting N, Tosi AJ, Li Y, Zhang Y-P, Disotell TR (2008) Phylogenetic incongruence between nuclear and mitochondrial markers in the Asian colobines and the evolution of the langurs and leaf monkeys. Mol Phyl Evol 46:466–474
Trigo TC, Freitas TRO, Kunzler G, Cardoso L, Silva JCR, Johnson WE et al (2008) Inter-species hybridization among Neotropical cats of the genus *Leopardus*, and evidence for an introgressive hybrid zone between *L. geoffroyi* and *L. tigrinus* in southern Brazil. Mol Ecol 17:4317–4333
Wakeley J (2008) Complex speciation of humans and chimpanzees. Nature 452:E3–E4
Wall JD, Lohmueller KE, Plagnol V (2009) Detecting ancient admixture and estimating demographic parameters in multiple human populations. Mol Biol Evol 26:1823–1827
Warmuth V, Eriksson A, Bower MA, Barker G, Barrett E, Hanks BK et al (2012) Reconstructing the origin and spread of horse domestication in the Eurasian steppe. Proc Natl Acad Sci USA 109:8202–8206
Weckworth BV, Musiani M, McDevitt AD, Hebblewhite M, Mariani S (2012) Reconstruction of caribou evolutionary history in western North America and its implications for conservation. Mol Ecol 21:3610–3624
Wegmann D, Excoffier L (2010) Bayesian inference of the demographic history of chimpanzees. Mol Biol Evol 27:1425–1435
White TA, Bordewich M, Searle JB (2010) A network approach to study karyotypic evolution: the chromosomal races of the common shrew (*Sorex araneus*) and house mouse (*Mus musculus*) as model systems. Syst Biol 59:262–276
Wood B, Baker J (2011) Evolution in the genus *Homo*. Annu Rev Ecol Evol Syst 42:47–69
Wright S (1931) Evolution in Mendelian populations. Genetics 16:97–159
Wu Y, Xia L, Zhang Q, Yang Q, Meng X (2011) Bidirectional introgressive hybridization between *Lepus capensis* and *Lepus* yarkandensis. Mol Phyl Evol 59:545–555
Yamamichi M, Gojobori J, Innan H (2012) An autosomal analysis gives no genetic evidence for complex speciation of humans and chimpanzees. Mol Biol Evol 29:145–156
Yang MA, Malaspinas A-S, Durand EY, Slatkin M (2012) Ancient structure in Africa unlikely to explain Neanderthal and non-African genetic similarity. Mol Biol Evol 29:2987–2995
Yannic G, Basset P, Hausser J (2008a) A hybrid zone with coincident clines for autosomal and sex-specific markers in the Sorex araneus group. J Evol Biol 21:658–667
Yannic G, Basset P, Hausser J (2008b) A new perspective on the evolutionary history of western European Sorex araneus group revealed by paternal and maternal molecular markers. Mol Phyl Evol 47:237–250
Yannic G, Dubey S, Hausser J, Basset P (2010) Additional data for nuclear DNA give new insights into the phylogenetic position of *Sorex granarius* within the *Sorex araneus* group. Mol Phyl Evol 57:1062–1071
Zinner D, Arnold ML, Roos C (2009) Is the new primate genus *Rungwecebus* a baboon? PLoS ONE 4:e4859

A Multiset Model of Multi-Species Evolution to Solve Big Deceptive Problems

Luís Correia and António Manso

Abstract This chapter presents symbiogenetic multiset genetic algorithm (SMuGA), an integration of symbiogenesis with the multiset genetic algorithm (MuGA). The symbiogenetic approach used here is based on the host–parasite model with the novelty of varying the length of parasites along the evolutionary process. Additionally, it models collaborations between multiple parasites and a single host. To improve efficiency, we introduced proxy evaluation of parasites, which saves fitness function calls and exponentially reduces the symbiotic collaborations produced. Another novel feature consists of breaking the evolutionary cycle into two phases: a symbiotic phase and a phase of independent evolution of both hosts and parasites. SMuGA was tested in optimization of a variety of deceptive functions, with results one order of magnitude better than state-of-the-art symbiotic algorithms. This allowed to optimize deceptive problems with large sizes and showed a linear scaling in the number of iterations to attain the optimum.

Keywords Genetic algorithm · Multisets · Symbiogenesis · Deceptive optimization problems

1 Introduction

Computational models of coevolution can be used to study both natural settings and artificial scenarios. Moreover, they can solve optimization problems. Computational models are an effective tool configurable to model different types

L. Correia (✉) · A. Manso
BioISI—Biosystems and Integrative Sciences Institute, Faculty of Science, University of Lisbon, Campo Grande, 1749-016 Lisbon, Portugal
e-mail: Luis.Correia@ciencias.ulisboa.pt

A. Manso
Polytechnic Institute of Tomar, Quinta do Contador, Estrada da Serra, 2300-313 Tomar, Portugal
e-mail: manso@ipt.pt

N. Gontier (ed.), *Reticulate Evolution*, Interdisciplinary Evolution Research 3,
DOI 10.1007/978-3-319-16345-1_11

of multi-species evolution: parasitism, commensalism, mutualism, and cooperative interactions. Competitive multi-species evolution has been useful in optimization applications as it provides better results when compared to a single problem-solver population. Coevolution of a solver population with a problem creator population pushes both populations to increasingly better solutions, a phenomenon called arms-race (Rosin and Belew 1997).

On the other hand, symbiosis is a form of cooperative coevolution, which has been gaining relevance in biology (Daida et al. 1996). In artificial systems, symbiogenetic coevolution has been shown to improve evolutionary optimization algorithms by a specialization of the different components of the symbiotic collaboration (Wallin et al. 2005). In this case of cooperative coevolution, there is a kind of division of labor between the different types of symbionts. Each host is combined with a set of parasites forming a collaboration. Each collaboration is evaluated as a solution to the optimization problem. This is repeated for different hosts and parasites. Artificial symbiogenetic evolution is proving useful in solving deceptive problems (Wallin et al. 2005), a class of functions that is especially difficult to optimize due to the fact that the optimum is surrounded by regions of low-quality solutions.

Artificial evolutionary models are inspired by nature, but when used as engineering tools they do not need to maintain a strict correspondence with their natural counterparts. The main goal of engineering was to obtain efficient tools, in this case designed to solve optimization problems. Taking this into account, we further explore different approaches of evolutionary algorithms (EA) and their operators that one may consider unrealistic by comparison to nature. The multiset representation of populations is one of those examples and in previous work we have used that representation to support the EA populations (Manso and Correia 2009). That algorithm is called multiset genetic algorithm (MuGA) and is successful in the optimization of various kinds of problems. The populations are represented by multisets and the operators that are used explore the representation in order to make the evolutionary process more efficient and effective in the optimization of difficult problems.

In this work, we present the symbiogenetic multiset genetic algorithm (SMuGA), which uses natural inspiration of symbiogenesis to solve large deceptive problems that are not solved by the common version of MuGA.

In the next section, we present the base algorithm of MuGA. In the following section, the symbiogenetic approach used is detailed. In particular we have two different evolutionary processes, one for the hosts and another for the parasites, and we describe each one separately and then aggregated. Next, we present results obtained in several types of deceptive functions. The final section of this chapter presents conclusion and proposes future work.

2 MuGA—A Multiset Genetic Algorithm

MuGA is a genetic algorithm that explores the features of a multiset to represent populations of EA and to improve their performance. The traditional representation of populations used in EA raises two types of problems: the loss of genetic diversity during the evolutionary process and the evaluation of redundant individuals. These problems can be alleviated when using multisets to represent populations.

Multiset population is not a representation that can be found in the natural world, but it works well for optimization of difficult engineering problems.

2.1 Populations Represented by Multisets

A multiset (or multiple memberships set) is a collection of objects, called elements, which are allowed to repeat. We can define the multiset as a set of ordered pairs ⟨***copies, element***⟩ where copies are the cardinality associated with the element. MuGA is a genetic algorithm in which populations represented by multisets are called Multipopulations (MP) and individuals represented by pairs ⟨***copies, genotype***⟩ are called Multi-individuals (MI).

Figure 1 shows a simple population (SP) with eight individuals of OnesMax problem and the equivalent MP with four MI. A multiset representation of populations contains characteristics that make it a good alternative to the collections that are usually used:

- MI has always different genotypes and the size of MP corresponds to the genotype diversity at the genotypic level;
- The number of copies of MI may be used to control the selection pressure in favor of the best fit individuals;
- The compact representation needs less computational effort to store the population and avoids evaluation of identical individuals.

The introduction of individuals in a MP is done either by incrementing the number of copies of corresponding MI if the genotype exists in the population or by introducing a new pair ⟨***1, genotype***⟩. The elimination is done by decrementing the number of copies of corresponding MI if the number of copies is greater than one, or otherwise by removing the MI.

(a)

Individual	Fitness
11111110	7
11111110	7
11111110	7
11110000	4
11110000	4
10001001	3
10001001	3
10000000	1

(b)

MultiIndividual	Fitness
< 3, 11111110 >	7
< 2, 11110000 >	4
< 2, 10001000 >	3
< 1, 10000000 >	1

Fig. 1 **a** Simple population of 8 individuals; **b** Multipopulation of 4 multi-individuals

2.2 MuGA Algorithm

In EA, populations are traditionally represented as a collection of individuals. To minimize the issues such models raise, we developed MuGA (Algorithm 1), whose most distinctive feature is that it represents populations by multisets.

The algorithm starts by randomly generating and evaluating ***n*** individuals of the ***problem*** to be optimized, while assuring that the base population, MP0, contains ***n*** different genotypes. The design of the MuGA is prepared to preserve the genetic diversity by maintaining the dimension of MP0 across generations. The evolutionary process starts by selecting ***m*** individuals from MP0. These ***m*** individuals are stored in MP1 and the number of MI is less than or equal to ***m***. The process continues with the recombination of MP1 and subsequent mutation of MP2, generating MP3. MP4 is produced by the application of the replacement operator on MP0 and MP3 to select ***n*** MI from the two populations. This operator maintains the number of MI as a constant across generations. The evolutionary process tends to produce many copies of good individuals. To reduce the number of copies in MP4, the rescaling operator is applied and produces a new population (MP0) for the iterative evolutionary process.

Algorithm 1 MuGA—Multiset Genetic Algorithm

```
MuGA (n , m , problem)
  MP0 = generate n MultiIndividuals from problem
  Evaluate MP0
  Repeat
     MP1 = Select m Individuals from MP0
     MP2 = Recombine the Individuals of MP1
     MP3 = Mutate the Individuals of MP2
     Evaluate MP3
     MP4 = Select n MultiIndividuals from MP3 and MP0
     MP0 = Rescale the number of copies of MP4
  Until stop criteria
End Function.
```

Multipopulations enable the execution of traditional genetic operators and allow the design of new operators using the extra information, a set of unique genotypes and associated number of copies, to extend operators that benefit from such information. Next, we briefly describe the behavior of genetic operators using MPs.

2.3 Multiset Selection

This operator chooses, from the base population, the parents that will be reproduced to generate new individuals. We first expand the MP to an SP, Fig. 1, so that MI with multiple copies has higher probability of being selected. We can then use traditional selection operators (tournament selection, proportional selection, or ranking selection) or any improved selection operator (Sivaraj and Ravichandran 2011). When the operator allows the selection of the same individual several times over, the mating population will contain MI and the number of copies will reflect the degree of fitness of the genotype. The number of copies of the fittest individuals tends to be larger than the remaining elements and can be explored by the subsequent genetic operators.

2.4 Multiset Recombination

The recombination operator is responsible for the combination of chromosomes to produce offspring that share genetic material of both parents. There is a great variety of recombination operators in accordance with the representation of the genes and chromosomes (e.g., binary strings, vectors of real numbers or trees) of individuals and the type of problem to be solved, e.g., optimization of real functions (Herrera et al. 2003), permutations (Otman and Jaafar 2011), or combinatorial (Spears and Anand 1991). All these operators can be used in MuGA through equivalence between MP and SP in terms of genotype representation. Nevertheless, we can design new operators using the number of copies to make a genotype associated with the various parameters of the genetic algorithm such as the probability of application, the number of cutting points, and the strength of individuals to spread their genes. A wide range of possibilities is available to explore the usefulness of this information, and (Manso and Correia 2011) presents a multiset recombination operator applied to the optimization of real-coded functions.

2.5 Multiset Mutation

The mutation operator in EA mimics what occurs in nature and randomly changes a (usually small) part of the genome. The main function of this operator is the introduction of new genes, enabling exploration of new areas in the search space that are not attainable by the recombination of parental characteristics. Like the recombination operator, mutation is also dependent on the type of problem and representation of the individuals (Abdoun et al. 2012; Droste et al. 2002). A new operator that uses multiset information to optimize deceptive binary functions, called multiset wave mutation (MWM), is presented in Manso and Correia (2013) and another one used to optimize real-coded functions is presented in Manso and Correia (2011).

2.6 Multiset Replacement

After recombination and mutation, the evolutionary algorithm has two populations of individuals: the main population and the offspring generated by genetic operators. The replacement operator selects which individuals will continue in the evolutionary process. The generational strategy replaces the parents with their children and the steady-state strategy replaces only a few parents with offspring (Lozano et al. 2008). The operator must maintain the genetic diversity in the main population so that the genetic operators can circumvent local optima and avoid premature convergence (Yu and Suganthan 2010; Jayachandran and Corns 2010). A new operator that uses multiset information to replace populations in a steady-state strategy, called multiset decimation replacement (MDR), is presented in Manso and Correia (2013).

2.7 Multiset Rescaling

The introduction of repeated elements in the MP tends to increase the number of copies of the best fit MI if nothing is done to oppose it.

The rescaling operator was proposed to avoid that the best individuals get too many copies (Manso and Correia 2009). In order to control the number of repeated elements, the rescaling operator divides the number of copies of each MI by a factor, controlling in this way the pressure exhibited by the fittest individuals. The operator ensures that each MI has at least one copy and that the total number of individuals in the MP is not greater than a constant. An adaptive form of this operator, called adaptive rescaling (AR), calculates in each iteration the value of the reduction factor to maintain approximately the desired number of individuals.

3 SMuGA—A Symbiogenetic Multiset Genetic Algorithm

Symbiosis is set of natural theories that try to explain the natural relationship between individuals that live together and how that relationship is vital to the survival of the group. In nature, symbiosis occurs and involves a relationship that is constant and intimate between dissimilar species (Daida et al. 1996). That relationship is more than the ecological interaction and includes mutualism, where both individuals gain advantages from the alliance; commensalism, in which one individual gains advantages and the other does not have any inconvenience; and parasitism, where one individual gains advantages and the other is harmed by the relation.

Symbiosis theory provides an additional genetic operator to the artificial evolutionary process and is successfully applied to solve a wide range of hard problems. See Heywood and Lichodzijewski (2010) for a review of symbiogenesis as a mechanism to build complex adaptive systems.

The SMuGA is inspired by the symbiogenetic coevolutionary algorithm (SCA), proposed by Wallin et al. in 2005, which explores a host–parasite relationship for optimization of concatenated deceptive functions. Although the names "hosts" and "parasites" suggest a parasitic relationship, the interaction between two species is benign and the gains of parasites are not harmful to the hosts. SCA is successfully used to optimize concatenated deceptive functions and MuGA by itself has proved to be an efficient algorithm in the optimization of such functions with a moderate size (Manso and Correia 2013).

However, when the size of the problems increases, MuGA experiences difficulties in its optimization. In this chapter, we apply the concept of symbiosis to increase the efficiency of the MuGA. SMuGA is an algorithm that uses two cooperative species, hosts and parasites, which evolve together in a mutualistic relationship. The parasites are composed of a tuple ⟨*position, genome*⟩, where the position represents the parasite genome location where the parasite acts, and the genome represents the genetic material of the parasite. In SMuGA, the host genome is replaced by the genome of the parasite in the location defined by the position attribute (Fig. 2). The parasite considers the host genome as a circle, which means that when the copy of the parasite genome to the host reaches the limit of the host genome, the copy continues in the beginning. In Fig. 2, parasite ***p1*** is applied in host genome alleles 4, 5, and 6 and parasite ***p2*** is applied in the host genome alleles 9 and 0. The collaboration is the combination of host genes and the genes introduced by parasites p1 and p2.

SCA has some deficiencies identified by the authors. The size of the parasites is static and defined as a parameter, and collaboration is from one parasite to one host, where each host can only be infected by a parasite at a time. The best results obtained by the algorithm are when the parasite genome size is similar to the size of the functions to be optimized, the building blocks (BB), and the performance degrades quickly as the size of the parasites deviates from the size of the BB. Another weakness of the SCA is that the collaboration is one to one, which limits its applicability to separable problems.

The SMuGA was designed to suppress these two shortcomings by combining the concept of symbiosis with the potential that the populations based on multisets present on the optimization of this kind of functions. In the next section, we present the representation and evolution of parasite populations, the evolution of host populations, and the interaction between them with SMuGA. In the design of the SMuGA, some choices are made with the objective of enhancing the success of the algorithm in the optimization of problems and contouring the shortcomings that SCA presents.

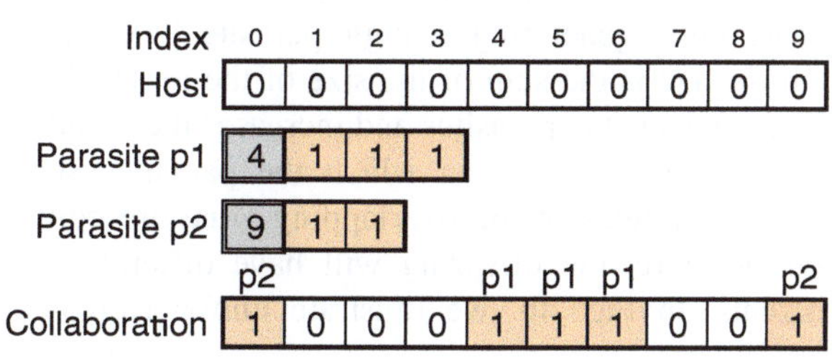

Fig. 2 Collaboration formed by the symbiosis of a host and a parasite

3.1 Evolution of Parasites

In order to avoid having a human choice interfere significantly in performance, we eliminate the need to specify the size of the parasites. As mentioned earlier, the work of Wallin et al. (2005) showed that there was a very strong dependence of performance relative to the size of the parasite. When the size of parasites approaches BB size, the performance is good; however, it decays very quickly with deviations from the ideal dimension.

In our approach, the user does not have to know the size of BB because the algorithm adapts the parasite's length as necessary. To the best of our knowledge, this is the first model of parasites that may vary their length along the evolutionary process. This system is important for solving problems in which the size of BB is not known or the BB has a variable size. The size of the parasites is changed by genetic operators of recombination and mutation. The selection operator gives opportunity to parasites that have a good performance in the host population to reproduce and to pass on their genetic material and position to their descendants. According to the theory of survival of the fittest, the parasites with a good genome, which includes the position of application and the genetic material, will spread their genes to subsequent generations, discovering and optimizing simultaneously the position, the size, and alleles of the parasites.

3.1.1 Parasite Recombination

The following four situations can occur when two parasites recombine:

1. The parasites do not share positions in the genome of the host;
2. The parasites occupy consecutive positions in the genome of the host;
3. The parasites share some positions in the host; and
4. All positions of one of the parasites occupy positions of the other.

In the first case, as the parasites infect different regions of the host genome, recombination between the two parasites cannot take place. In all other cases, the idea underlying this operator is not only to recombine genetic material but also to introduce different genome lengths. We selected the recombination of parasite genomes as the principal operator to grow and shrink the length of the parasites.

In the second case, Fig. 3, in which the parasites occupy consecutive locations in the host genome, we determine that recombinant parasites are the union of genomes generating a single parasite. The offspring ***o1***, Fig. 3c, has a genome whose size is the sum of the size of the parental genomes. This type of reproduction connects the parasites and increases the length of the parasite genome.

In case 3, Fig. 4, in which the parasites share some positions in the host genome, alleles in the overlapping zone are combined using uniform crossover. Furthermore, the offspring will have different genome sizes compared to their parents. In Fig. 4b, we illustrate uniform crossover. A recombination mask is

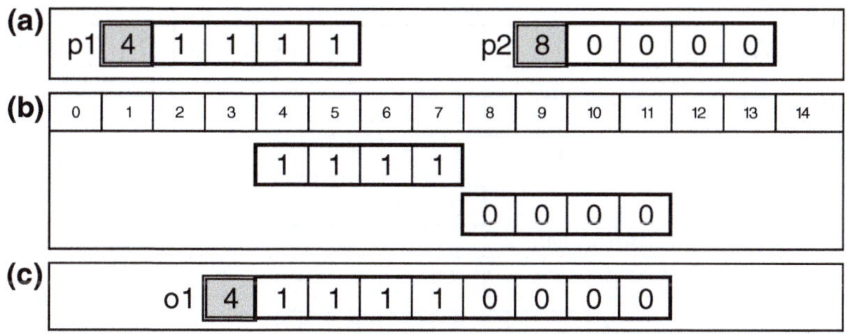

Fig. 3 Recombination by the union of consecutive parasites: **a** selected parasites; **b** positions occupied by parasites in the genome; **c** result of p1 and p2 recombination

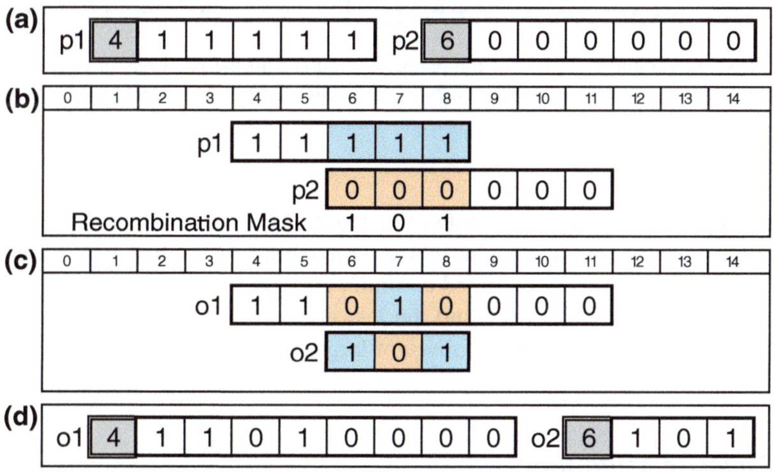

Fig. 4 Recombination by the share of some positions in the host: **a** selected parasites; **b** positions occupied by parasites in the genome; **c** result of p1 and p2 recombination; **d** offspring parasites

randomly obtained to perform an exchange of the parental alleles in the overlapping zone. The symbol ***1*** in the mask means that there is an exchange of alleles in the overlapping zone and the symbol ***0*** means the opposite. Figure 4c, d shows the recombination result of parents ***p1*** and ***p2***. The offspring ***o1*** inherits from both parents the parts that are not common between them, as well as the recombined genome produced by the recombination mask. The offspring ***o2*** inherits only the recombined common part with a dual mask. The offspring ***o1*** is longer than the parents and ***o2*** is shorter.

In case 4, Fig. 5, where one of the parasites, ***p1***, occupies all the positions of the other, ***p2***, in the genome of the host, the overlapping zone is also recombined using uniform crossover. As in the previous case, the genetic material is exchanged in the overlapping zone through a recombination mask, Fig. 5b generated from a

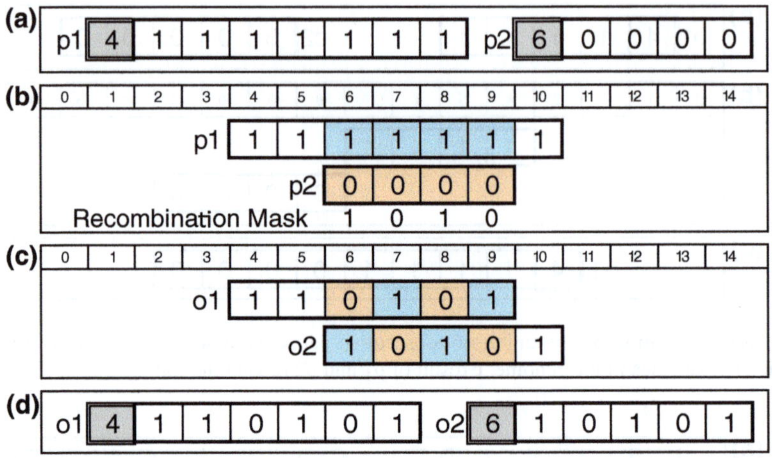

Fig. 5 Recombination when one of the parasites occupies all the positions of other: **a** selected parasites; **b** positions occupied by parasites in the genome; **c** result of p1 and p2 recombination; **d** offspring parasites

uniform distribution. Figure 5d shows the result of the recombination and the complete offspring. Individual ***o1*** inherits from the parent ***p1*** the first part not common to both parents, and the recombined common part, and the offspring ***o2*** inherits the dual recombined common part, and the last not common part of ***p1***. In this case, the small parasites act as cutting knives of larger parasites.

3.2 Parasite Mutation

The mutation operator randomly changes features of a parasite. These features include the position, length, and their genetic material. We use three types of parasite mutation:

1. Change in anchoring position;
2. Change in the genome; and
3. Parasite genome splitting whereby two new parasites are formed.

In the first situation, parasites change the position of host infection. In Fig. 6a, the parasite ***p1*** that infects the fourth position generates the mutant ***m1*** infecting position 10 with the same genotype. Note that the parasite ***m1*** affects the host genome in a circular way where the last three bits of the parasite infect the first three positions of the host.

In the second case, the value of the alleles is changed by a probability distribution that generates the mutation mask shown in Fig. 7b. At the positions where the mask has the value ***1***, the bit value of parasite is flipped. In this situation, only

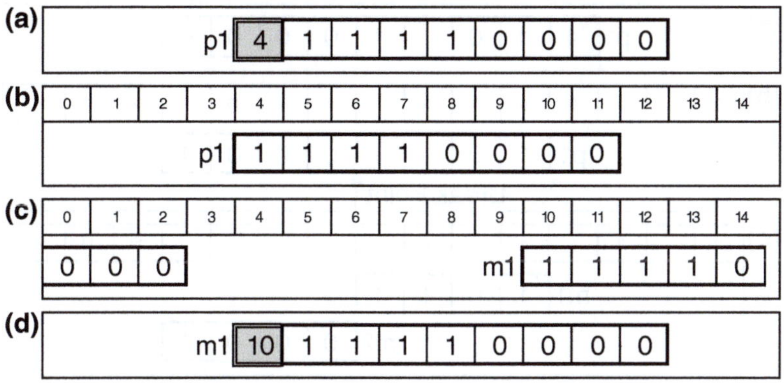

Fig. 6 Mutation by changing position: **a** original parasite; **b** positions occupied by original; **c** positions occupied by mutant parasite; **d** mutant parasite

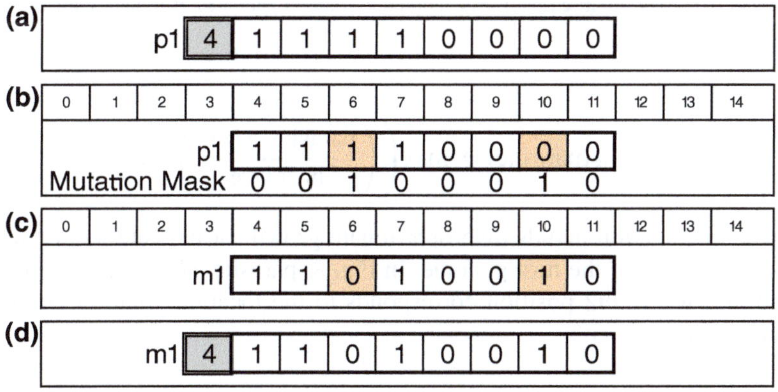

Fig. 7 Mutation by changing genome: **a** original parasite; **b** positions occupied by original and mutation mask; **c** positions occupied by mutant parasite; **d** mutant parasite

the value of the parasite's genome is modified, which enhances the appearance of parasites in the population with new genomes.

In the latter situation, the parasite genome is split into two parts, originating into two new parasites. The probability to split a genome is proportional to its length in bits.

Equation (1) shows the formula to calculate the probability of parasite splitting. Parameter k controls the dimension from which the splitting of a parasite is inevitable, i.e., when the ratio is greater than one; Parameter n controls the shape of distribution probability of splitting in other cases. The genotype split point is selected by a uniform probability distribution over the genotype of the parasite.

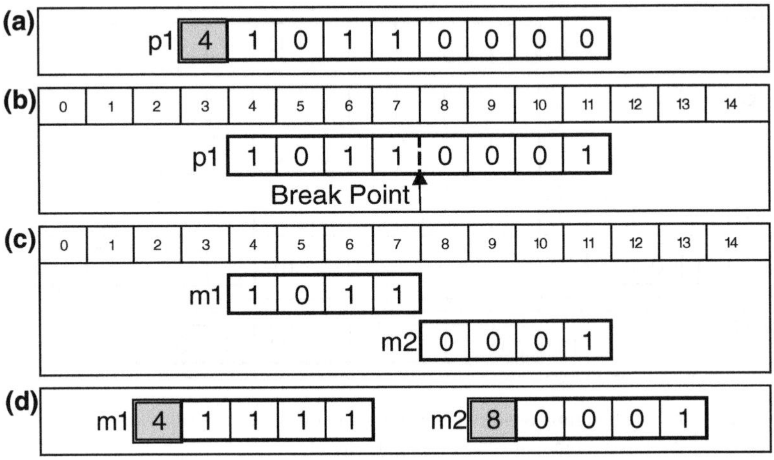

Fig. 8 Mutation by breaking genome: **a** original parasite; **b** positions occupied by original parasite and the break point; **c** positions occupied by mutant parasites; **d** mutant parasites

$$p_{\text{break}}(\text{parasite}) = \max\left(\left(\frac{\text{parasite} \cdot \text{size}^k}{\text{host.size}}\right)^n, 1\right) \tag{1}$$

This type of mutation avoids disproportionate growth of parasite length and possible subsumption of the host genome. In Fig. 8, parasite ***p1*** creates two parasites, ***m1*** and ***m2***, where ***m2*** position corresponds to the location splitting point of the parasite genome p1.

3.3 Evaluation of Parasites

The evaluation of the population of parasites is obtained indirectly through the genomes present in the population of hosts. This feature allows the parasites to be evaluated without the need to apply them to the hosts and then call the fitness function to evaluate the collaboration. In this way, we replace fitness function calls by a proxy consisting of simply checking whether the parasite genome is present in the host genome and using the host fitness rank. Therefore, we significantly save function fitness calls as well as computational resources that would be spent on testing and generating collaborations.

We defined three goals for the parasites:

1. Promoting the emergence of parasites with new genetic material, necessary for the evolution of the combined population and prevention of its stagnation;
2. Promoting the dissemination of parasites with good genotypes in the host population so that all individuals have the parasite;

3. Promoting the variability of the anchoring point of good parasites in the host genome in order to allow different regions to be infected.

The last two goals are incompatible with the first, since it involves the destruction of the original genetic material. Also, the evaluation function should promote growth of the parasite length to speed up the evolutionary process to discover large BB, and therefore we made the value of parasite fitness directly proportional to its size.

In addition, the evaluation function of the parasites must be independent from the scale of the fitness values in the hosts. To accomplish this, hosts are sorted with a descending rank and parasites use those ranks to compute their evaluation. The parasite evaluation algorithm sums the ranks of the hosts that have the parasite in their genome. If the host rank is defined in the interval $[1, n]$, where n is the rank of fittest host and 1 the worst, parasites that infected the entire population have maximum fitness value. Their contribution to diversification of the population is zero, contrary to goal 1, nevertheless they are good candidates for dissemination, goals 2 and 3. To circumvent this obstacle, we shifted the rank of the hosts to the interval $[-n/2 - 1, n/2]$ where n is the size of the population. This shift in ranking of the population provides a number of significant advantages. First of all, the fitness of parasites that infect the entire population is zero; parasites present only in the best individuals have positive fitness, and by opposition, parasites that are present only in worst individuals have negative fitness.

In order to reward individuals with a large genome, the value of the sum of ranks is multiplied by the size of the parasite. Thus, if a parasite has a positive sum of ranks, its size is rewarded; otherwise, its size contributes to the decrease of its fitness. Such evaluation makes the discovery of a good parasite to be valuable at the beginning, thereby promoting its spreading, and as it infects the population through successive generations, its interest fades because the population has already assimilated its genome. This parasite evaluation is very efficient because it does not use a single call to the fitness function.

When evolution discovers a new parasite, whose genotype does not exist in the population, the evaluation function should reward its discovery with a fitness that allows it to survive and reproduce if it is a good parasite. On the other hand, the length of a new parasite should decrease its fitness to prevent the emergence of large parasites with random genomes that contrast with large parasites evolved from good BB. We decided to assign the new parasite a fitness value equal to the population size divided by its length in bits, as a reward for the discovery of new parasite genomes. The evaluation function allows small parasites with new genotypes to appear in the population and to recombine themselves with existing ones, thereby promoting their growth if they contain useful genetic material for evolution.

3.4 Algorithm of Parasites Evolution

The evolution of parasites is done by Algorithm 2. The algorithm receives as parameters the population of parasites to evolve, ***pPop***, the population of hosts to perform the evaluation of the parasites, ***hPop***, and the number of parasites that will be selected to evolve, ***n***.

Algorithm 2 Parasite Evolution Algorithm

```
ParasiteEvolution (pPop, hPop, n)
  selectPop = select n parasites from pPop
  offspringPop = recombine selectedPop
  while offspringPop.size < pPop.size
       Select random parasite from offspringPop
       Mutate a clone of parasite
       Insert mutated parasite clone in offspringPop
  End while
  Evaluate offspring in hPop
  pPop = select pPop.size parasites from pPop and
                                             offspringPop
End Function.
```

The algorithm starts by selecting ***n*** parasites from ***pPop***. It continues with the recombination of the selected population giving rise to the population ***offspring-Pop***. This step recombines genetic material of selected parasites and changes the length of the offspring with the rules described above. The population ***offspring-Pop*** is constructed by removing a pair of individuals from the selected population, applying the recombination algorithm to the parents and inserting the offspring in ***offspringPop*** population. The algorithm continues completing ***offspringPop*** through successive mutations of clones of randomly selected individuals in off-springPop. One of the three types of mutation described above, genomic mutation, position mutation, and genome splitting, is randomly applied with uniform probability. This way of completing a population allows a parasite to undergo several mutations in a single generation, because a mutant parasite can be selected and cloned several times.

The population ***offspringPop*** is evaluated through the genes of individuals of the population ***hPop***. The algorithm terminates with the calculation of a new population through replacement operator applied to the original ***pPop*** and to the population of its descendants, the ***offspringPop***.

3.5 Evolution of Hosts

A population of hosts is evolved with a MuGA, Algorithm 1, that uses some genetic operators adapted to MP. The adaptation of genetic operators to use the number of copies is critical to MuGA being able to solve difficult problems. MuGA uses standard operators of selection and recombination and an adapted form of mutation and replacement operators. In the next section, we describe the adaptations made in operators to take advantages of the number of copies present in MI of MuGA populations.

4 MWM—Multiset Wave Mutation

To solve problems where the solution cannot be found by a recombination of parent genes, the mutation operator performs a critical mission to introduce new genes into the population. Mutation introduces random changes in the genome of the individuals. Usually the operator introduces small changes in the genome of the individual and the new features acquired are propagated in the population through generations. A high rate of mutation is required if the changes to escape from local maxima include many alleles but it is harmful if this assumption does not happen. MI in multiset populations represents a set of clones of the same genotype on which we apply different mutation rates.

$$\text{waveFunction}(\text{copy}) = \left\lceil \frac{\sin\left(\frac{\pi}{2} + \frac{\text{copy}-1}{\text{roughness}}\right)}{2} \right\rceil^{\text{thinness}} \tag{2}$$

Equation (2) presents a waveFunction formula to calculate the probability of mutation from each clone of the MI that produces values between 0 and 1 (Fig. 9). When the mutation value reaches the value 1, all the bits are changed and that feature is very important to optimize deceptive functions where, usually, the optimum is the complement of the local maxima.

Algorithm 3 Multiset Mutation Algorithm

```
MultisetWaveMutation (MI, minProb , mutOperator)
  mutants = empty Multipopulation
  For copy = 1 to MI.copies
    probabilty = min ( minProb + waveFunction(copy), 1)
    individual = MI.genotype
    mutOperator( individual , probability )
    mutants.add( individual )
  next copy
  return mutants
End Function.
```

MWM Algorithm 3, fully explained in Manso and Correia (2013), was designed to apply a traditional mutation operator, ***mutOperator,*** to a multi-individual, ***MI***, using the waveFunction to calculate the probability of mutation of each clone. The probability is calculated adding a minimal probability, ***minProb***, to the result of waveFunction and truncating the result to 1 if the sum is greater than 1. Mutation in the offspring population is brought about by applying Algorithm 3 to every MI present in the population.

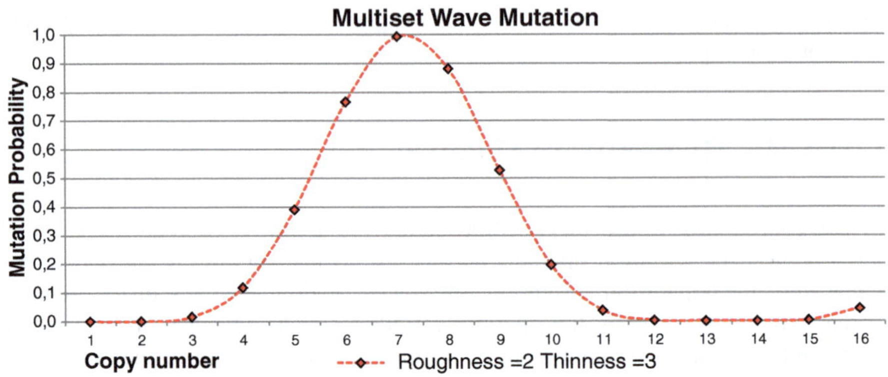

Fig. 9 Graph of wave function with roughness = 2 and thinness = 3

5 MDR—Multiset Decimation Replacement

The replacement operator has the task of forming the population that will continue the evolutionary process. This operator selects from parents and offspring MP which individuals are selected to continue the evolutionary process.

Algorithm 4 Multiset Decimation Algorithm

```
MultisetDecimation (parentsPop,offspingPop , n)
  parentsSize = parentsPop.size
  parentsPop = parentsPop + offspringPop
  while parentsPop.size > parentsSize
      tournament = select n random MultiIndividuals
                                    from parentsPop
      selected = weakest MultiIndividual in tournament
      remove selecteded from parentsPop
  end while
End Function.
```

MDR operator, Algorithm 4, was designed to replace the parent population with an offspring population in a steady-state approach maintaining the multiset characteristics of MI present in both populations. MDR joins the offspring population with the parent population and the individuals with the same genotype increase their number of copies. The algorithm then selects a group of random MI and removes the weakest. This procedure is repeated until the parent population is reduced to the same number of MI of the original population.

5.1 Co-evolution of Hosts and Parasites

The SMuGA is an evolutionary algorithm that uses two cooperating populations to solve difficult problems: The host population that contains solutions of the problem, and the parasite population that helps the first to reach the best solution. Parasite populations evolve to achieve good genes that represent partial solutions, and infect hosts through the incorporation of those genes.

The interaction between hosts and parasites produces a new population using symbiosis that mimics what occurs in the natural world. We define collaboration as the result of a host infected by one or more parasites using symbiosis.

5.1.1 Collaboration

Collaboration is obtained by copying the alleles of the parasite into the host. In this case, the alleles of the host are replaced by those of the parasite.

A collaboration of a parasite with a host is only allowed if the host does not have all the bits of the parasite, Fig. 10a. This means that a parasite can infect a host only once, Fig. 10b. This detail allows the elimination of collaborations that do not add anything new and clears space for collaborations that do modify something in the host.

We restrict the application of multiple parasites to cases where parasites do not have incompatible bits. This means that the parasites may overlap, provided that the overlapping segment does not contain different bits.

In Fig. 11a, parasites ***p1*** and ***p2*** can infect host ***h*** because they infect disjoint regions. In Fig. 11b, parasites ***p1*** and ***p3*** can infect the host ***h*** because, although they share two genes, they have the same value and therefore the infection causes no ambiguity. In Fig. 11c, parasites ***p1*** and ***p2*** cannot be used simultaneously because they overlap in two genes, one of which has distinct alleles. In this case, the host can be infected by any of them but not by both simultaneously.

Algorithm 5 controls the formation of collaborations among a population of hosts and a population of parasites. Algorithm 5 takes as parameters a host multi-population, ***sortedHostPop***, sorted in descending order, a parasite population, ***parasitePop,*** and a parameter ***n*** that controls the probability of infection. The order of the population is important because the index of the host in a population determines the probability of the host receiving parasites. The algorithm continues with the definition of the population resulting from the collaboration, ***symbPop***, among populations that are passed as a parameter. Afterward, the hosts are selected sequentially and the probability of infection is calculated. As hosts are MI, the algorithm proceeds to expand into clones and applies parasites to each one of them

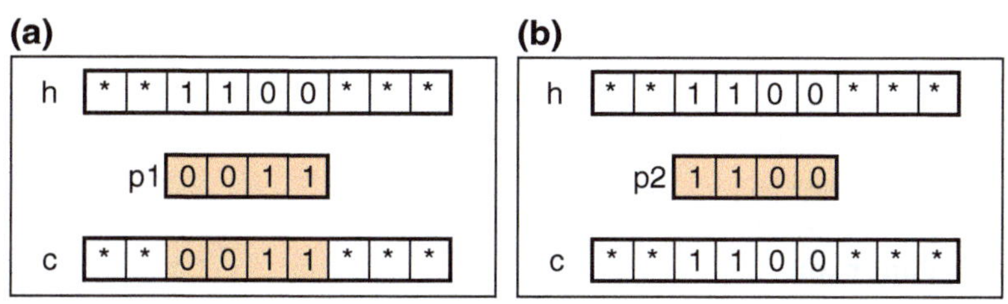

Fig. 10 Collaboration between one host and one parasite: **a** successful collaboration; **b** collaboration rejected

independently. Individuals with a higher ranking are those that usually make more copies and thus may suffer various combinations of parasites.

After selecting a host and calculating a probability of infection, the algorithm continues with the application of parasites to each of its clones. The parasites are randomly arranged within the population of parasites to ensure no preference in its application. In the next step, the algorithm tries to apply each parasite to the host selected using the compatibility rules of Fig. 11. In order to preserve the good individuals of the population from a generalized infection, and hence the sudden change of its genome, parasites are applied in a probabilistic manner. A host is particularly vulnerable to parasites when its rank in the population is smaller. This allows the fittest individuals to receive few parasites, thereby preserving their genes, and lower ranked individuals are subject to a generalized infection accommodating several parasites. This process is similar to that described in (Dumeur 1996).

$$p_{\text{infection}}(h, \text{pop}) = \left(\frac{\text{rank}(h, \text{pop})}{\text{pop.size}}\right)^n \tag{3}$$

Equation (3) shows the formula to calculate the probability of a parasite infecting a host, ***h,*** contained within a population, ***pop***. The ***rank*** function returns the rank of the individual within the population, in descending order of fitness and ***pop*** size represents the number of hosts that the population has. Parameter ***n*** controls the shape of the ratio described above.

Algorithm 5 Collaboration between Hosts and Parasites

```
Collaboration (parasitesPop, sortedHostsPop, n)
  symPop = empty MultiPopulation
  for index = 1 to sortedHostsPop.size
    host = sortedHostsPop.get(index)
    pInfection = (index / hosts.size)^n
    for copy = 1 to host.numberOfCopies
       symbiosis = host.genotype
       randomize parasites in parasitesPop
       foreach parasite in parasitesPop
          if compatible(parasite, symbiosis) and
                 uniformRandom(0,1) < pInfection
             symbiosis = symbiosis + parasite
             add symbiosis.clone to symbPop
          end if
       next parasite
    next copy
  next index
  return symbPop
End Function.
```

The symbiosis population is built by the infection of selected parasites into the host genomes. When a parasite is applied to the host, the genome of the parasite is

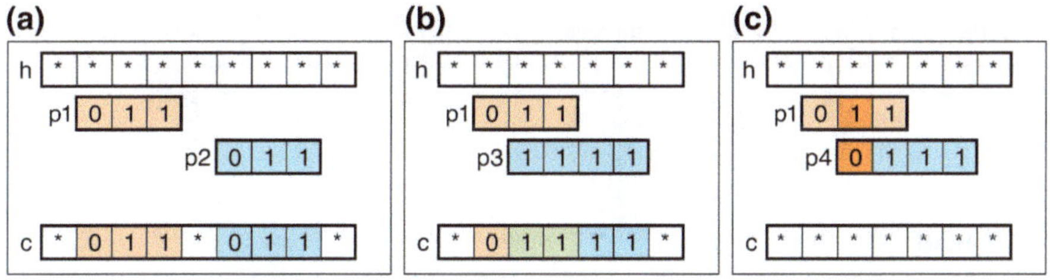

Fig. 11 Infection of a host by two parasites: **a** non-overlapping parasites; **b** compatible overlapping parasites; **c** incompatible overlapping parasites

copied to the genome of the host generating a new individual through symbiosis. A clone of that collaboration is added to the population of symbiosis, and the symbiosis continues the process of being infected by other parasites.

6 SMuGA—Symbiogenetic Multiset Genetic Algorithm

SMuGA, Algorithm 6, uses multipopulations to represent the populations of hosts and parasites. This representation enables the use of multiset-adapted genetic operators in both populations to help the evolutionary process. The use of multipopulations is required to optimize deceptive problems, and every challenging problem has a degree of deception (Whitley 1991). This algorithm has two phases: the collaboration phase, where the parasites infect the hosts; and the evolution phase, where hosts and parasites evolve using coevolution.

Algorithm 6 SMuGA—Symbiogenetic Multiset Genetic Algorithm

```
SMuGA (h, p, problem, iterations, k, n)
  hPop = generate h MultiIndividuals from problem
  Evaluate hPop
  pPop = generate p MultiParasites from problem
  Evaluate pPop with hPop
  Repeat
     /* Collaboration phase */
     selPop = Select k hosts from hPop
     symbPop = collaboration( pPop, selPop, n)
     hPop = Select h hosts from symbPop and hPop
     /* Evolution phase */
     Repeat iterations times
        Evolve hPop
        Evolve pPop
     End repeat
  Until stop criteria
End Function.
```

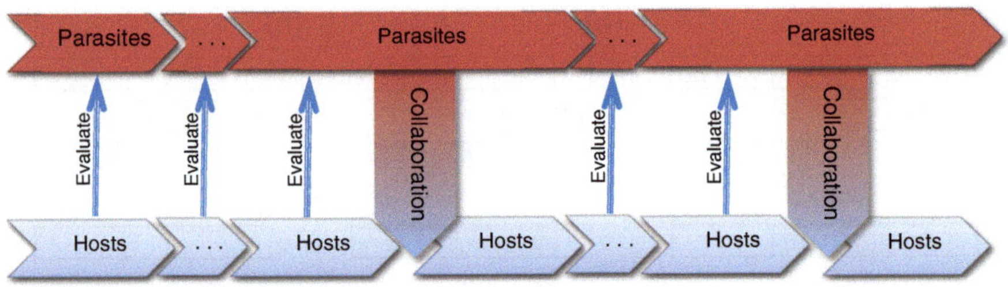

Fig. 12 Interaction between populations in SMuGA algorithm

The algorithm has six parameters: ***h*** represents the size of the host population; ***p*** the size of the parasite population; ***problem*** the problem to be solved; ***iterations*** the number of iterations that hosts and parasites evolve without collaboration; ***k*** the number of hosts selected to participate in the collaboration; and ***n*** that controls the probability of hosts' infection.

The algorithm starts by generating and evaluating a host population, ***hPop***, with ***h*** hosts of ***problem***, and a parasite population, ***pPop***, with ***p*** parasites. The only information about the problem needed by parasites is the size of the host to perform mutations. The evaluation of ***pPop*** is donc using ***hPop***. Figure 12 shows the interaction between ***sPop*** and ***hPop***.

The evolutionary iterative process starts with the collaboration phase followed by the evolution phase until a stop criterion is reached.

Collaboration phase is performed by Algorithm 5 between populations of parasites, ***pPop***, and the ***k*** selected hosts in the host population, ***selPop***, using the parameter ***n*** to control the infection probability of hosts. The result of Algorithm 5 is a symbiosis population, ***symbPop,*** that contains the selected host clones infected by the parasites. Because one host may be infected by many parasites and the algorithm saves clones when a host is infected by one parasite, the number of symbiosis is huge when compared to the number of parasites and number of hosts. This phase is computationally expensive. That effort is relieved by the use of multipopulations since the collaboration algorithm produces symbiosis with repeated genotypes and the multiset representation helps in its storage and evaluation. The collaboration phase ends with the selection of ***h*** hosts from the union of host population, ***hPop,*** and symbiosis population, ***symbPop***.

The evolution phase starts with the evolution of ***hPop*** using MuGA, Algorithm 1, and the evolution of ***pPop*** using Algorithm 2. Both populations evolve for ***iteration*** generations without establishing new collaborations. This phase is used to stabilize the individuals in the populations and to assimilate, in the hosts, genetic material introduced by the collaboration phase. The host population evolves on its own; however, the parasite population still uses hosts, since parasites are evaluated using the genes of the host population as a proxy for fitness evaluation. When hosts evolve and change their genes, the fitness value of parasites may change too.

7 Experimental Study

To examine the influence of symbiosis in the solutions of hard problems, we conducted a set of experiments with the SMuGA and compared the results with the standard MuGA. We compared, also, the results of SMuGA with SCA in order to assess the scalability of the algorithm to big deceptive problems.

7.1 Experimental Setup

MuGA was configured with 128 MI in the main population. Selection is made by tournaments with size 3. The operator selects 256 individuals for the mating pool, and in this way MI with copies is guaranteed for the following operators. Recombination is made by one-point crossover operator with probability 0.6. Mutation is made by the multiset wave mutation, MWM, configured with roughness = 2 and thinness = 3 (Fig. 9). The minimal mutation probability, parameter ***minProb*** of Algorithm 3, is equal to ***1/l***, where ***l*** represents the size in bits of the genome of the individual. Rescaling was applied to maintain a maximum total of copies in the main population of twice the number of MI.

SMuGA is configured with 32 MI in the host population and 32 MI in the parasite populations. In this case, we can use a smaller population than with MuGA, due to the increased genetic variety introduced by parasites. The size of the population selected to make collaboration is 16 MI, and the parameter that controls the probability of infection, parameter ***n*** in Algorithm 5, has value 1. The number of iterations of the evolution phase in Algorithm 6 is set to 16. The evolution of hosts uses tournament selection with tournament size 3 and selects 32 individuals. Recombination is done by uniform crossover with probability 0.6, and mutation, replacement, and rescaling are performed in the same way as in MuGA. Table 1 shows evolutionary parameters of MuGA and SMuGA.

To obtain statistical confidence, we performed 128 independent runs for each experiment. In each run, random initial populations were generated for individuals,

Table 1 Configuration of MuGA and SMuGA

	MuGA		SMuGA	
	Parameter	Settings	Parameter	Settings
Size of population	Individuals	128	Hosts	32
			Parasites	32
Selection	Tournament size 3	256	Tournament size 3	32
Recombination	Crossover 1 cut	0.6	Uniform Crossover	0.6
Mutation	MWM	2, 3	MWM	2, 3
Replacement	Decimation	2	Decimation	2
Rescaling	Adaptive	2	Adaptive	2

hosts, and parasites. The stop criteria used in the simulations are the number of evaluation function calls and, due to the varied difficulty of the problems that limit, are adjusted to allow the success of the evolutionary process. For each experiment, we compute the average of the number of evaluations to find the optimum. We assign the maximum number of evaluations to the experiments where the optimum is not found. We also compute what we consider a more revealing result, which is the success rate, meaning the percentage of runs that reach the optimum.

To compare the algorithms, we use pair-wise Student's *t* tests with 95 % confidence interval for the means. Due to the large number of simulations, we assume the normality of the variables. For each problem, we also compare results with other previously referred algorithms, when available, which means only for smaller genome lengths. However, results published for these problems are not always precise. In some cases, only logarithmic graphs are printed and the results here presented are best effort readings. And they never present the percentage of runs that reach the optimum.

7.2 Experimental Results with Deceptive Functions

The key to the success of EA is their combination of low-order BB to form high-order BB, which eventually leads to the optimum. When the solution cannot be built through this incremental combination of BB, we are in the presence of deceptive problems and we need to improve the artificial evolutionary process in order to solve those problems. The concept of deception was first introduced by Goldberg (1987) and much work has been done in addressing this class of problems. MuGA and SCA are two EA that are able to optimize deceptive functions. In the next sections, we present experimental results on different deceptive benchmark functions, for SMuGA, MuGA, and SCA.

7.2.1 Fully Deceptive F3 Function

Goldberg (1989) devised a 3-bit function, F3, presented in Eq. (4), that is fully deceptive since BB of order n are deceptive to build blocks of order $n + 1$.

$$\begin{array}{llll} \text{F3}(000) = 28, & \text{F3}(001) = 26, & \text{F3}(010) = 22, & \text{F3}(011) = 0 \\ \text{F3}(100) = 14, & \text{F3}(101) = 0, & \text{F3}(110) = 0, & \text{F3}(111) = 30 \end{array} \quad (4)$$

Fully deceptive function ***F3*** is easily solved by EA because of its size, i.e., three bits. To get a changeling problem, we define the function ***F3 10*** as ten consecutive copies of ***F3***. This procedure is usual in the optimization in this kind of deceptive problems and is adequate to be solved using symbiogenesis present in SMuGA.

Optimization of ***F3 10*** was successfully solved by the two algorithms (SMuGA and MuGA), Table 2, and the symbiotic approach speeds up the evolutionary process. Figure 13 shows the collaboration events between hosts and parasites in SMuGA, Fig. 15 shows the evolution of the success rate of the algorithms in the first 30,000 evaluation function calls, and Fig. 14 presents a statistical view of the

Table 2 Statistics of SMuGA and MuGA result in F3 10 function

F3 10	SMuGA		MuGA	
	Mean	Std	Mean	Std
Evals. to find best	3309.79	1273.36	6074.30	2516.68
Best value found	300.00	0.00	300.00	0.00
Success rate (%)	100.00	0.00	100.00	0.00

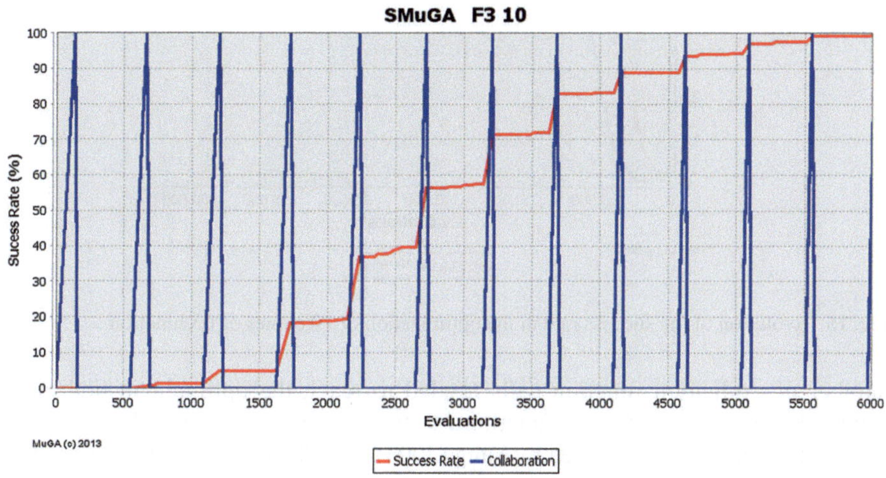

Fig. 13 Detail of the evolution of the success rate in the optimization of 10 copies of F3 function with SMuGA solver. The *blue line* represents collaboration events between hosts and parasites

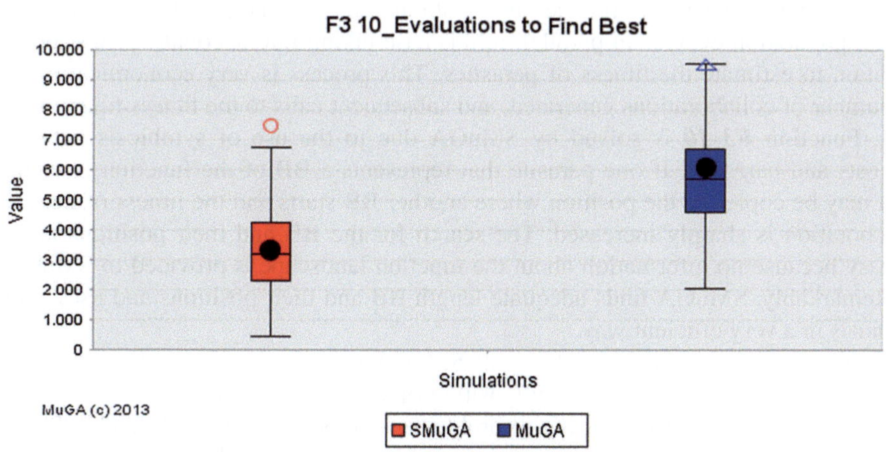

Fig. 14 Box-plots of the evaluation function calls to find the best of 10 copies of F3 function

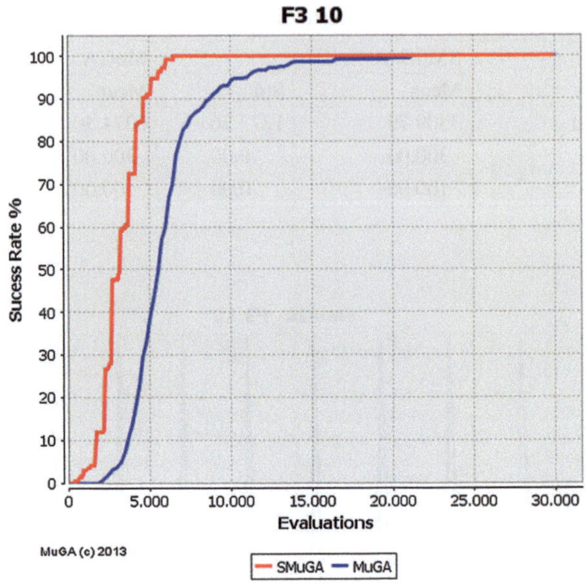

Fig. 15 Evolution of the success rate in the optimization of 10 copies of F3 function

number of evaluation function calls needed to reach the optimum in both algorithms. Results of SMuGA in function ***F3 10*** are more than one order of magnitude better than those presented by Yang (2004) and Chen et al. (2008).

Figure 13 shows in more detail the evolution of the success rate, observing only the first 6000 evaluation function calls. In that figure we can clearly see, in the major steps, the effect of the periodic incorporation of parasites in hosts, when new collaborations are formed and integrated into the host population. The evolution of the isolated host population over a few generations allows spreading of good genetic material introduced by symbionts through the population. The parasite population evolves in parallel, in this case taking into account the host population to estimate the fitness of parasites. This process is very economical in the number of collaborations generated, and subsequent calls to the fitness function.

Function ***F3 10*** is solved by SMuGA due to the use of symbiosis between hosts and parasites. If one parasite that represents a BB of the function is found, it may be copied to the position where another BB starts and the fitness of the collaboration is sharply increased. The search for the BB and their positions is not easy because no information about the function landscape is provided to SMuGA. Remarkably, SMuGA finds adequate length BB and their positions and uses symbiosis in a very efficient way.

In order to verify the scalability of SMuGA to big genome problems, we performed a set of tests with the composition of 10, 20, 40, 80, and 160 fully deceptive F3 functions corresponding to problems with 30, 60, 120, 240, and 480 bits, respectively. For these tests, we only present results for SMuGA since, in large problems, MuGA does not achieve solutions in reasonable time, and other algorithms do not present results.

Table 3 Statistics of SMuGA evolution result in optimization of concatenated F3 function with different lengths

SMuGA F3	Evals. to find best		Success (%)	
	Mean	Std	Mean	Std
30 bits	3088.69	1363.68	100.00	0.00
60 bits	5054.30	1495.08	100.00	0.00
120 bits	8457.08	3021.80	100.00	0.00
240 bits	17,500.25	12,182.22	98.44	12.40
480 bits	24,960.44	13,143.26	95.31	21.14

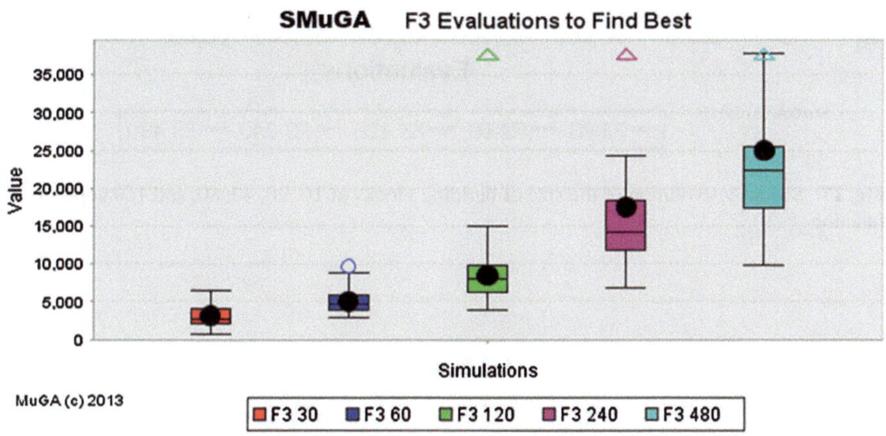

Fig. 16 SMuGA: Box-plots of the evaluation function calls to find the best value in 10(30), 20(60), 40(120), 80(240), and 160(480) copies(bits) of F3 function. Notice that *vertical axis* is linear while *horizontal axis* is exponential

Table 3 and Fig. 16 show the evolution statistics in the optimization of the concatenated F3 function with different lengths using SMuGA after 75,000 function evaluations calls. The algorithm scales in a linear way in this kind of functions due to its ability in finding good BB, assembling them with recombination, Fig. 3, and thus forming larger BB which can be moved to other locations in the genome, Fig. 6. This feature allows the solution of problems with long genomes of concatenated functions in a very efficient way. Figure 17 shows the evolution of the size of BB in that experiment. As we can see, problems with long genomes are solved by parasites also with long genomes, which will eventually incorporate a collaboration, speeding up the evolution of hosts. Again we note that the algorithm does not receive any information about BB.

Figure 18 shows the evolution of the success rate. The decrease of success in optimization of F3 with 240 bits, 98 %, and 480 bits, 95 %, can be explained by the small size of the parasite population (32 individuals) for a very large genome of the hosts. In that case, the probability of assembling useful BB in parasites decreases due to the large space that they explore.

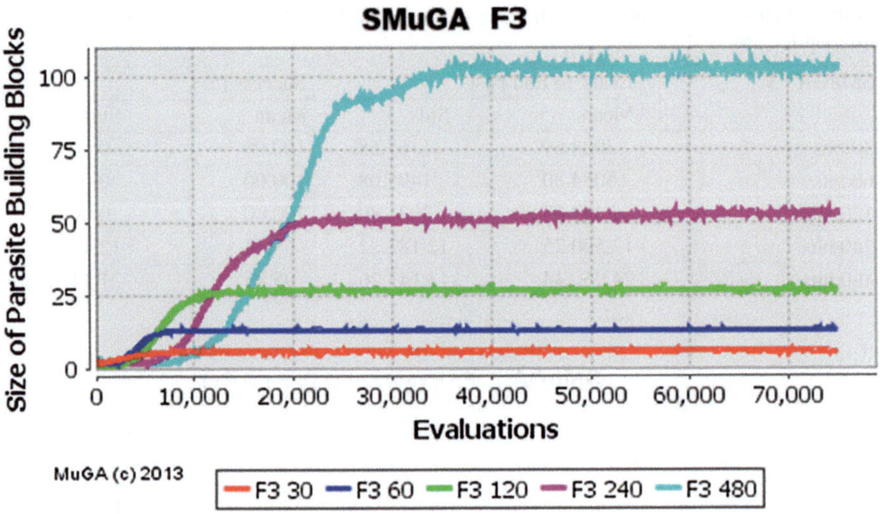

Fig. 17 SMuGA: evolution of the size of building blocks of 10, 20, 40, 80, and 160 copies of F3 function

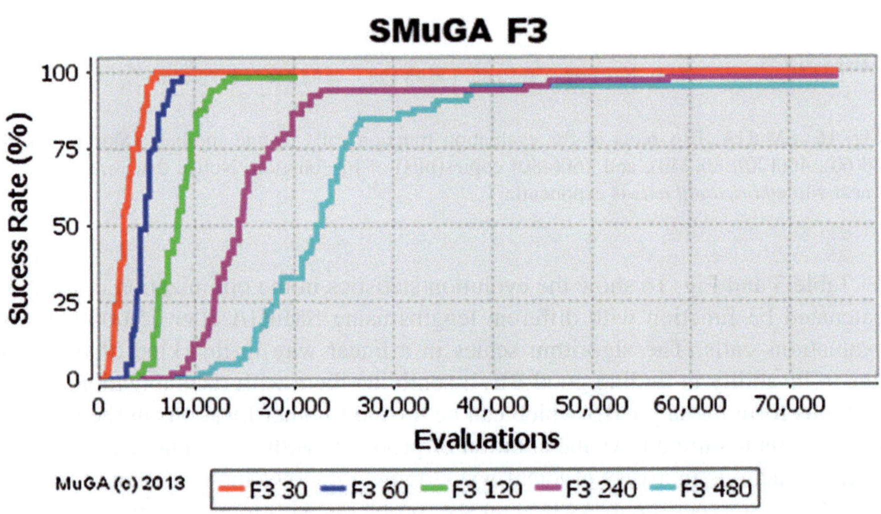

Fig. 18 SMuGA: evolution of the success rate in the optimization of 10, 20, 40, 80, and 160 copies of F3 function

For the functions analyzed next, we notice similar behavior to the one depicted in Fig. 13 in the step growth of success rate, and also a similar result to the one depicted in Fig. 17, regarding the evolution of the length of parasites as a function of the size of the problem. Therefore, we do not present such graphs for the remaining functions.

7.2.2 Maximally Separated Fully Deceptive F3 Function

The composition of functions in a sequential way is solved by SMuGA using the mobility property of parasites present in the algorithm. The application of one good parasite, which represents a BB, in a position where other BB starts, contributes to the success of the algorithm due to the nature of the function composition.

The problem becomes difficult when the bits of each function are separated. The most difficult case of separation is when they are uniformly and maximally distributed in the chromosome. We call these functions ***F3S N***, where ***N*** represents the number of F3 functions in the chromosome. In case of ***F3S 10,*** each bit of one function is located in positions $\boldsymbol{i}, \boldsymbol{i+10}$, and $\boldsymbol{i+20}$.

These functions are difficult because the problem is not separable and the formation of BB is not possible with a naïve strategy. In this way, the bits of the functions are spread and the application of one parasite in different positions is not enough to solve the problem. SMuGA escapes this situation by combining several parasites in a single host. With this experiment, we verify SMuGA's effectiveness in non-separable problems as well.

Table 4 presents the results of the optimization of ***F3S 10***. Both SMuGA and MuGA solve the function with notable efficacy and, again, symbiogenesis speeds up the evolutionary process. Figure 19 shows the evolution of the success rate of the algorithms in the first 100,000 evaluation function calls, and Fig. 20 presents a statistical view of the number of function evaluations needed to reach the optimum in both algorithms.

Figure 21 and Table 5 show the statistics of the optimization of the composition of 10, 20, 40, 80, and 160 F3S function in the chromosome after 500,000 evaluation function calls. As previously stated, the bits of F3S N functions are maximally spread over the chromosome, and big genomes separate the bits of one function with large distances. SMuGA fully succeeds in the optimization of 10 and 20 F3S N functions. In the optimization of 40 F3S, whose chromosome has 120 bits and the bits of each F3S function are separated by 40 bits, SMuGA

Table 4 Statistics of SMuGA and MuGA result in F3S 10 function

F3 separated	SMuGA		MuGA	
	Mean	Std	Mean	Std
Evals. to find best	9419.20	5604.92	34,063.77	18,018.62
Best value found	300.00	0.00	299.95	0.30
Success rate (%)	100.00	0.00	99.22	8.80

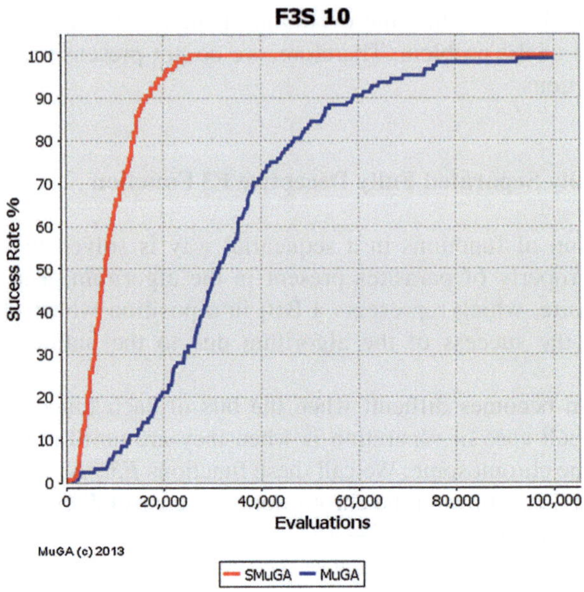

Fig. 19 Evolution of the success rate in the optimization of F3S 10 function

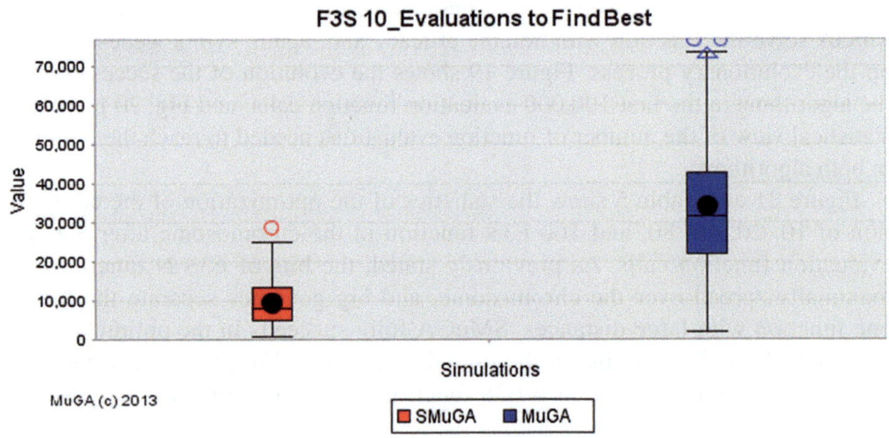

Fig. 20 Box-plot of the evaluations to find the best in the optimization of F3S 10 function

succeeds in 95 % of simulations and needs more generations to fully succeed. In the larger simulations, the small population of parasites and the large genome of the hosts hinders the optimization, and the parameters must be adjusted (Fig. 22).

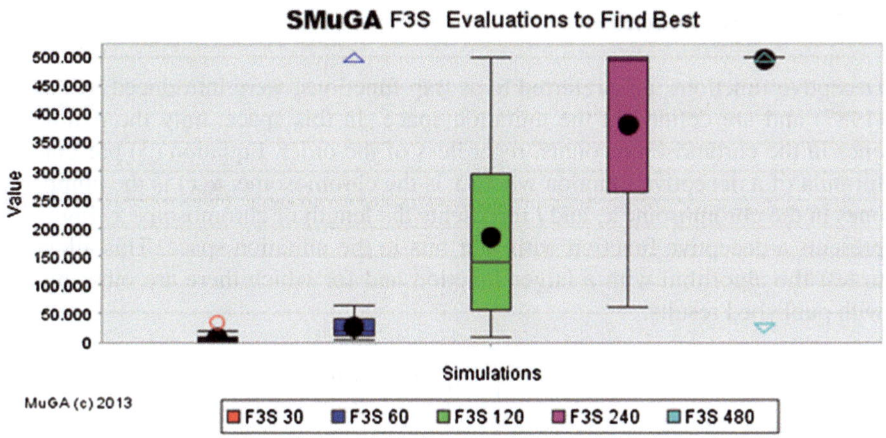

Fig. 21 SMuGA: box-plots of the evaluations to find the best value in 10, 20, 40, 80 and 160 copies of F3S function

Table 5 Statistics of SMuGA evolution result in optimization of separated F3S with different lengths

SMuGA F3S	Evals. to find best		Success (%)	
	Mean	Std	Mean	Std
30 bits	7651.08	5049.29	100.00	0.00
60 bits	26,429.98	19,408.04	100.00	0.00
120 bits	184,549.02	145,359.55	95.31	21.14
240 bits	381,741.92	148,172.39	51.56	49.98
480 bits	496,565.75	27,516.57	1.56	12.40

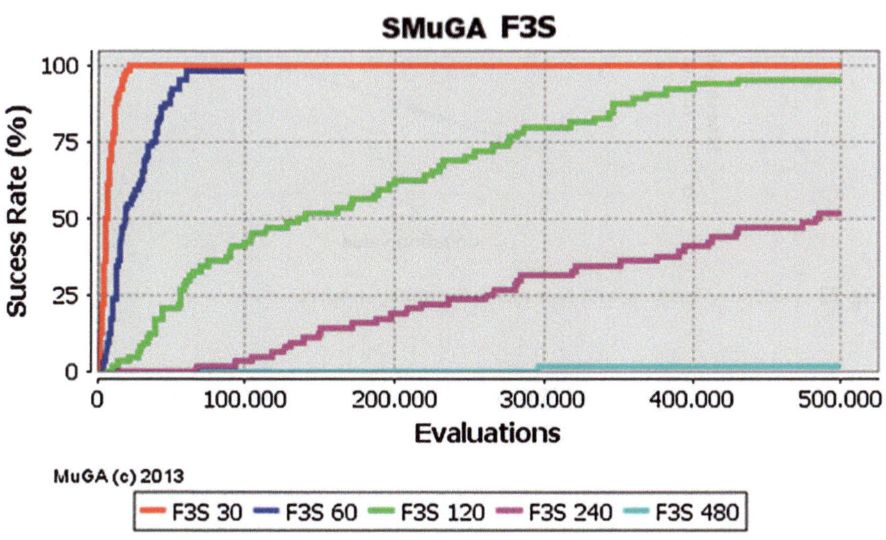

Fig. 22 SMuGA: evolution of the success rate in the optimization of 10, 20, 40, 80, and 160 copies of F3 function

7.2.3 Deceptive Functions

Deceptive functions, also referred to as trap functions, were introduced by Ackley (1987) and are defined in the unitation space. In this space, only the number of ones in the chromosome counts, regardless of the order. Equation (5) presents the formula of a deceptive function where $\boldsymbol{x}$ is the chromosome, $\boldsymbol{u}(\boldsymbol{x})$ is the number of ones in the chromosome $\boldsymbol{x}$, and $\boldsymbol{l}$ represents the length of chromosome $\boldsymbol{x}$. Figure 23 presents a deceptive function with four bits in the unitation space. This allows us to test the algorithm with a larger function and for which there are other models with published results.

$$\text{deceptive}(x) = \begin{cases} u(x) & \text{if } u(x) > 0 \\ l+1 & \text{if } u(x) = 0 \end{cases} \tag{5}$$

In this experiment, we use a concatenated 16 blocks of four bits deceptive function, Fig. 23, representing a chromosome with 64 bits. Table 6 shows the results of MuGA and SMuGA in the optimization of the function after 100,000 evaluation function calls. SMuGA optimizes all the experiments with very little evaluation function calls when compared to MuGA. Figure 24 shows the evolution of the success rate of both algorithms in evolution. MuGA experiences several difficulties in optimizing deceptive functions with large genomes.

Comparing the results with SCA presented in Wallin et al. (2005), where SCA needs hundreds of thousands of function evaluations, we conclude that SMuGA is

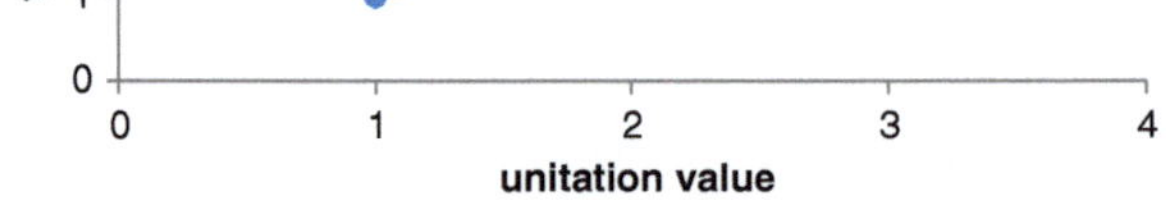

Fig. 23 Deceptive function with four bits in the unitation space

Table 6 Statistics of SMuGA and MuGA results in deceptive 4 functions with 16 copies

Deceptive 16 4	SMuGA		MuGA	
	Mean	Std	Mean	Std
Evals. to find best	5431.04	3724.07	84,749.20	30,440.50
Best value found	80.00	0.00	78.34	1.43
Success rate (%)	100.00	0.00	27.34	44.57

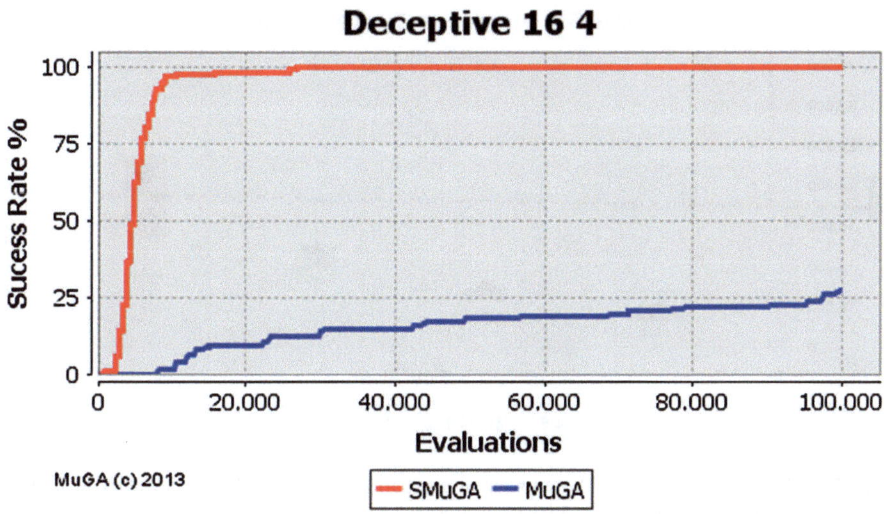

Fig. 24 Evolution of the success rate in the optimization of 16 copies of deceptive 4 functions

significantly better. The ability of SMuGA to manipulate the size of the parasite genomes is the key to solve this kind of problems. SCA do not have that property, and the static size of the parasites slows down the evolution.

Table 7 and Fig. 25 show the statistics of evolution after 75,000 function evaluation calls for the problems composed by 16, 32, 64, and 128 deceptive 4 functions that represent genomes with 64, 128, 256, and 512 bits. SMuGA was successful in all the simulations. However in a simulation with problems composed by 512 bits, SMuGA experiments some difficulties in the optimization due to the large genome of the host and more generations are needed to optimize all the problems as shown in Fig. 26.

SMuGA scales up very well to optimize large deceptive 4 problems, and results are again over one order of magnitude better than those presented in Wallin et al. (2005) using SCA and Thierens (2010) using linkage tree genetic algorithm (LTGA). Table 8 shows the number of function evaluations to solve deceptive 4 functions with different lengths provided by our best effort to read the graphics supplied in the papers.

Table 7 Statistics of SMuGA evolution result in optimization of deceptive 4 functions with different lengths

SMuGA deceptive 4	Evals. to find best		Success (%)	
	Mean	Std	Mean	Std
64 bits	4243.00	1645.40	100.00	0.00
128 bits	7061.89	3215.18	100.00	0.00
256 bits	11,180.36	4897.55	100.00	0.00
512 bits	21,458.61	13,696.15	96.88	17.40

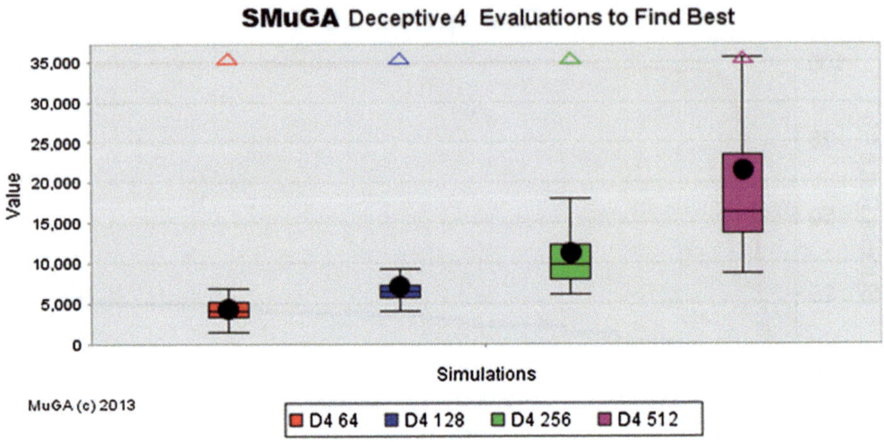

Fig. 25 SMuGA: box-plots of the number of evaluation function calls for SMuGA to find the best value in 16, 32, 64, and 128 copies of deceptive 4 functions

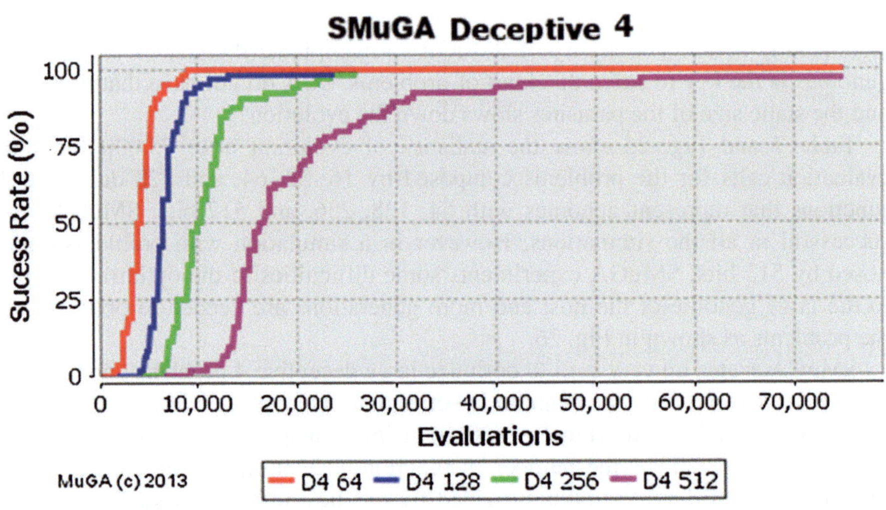

Fig. 26 SMuGA: evolution of the success rate in the optimization of 16, 32, 64, and 128 copies of deceptive 4 functions

Table 8 Number of functions evaluation calls to solve deceptive 4 functions using SCA and LTGA algorithms (approx.)

Algorithm	Size	Evals.
SCA	64	100,000
	128	200,000
LTGA	60	40,000
	100	75,000

7.2.4 Intertwined Deceptive Functions

The pair-intertwined function proposed by Wallin et al. (2005) is defined as two deceptive functions where the bits are intertwined in the same function, Fig. 27. The pair-intertwined function was, many local optima, introduced by the combination of the pair of deceptive functions. In this experiment, we use as building block two deceptive functions of four bits each composing a deceptive intertwine function, ***D4PI***, with eight bits.

Table 9 presents the statistics of the optimization of 8 D4PI functions, amounting to 64 bits, after 100,000 evaluation function calls. SMuGA optimizes all the problems with a small number of evaluation function calls due to the capability, provided by the parasites, to discover the BB of the ***D4TI*** function and the ability to concatenate BB and move them along the chromosome. The success of MuGA in this experiment is very limited due to the large length of the BB and the long genome of the individuals, Fig. 28. Comparing results with SCA presented in Wallin et al. (2005), Table 10, we notice SMuGA is more than one order of magnitude faster (in number of evaluations).

Figure 29 and Table 11 show the statistics of evolution after 500,000 function evaluation calls for the problems composed by 8, 16, 32, and 64 D4PI functions which represent chromosomes with 64, 128, 256, and 512 bits. SMuGA was successful in all the simulations. However, in the 512-bit problems, SMuGA experiments some difficulties in the optimization due the large genome of the host. More generations would allow to optimize these problems as we can infer from Fig. 30, but adjusting the parameters for the 512-bit problem would supposedly increase convergence.

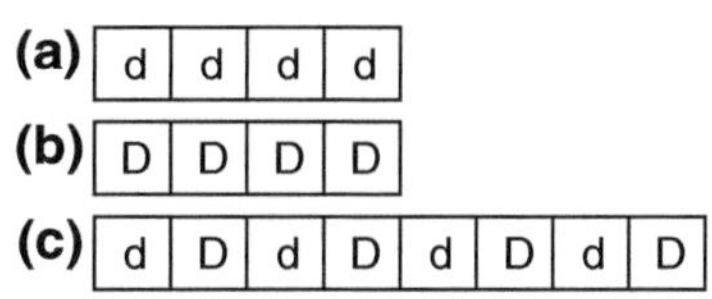

Fig. 27 Intertwined pair deceptive functions: **a** deceptive function d; **b** deceptive function D; **c** intertwined deceptive function dD

Table 9 Statistics of SMuGA and MuGA result in deceptive 4 pair-intertwined functions with 8 copies

D4PI 8	SMuGA		MuGA	
	Mean	Std	Mean	Std
Evals. to find best	12,401.04	12,114.49	97,294.48	12,106.18
Best value found	80.00	0.00	76.88	1.61
Success rate (%)	100.00	0.00	6.25	24.21

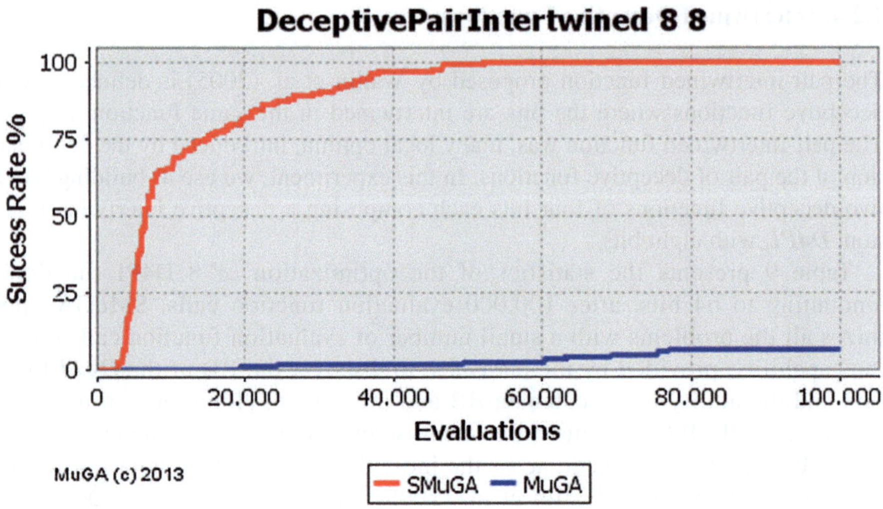

Fig. 28 Evolution of the success rate in the optimization of 8 copies of D4PI function

Table 10 Number of functions evaluation calls to solve deceptive pair-intertwined function using SCA (approx.)

Algorithm	Size	Evals.
SCA	64	150,000
	128	250,000

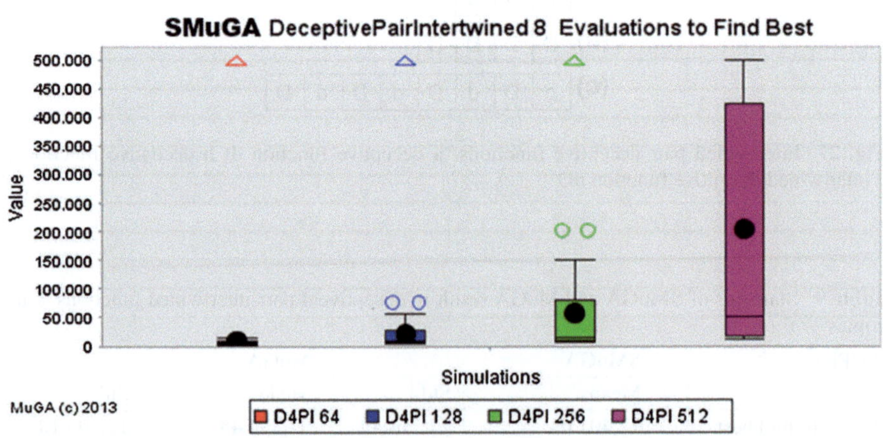

Fig. 29 SMuGA: box-plots of the evaluation function calls to find the best value in 8, 16, 32, and 64 copies of D4PI function

Table 11 Statistics of SMuGA evolution result in optimization of D4PI with different lengths

SMuGA D4PI	Evals. to find best		Success (%)	
	Mean	Std	Mean	Std
64 bits	9246.20	9045.93	100.00	0.00
128 bits	20,050.59	21,544.04	100.00	0.00
256 bits	56,673.89	84,180.82	100.00	0.00
512 bits	204,222.19	206,614.44	79.69	40.23

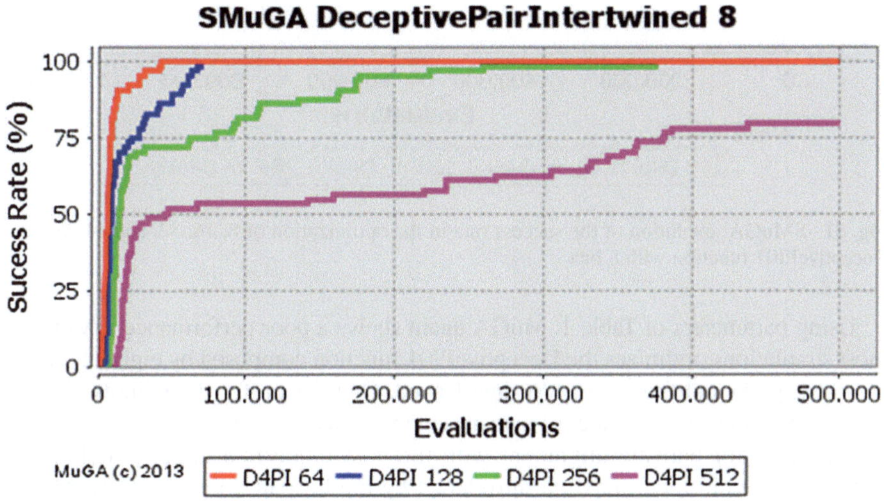

Fig. 30 SMuGA: evolution of the success rate in the optimization of 8, 16, 32, and 64 copies of D4PI function

7.2.5 Deceptive Intertwined Pair 0–1 Function

To assess the ability of SMuGA to evolve BB with optima that are not all ones or all zeroes, we defined a new intertwined function, DeceptivePI01, where one function is evaluated by Eq. (5) and other by Eq. (6). In the ***deceptiveZ*** function, Eq. (6), $z(\boldsymbol{x})$ counts the number of zeroes in the string $\boldsymbol{x}$. The optimum of function DeceptivePI01 is composed by a string with alternating zeros and ones and the translocations of the BB done by the parasites need alignment in the host.

$$\text{deceptive}Z(x) = \begin{cases} z(x) & \text{if } z(x) > 0 \\ l+1 & \text{if } z(x) = l \end{cases} \tag{6}$$

Figure 25 show the evolution of success rate along the 1,000,000 function evaluation calls for the problems composed by 16, 32, 64, and 128 DeceptivePI01 functions that represent genomes with 64, 128, 256, and 512 bits.

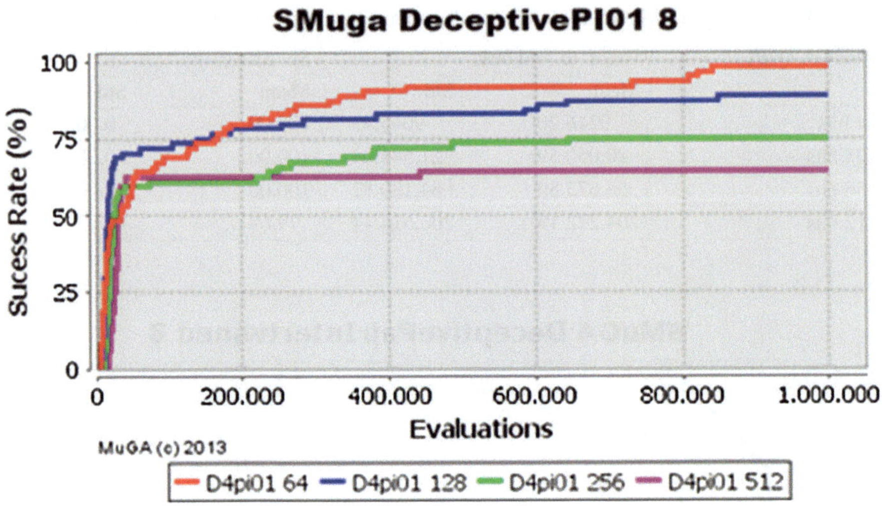

Fig. 31 SMuGA: evolution of the success rate in the optimization of 8, 16, 32, and 64 copies of DeceptivePI01 function with 8 bits

Using parameters of Table 1, MuGA again shows a poor performance. SMuGA in most simulations optimizes the DeceptivePI01 function composed by eight-bit blocks, four of Eq. (5) and four of Eq. (6) interleaved. One reason for the failures could be explained by the small number of parasites in the parasite population (Fig. 31).

The need for parasite alignment with the host requests a larger population of parasites to avoid local maxima introduced by the bit pattern of the DeceptivePI0 functions. The two local maxima, all ones and all zeroes, are more attractive to the parasites because that pattern does not need alignment and that parasites are easily assimilated by the hosts.

Figure 32 shows the effect of the size of parasite population in the optimization of 8 copies of DeceptivePI01 with 8 bits. As can be seen, the increase of the number of parasites in the symbiotic system increases the robustness of the solver. The increase of parasite population increases the computational complexity of the algorithm, but parasite population can evolve in parallel to the host population exploring the multicore resources of the computers.

Figure 33 and Table 12 present the same situation of Fig. 31 but now with 128 elements in the parasite population, instead of 32. The success of the algorithm is increased and simulations evolving functions with 64, 128, and 256 bits are always successfully optimized. The rate of success of simulation with 512 bits also increases although not attaining 100 % success. Further parameter tuning is one possible solution to achieve perfect score.

These results show that a large size of the parasite population makes SMuGA more robust in the evolution of difficult functions. Complex bit patterns impose difficulties to SMuGA in the alignment of parasites but these seem to be circumvented by larger parasite populations.

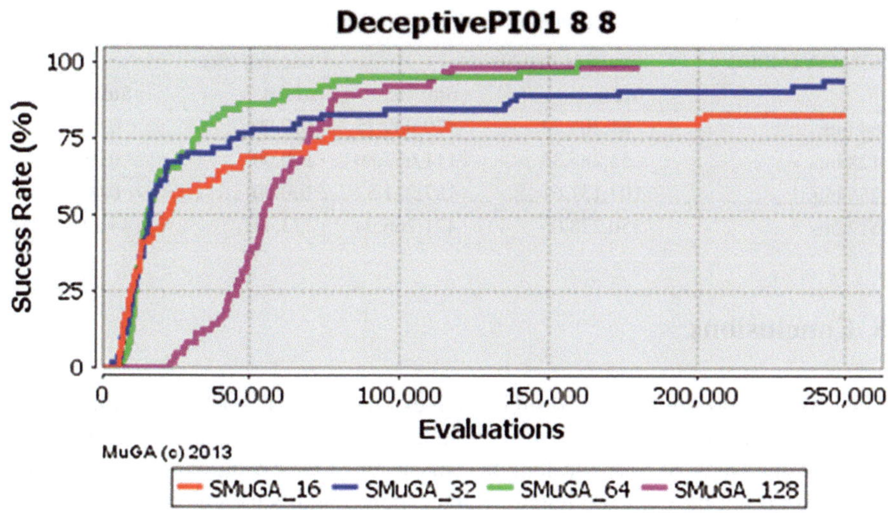

Fig. 32 SMuGA: evolution of the success rate in the optimization of 8 copies of DeceptivePI01 function with 8 bits with solver with 16, 32, 64, and 128 parasites in the parasite population

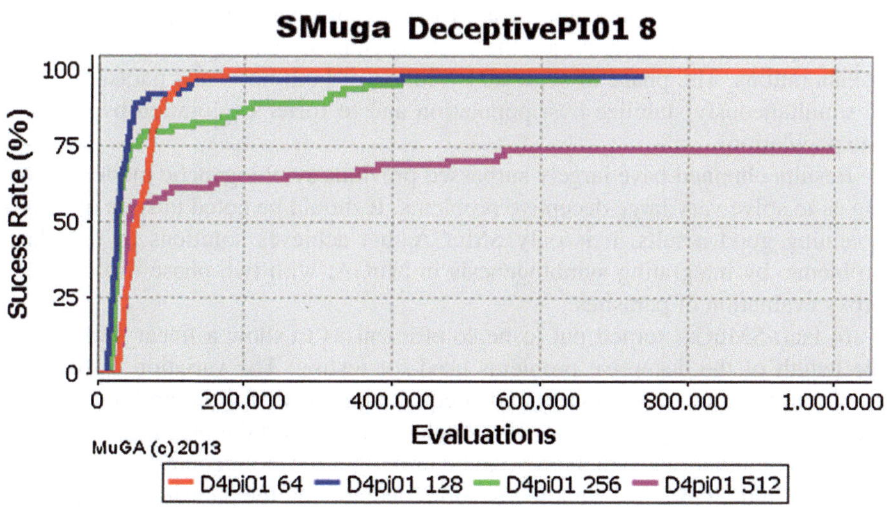

Fig. 33 SMuGA: evolution of the success rate in the optimization of 8, 16, 32, and 64 copies of DeceptivePI01 function with 8 bits intertwined with 128 parasites solvers

Table 12 Statistics of SMuGA evolution result in optimization of D4PI01 with different lengths

SMuGA D4PI01	Evals. to find best		Success (%)	
	Mean	Std	Mean	Std
64 bits	61,204.72	29,939.75	100.00	0.00
128 bits	54,285.36	111,617.89	100.00	0.00
256 bits	107,131.61	187,231.57	100.00	0.00
512 bits	350,228.05	421,266.94	73.44	44.17

8 Conclusions

This chapter presented the SMuGA, an extension of the MuGA with a novel approach to artificial symbiogenesis where a host receives genetic material from multiple parasites of variable length. This is the first evolutionary model where parasites do not have a fixed length. Rather their length varies along the evolutionary process.

The model proposed also introduced a two-phased step of evolution. In one phase, symbiotic collaborations are generated and compete with previous hosts to form the next generation host population. In the other phase, host and parasite populations evolve on their own for a few generations, but parasites use hosts' fitness as proxies to compute their own. Proxy parasite evaluation significantly saves fitness function calls and avoids the need to generate an exponential number of collaborations. The phase of separate evolution of both hosts and parasites allows to simultaneously stabilize host population and to foster exploration by the parasite population.

Results obtained have largely surpassed previous symbiogenetic models, allowing us to solve very large deceptive problems. It should be noted in spite of MuGA obtaining good results, it is only SMuGA that achieves solutions to very large problems, by integrating symbiogenesis in MuGA, with two-phase evolution and proxy evaluation of parasites.

In fact, SMuGA turned out to be so efficient as to show a linear scaling with the length of the deceptive problems used for testing. The variation of the parasites' length allows evolution to find adequate length BB for the problem at hand. Accumulating multiple parasites in a single host provides the opportunity of using parasite combinations, which prove to be important for more complex problems.

In the future work, we want to test more operators in the parasites. In particular, inversion might be important to hierarchical deceptive problems. We also need to explore different types of problems with SMuGA. Those used in this chapter are repeated concatenations of the same function. Also, the flexibility of this model indicates that it is adequate for dynamic fitness functions, and we should test it on dynamic problems. The symbiotic system can also be taken as a new operator introducing new parameters in the evolutionary process. Consequently, the new parameters can be tuned to increase the effectiveness of SMuGA and in the future we will make an effort in optimization and automation of these parameters.

Acknowledgments The authors thank Mel Todd, Guida Manso, and Nathalie Gontier for the precious revisions that made this text more clear and readable.

Glossary

Crossover operator Genetic operator inspired by biological reproduction, where two or more parents exchange genetic information to produce offspring that inherits features from the parents

Coevolution Simultaneous evolution of two or more species that have a strong ecological relationships among them (predator-prey, mutualism, or parasitic)

Collaboration Process for creating symbionts through the interaction of two distinct species

Deceptive problems Problems where the combination of building blocks with low order to form high order building blocks lead to a solution that is not a global optimum

Evolutionary Algorithm Generic population-based metaheuristic, inspired by biological evolution, which uses genetic inspired operators to evolve solutions to optimization problems that are represented by chromosomes

Genetic Algorithm Subclass of evolutionary algorithms that evolve a population of individuals, representing solutions to optimization problems, using genetic operators that mimic natural evolution such as selection, crossover, and mutation. Bit string chromosome is the standard

MDR Multiset Decimation Replacement—multiset selection operator used by MuGA to merge parents and offspring multiset populations

MuGA Multiset genetic algorithm—evolutionary algorithm that uses multisets to represent populations and genetic operators that take advantage of this representation

Multi-individuals Set of identical individuals represented by a 2-tuple composed by the chromosome and the number of clones (copies)

Multiset Collection in which members are allowed to appear more than once. May be formally defined as a set of 2-tuples $\langle n, e \rangle$ where n is the number of copies of the element e

Mutation Operator Analogous to biological mutation, this operator introduces probabilistic random changes in the chromosomes of the individuals

MWM Multiset wave mutation—multiset mutation operator used by MuGA that applies different probabilities of mutation to clones present in a multi-individual

Rescaling Operator MuGA genetic operator used to control the number of copies present in Multi-individuals

Selection Operator Genetic operator that mimics the natural selection of the fittest individuals in the population. In the genetic algorithm context, selection operator is used to choose parents for reproduction and to introduce the offspring in the population

SMuGA Symbiogenetic multiset genetic algorithm—coevolutionary algorithm that uses symbiogenesis

Symbiogenesis Evolutionary theory according to which individuals of different species come together to form a new individual (symbiont)

References

Abdoun O, Abouchabaka J, Tajani C (2012) Analyzing the performance of mutation operators to solve the travelling salesman problem. arXiv:1203.3099

Ackley DH (1987) A connectionist machine for genetic hillclimbing. Kluwer Academic Publishers, Norwell

Chen Y, Hu J, Hirasawa K, Yu S (2008) Solving deceptive problems using a genetic algorithm with reserve selection. Presented at the IEEE congress on evolutionary computation, 2008 (CEC 2008). IEEE World Congress on Computational Intelligence, pp 884–889

Daida JM, Grasso CS, Stanhope SA, Ross SJ (1996) Symbionticism and complex adaptive systems I: implications of having symbiosis occur in nature. In: Proceedings of the fifth annual conference on evolutionary programming. The MIT Press, pp 177–186

Droste S, Jansen T, Wegener I (2002) On the analysis of the (1 + 1) evolutionary algorithm. Theor Comput Sci 276:51–81

Dumeur R (1996) Evolution through cooperation: the symbiotic algorithm. In: Alliot J-M, Lutton E, Ronald E, Schoenauer M, Snyers D (eds) Artificial evolution. Lecture notes in computer science. Springer, Berlin, pp 145–158

Goldberg DE (1987) Simple genetic algorithms and the minimal, deceptive problem. In: Davis L (ed) Genetic algorithms and simulated annealing. Morgan Kaufmann, San Mateo, pp 74–88

Goldberg DE (1989) Genetic algorithms and Walsh functions: Part I, a gentle introduction. Complex Syst 3:129–152

Herrera F, Lozano M, Sanchez AM (2003) A taxonomy for the crossover operator for real-coded genetic algorithms: an experimental study. Int J Intell Syst 18:309–338

Heywood MI, Lichodzijewski P (2010) Symbiogenesis as a mechanism for building complex adaptive systems: a review. In: Chio CD, Cagnoni S, Cotta C, Ebner M, Ekárt A, Esparcia-Alcazar AI, Goh C-K, Merelo JJ, Neri F, Preuß M, Togelius J, Yannakakis GN (eds) Applications of evolutionary computation., Lecture notes in computer scienceSpringer, Berlin, pp 51–60

Jayachandran J, Corns S (2010) A comparative study of diversity in evolutionary algorithms. IEEE, pp 1–7

Lozano M, Herrera F, Cano J (2008) Replacement strategies to preserve useful diversity in steady-state genetic algorithms. Inf Sci 178:4421–4433

Manso A, Correia L (2009) Genetic algorithms using populations based on multisets. In: Lopes LS, Lau N, Mariano P, Rocha L (eds) New trends in artificial intelligence, EPIA 2009. Universidade de Aveiro, pp 53–64

Manso A, Correia L (2011) A multiset genetic algorithm for real coded problems. In: Proceedings of the 13th annual conference companion on genetic and evolutionary computation—GECCO'11. Presented at the the 13th annual conference companion, Dublin, Ireland, p 153

Manso A, Correia L (2013) A multiset genetic algorithm for the optimization of deceptive problems. In: Proceeding of the fifteenth annual conference on genetic and evolutionary computation conference (GECCO'13). ACM, New York, pp 813–820

Otman A, Jaafar A (2011) A comparative study of adaptive crossover operators for genetic algorithms to resolve the traveling salesman problem. IJCA (0975-8887)

Rosin CD, Belew RK (1997) New methods for competitive coevolution. Evol Comput 5:1–29

Sivaraj R, Ravichandran T (2011) A review of selection methods in genetic algorithm. Int J Eng Sci Technol 3:3792–3797

Spears W, Anand V (1991) A study of crossover operators in genetic programming

Thierens D (2010) Linkage tree genetic algorithm: first results. In: Proceedings of the 12th annual conference companion on genetic and evolutionary computation (GECCO'10). ACM, New York, pp 1953–1958

Wallin D, Ryan C, Azad RMA (2005) Symbiogenetic coevolution. Presented at the 2005 IEEE congress on evolutionary computation, 2005, vol 2, pp 1613–1620

Whitley LD (1991) Fundamental principles of deception in genetic search. Found Genet Algorithms 1:221–241
Yang S (2004) Adaptive group mutation for tackling deception in genetic search. WSEAS Trans Syst 3:107–112
Yu EL, Suganthan PN (2010) Ensemble of niching algorithms. Inf Sci 180:2815–2833

MIX
Papier aus verantwortungsvollen Quellen
Paper from responsible sources
FSC® C105338

If you have any concerns about our products,
you can contact us on
ProductSafety@springernature.com

In case Publisher is established outside the EU,
the EU authorized representative is:
Springer Nature Customer Service Center GmbH
Europaplatz 3, 69115 Heidelberg, Germany

Printed by Libri Plureos GmbH
in Hamburg, Germany